Jens Hagen

Industrial Catalysis

Further Titles on Catalysis:

Wijngaarden, R. J., Kronberg, A., Westerterp, K. R.
Industrial Catalysis – Optimizing Catalysts and Processes
ISBN 3-527-28581-4

Thomas, J. M., Thomas, W. J.
Principles and Practice of Heterogeneous Catalysis
ISBN 3-527-29288-8 (Hardcover)
ISBN 3-527-29239-X (Softcover)

Ertl, G., Knötzinger, H., Weitkamp, J. (eds.)
Handbook of Heterogeneous Catalysis
5 Volumes
ISBN 3-527-29212-8

Niemantsverdriet, J. W.
Spectroscopy in Catalysis
An Introduction
ISBN 3-527-28593-8 (Hardcover)
ISBN 3-527-28726-4 (Softcover)

Jens Hagen

Industrial Catalysis

A Practical Approach

Weinheim · New York · Chichester
Brisbane · Singapore · Toronto

Prof. Dr. Jens Hagen
Fachhochschule Mannheim –
Hochschule für Technik und Gestaltung
Windeckstraße 110
D-68163 Mannheim

> This book was carefully produced. Nevertheless, author and publisher do not warrant the information contained therein to be free of errors. Readers are advised to keep in mind that statements, data, illustrations, procedural details or other items may inadvertently be inaccurate.

Library of Congress Card No. applied for

A catalogue record for this book is available from the British Library

Die Deutsche Bibliothek – CIP-Einheitsaufnahme
Hagen, Jens:
Industrial catalysis : a practical approach / Jens Hagen. –
Weinheim ; New York ; Chichester ; Brisbane ; Singapore ; Toronto :
Wiley-VCH, 1999
  ISBN 3-527-29528-3

© WILEY-VCH Verlag GmbH, D-69469 Weinheim (Federal Republic of Germany), 1999

Printed on acid-free and low chlorine paper

All rights reserved (including those of translation into other languages). No part of this book may be reproduced in any form – by photoprinting, microfilm, or any other means – nor transmitted or translated into a machine language without written permission from the publishers. Registered names, trademarks, etc. used in this book, even when not specifically marked as such, are not to be considered unprotected by law.
Cover design: Susanne Pauker
Composition: Prosatz, D-69469 Weinheim
Printing: betz-druck gmbh, D-64291 Darmstadt
Bookbinding: Großbuchbinderei J. Schäffer, D-67269 Grünstadt
Printed in the Federal Republic of Germany

Preface

Catalysts have been used in the chemical industry for hundreds of years, and many large-scale industrial processes can only be carried out with the aid of catalysis. However, it is only since the 1970s that catalysis has become familiar to the general public, mainly because of developments in environmental protection, an example being the well known and widely used catalytic converter for automobiles.

Catalysis is a multidisciplinary area of chemistry, in particular, industrial chemistry. Anyone who is involved with chemical reactions will eventually have something to do with catalysts.

In spite of years of experience with catalysts and the vast number of publications concerning catalytic processes, there is still no fundamental theory of catalysis. As is often the case in chemistry, empirical concepts are used to explain experimental results or to make predictions about new reactions, with greater or lesser degrees of success.

To date there has been no standard book that deals equally with both heterogeneous and homogeneous catalysis, as well as industrial aspects thereof. The books published up to now generally describe a particular area or special aspects of catalysis and are therefore less suitable for teaching or studying on one's own. For this reason, it is not easy for those commencing their careers to become familiar with the complex field of catalysis.

This book is based on my own lecture course for chemical engineers at the Fachhochschule Mannheim (Mannheim University of Applied Sciences M.U.A.S) and is intended for students of chemistry, industrial chemistry, and process engineering, as well as chemists, engineers, and technicians in industry who are involved with catalysts. Largely dispensing with complex theoretical and mathematical treatments, the book describes the fundamental principles of catalysis in an easy to understand fashion. Numerous examples and exercises with solutions serve to consolidate the understanding of the material. The book is particularly well suited to studying on one's own.

It is assumed that the reader has a basic knowledge of chemistry, in particular, of reaction kinetics and organometallic chemistry. Homogeneous transition metal catalysis and heterogeneous catalysis are treated on the basis of the most important catalyst concepts, and the applications of catalysts are discussed with many examples. The book aids practically oriented readers in becoming familiar with the processes of catalyst development and testing and therefore deals with aspects of test planning, optimization, and reactor simulation. Restricting the coverage to fundamental aspects made it necessary to treat certain areas that would be of interest to specialists in concise form or to omit them completely.

I wish to thank all those who supported me in producing this book. Special thanks are due to Dr. R. Eis for all the hard work and care he invested in preparing the figures and for his helpful contributions and suggestions. I am grateful to the following companies for providing photographic material: BASF, Ludwigshafen, Germany; Degussa, Hanau, Germany; Hoffmann-LaRoche, Kaiseraugst, Switzerland; Doduco, Sinsheim, Germany; and VINCI Technologies, Rueil-Malmaison, France. Interesting examples of catalyst development were taken from the Diploma theses of Fachhochschule graduates, of whom K. Kromm and T. Zwick are especially worthy of mention.

I was pleased to accept the publisher's offer to produce an English version of the book. The introduction of international study courses leading to a Bachelor's or Master's degree in Germany and other countries makes it necessary to provide students with books in English. I am particularly grateful to Dr. S. Hawkins for his competent translation of the German text with valuable advice and additional material.

I thank the publishers, Wiley-VCH Weinheim, for their kind support. Thanks are due to Dr. B. Böck, who directed the project, C. Grössl for production, and S. Pauker for the cover graphics.

Mannheim, January 1999   Jens Hagen

Contents

1	**Introduction**	1
1.1	The Phenomenon Catalysis	1
1.2	Mode of Action of Catalysts	4
1.3	Classification of Catalysts	8
1.4	Comparison of Homogeneous and Heterogeneous Catalysis	10
	Exercises for Chapter 1	14
2	**Homogeneous Catalysis with Transition Metal Catalysts**	17
2.1	Key Reactions in Homogeneous Catalysis	18
2.1.1	Coordination and Exchange of Ligands	18
2.1.2	Complex Formation	21
2.1.3	Acid–Base Reactions	24
2.1.4	Redox Reactions: Oxidative Addition and Reductive Elimination	26
2.1.5	Insertion and Elimination Reactions	33
2.1.6	Reactions at Coordinated Ligands	37
	Exercises for Section 2.1	41
2.2	Catalyst Concepts in Homogeneous Catalysis	44
2.2.1	The 16/18-Electron Rule	44
2.2.2	Catalytic Cycles	45
2.2.3	Hard and Soft Catalysis	47
2.2.3.1	Hard Catalysis with Transition Metal Compounds	48
2.2.3.2	Soft Catalysis with Transition Metal Compounds	50
	Exercises for Section 2.2	55
2.3	Characterization of Homogeneous Catalysts	57
	Exercises for Section 2.3	63
3	**Homogeneously Catalyzed Industrial Processes**	65
3.1	Overview	65
3.2	Examples of Industrial Processes	68
3.2.1	Oxo Synthesis	68
3.2.2	Production of Acetic Acid by Carbonylation of Methanol	72
3.2.3	Selective Ethylene Oxidation by the Wacker Process	73

3.2.4	Oxidation of Cyclohexane	75
3.2.5	Asymmetric Hydrogenation: Monsanto L-Dopa Process	77
3.2.6	Oligomerization of Ehtylene (SHOP Process)	78
	Exercises for Chapter 3	80
4	**Heterogeneous Catalysis: Fundamentals**	**83**
4.1	Individual Steps in Heterogeneous Catalysis	83
4.2	Kinetics and Mechanisms of Heterogeneously Catalyzed Reactions	86
4.2.1	The Importance of Adsorption in Heterogeneous Catalysis	86
4.2.2	Kinetic Treatment	91
4.2.3	Mechanisms of Heterogeneously Catalyzed Gas-Phase Reactions	93
	Exercises for Section 4.2	98
4.3	Catalyst Concepts in Heterogeneous Catalysis	100
4.3.1	Energetic Aspects of Catalytic Activity	100
	Exercises for Section 4.3.1	115
4.3.2	Steric Effects	117
	Exercises for Section 4.3.2	128
4.3.3	Electronic Factors	129
4.3.3.1	Metals	131
4.3.3.2	Semiconductors	140
4.3.3.3	Isolators: Acidic and Basic Catalysts	154
	Exercises for Section 4.3.3	163
4.4	Interaction of Catalysts with Supports and Additives	165
4.4.1	Supported Catalysts	165
4.4.2	Promoters	174
	Exercises for Section 4.4	179
4.5	Catalyst Deactivation and Regeneration	180
	Exercises for Section 4.5	192
4.6	Characterization of Heterogeneous Catalysts	193
4.6.1	Physical Characterization	193
4.6.2	Surface Analysis and Chemical Characterization	198
	Exercises for Section 4.6	205

Contents IX

5 Catalyst Shapes and Production of Heterogeneous Catalysts 207
5.1 Catalyst Production 207
5.2 Immobilization of Homogeneous Catalysts 214
   Exercises for Chapter 5 222

6 Shape-Selective Catalysis: Zeolites 225
6.1   Composition and Structure of Zeolites 225
6.2   Production of Zeolites 228
6.3   Catalytic Properties of the Zeolites 229
6.3.1 Shape Selectivity 230
6.3.2 Acidity of Zeolites 235
6.4   Isomorphic Substitution of Zeolites 239
6.5   Metal-Doped Zeolites 240
6.6   Applications of Zeolites 242
   Exercises for Chapter 6 245

7 Planning, Development, and Testing of Catalysts 249
7.1   Stages of Catalyst Development 249
7.2   An Example of Catalyst Planning: Conversion of Olefins to Aromatics 251
7.3   Selection and Testing of Catalysts in Practice 256
7.3.1 Catalyst Screening 258
7.3.2 Catalyst Test Reactors for Reaction Engineering Investigations . 260
7.3.3 Statistical Test Planning and Optimization 267
7.3.4 Kinetic Modeling and Simulation 281
   Exercises for Chapter 7 295

8 Heterogeneously Catalyzed Processes in Industry 299
8.1   Overview 299
8.2   Examples of Industrial Processes 304
8.2.1 Ammonia Synthesis 304
8.2.2 Hydrogenation 306
8.2.3 Methanol Synthesis 311
8.2.4 Selective Oxidation of Propene 314
8.2.5 Selective Catalytic Reduction of Nitrogen Oxides 318
8.2.6 Olefin Polymerization 320
   Exercises for Chapter 8 322

9 Catalysis Reactors 327
9.1   Two-Phase Reactors 329
9.2   Three-Phase Reactors 333

9.2.1	Fixed-Bed Reactors	334
9.2.2	Suspension Reactors	336
9.3	Reactors for Homogeneously Catalyzed Reactions	340
	Exercises for Chapter 9	342
10	**Economic Importance of Catalysts**	345
11	**Future Development of Catalysis**	351
11.1	Homogeneous Catalysis	351
11.2	Heterogeneous Catalysis	353

Solutions to the Exercises 359

References 397

Textbooks and Reference Books on Homogeneous Catalysis 397

Textbooks, Reference Books, and Brochures on Heterogeneous Catalysis 397

References to Chapters 1–11 398

Index 405

Abbreviations

A	area [m^2]
A^*	adsorbed (activated) molecules of component A
a	catalyst activity
a_s	area per mass [m^2/kg]
ads	adsorbed (subscript)
AES	Auger electron spectroscopy
aq	aqueous solution (subscript)
bcc	body-centered cubic
bipy	2,2′-bipyridine
Bu	butyl C$_4$H$_9$-
c_i	concentration of component i [mol/L]
C.I.	constraint index
Cp	cyclopentadienyl C$_5$H$_5$-
D	diffusion coefficient [m^2/s]
d	deactivation (subscript)
E_a	activation energy [J/mol]
E_F	Fermi level
eff	effective (subscript)
E_i	ionisation energy
Et	ethyl C$_2$H$_5$-
ESCA	electron spectroscopy for chemical analysis
ESR	electron spin resonance spectroscopy
e	electrons
fcc	face-centered cubic
ΔG	Gibb's free energy [J/mol]
G	gas (subscript, too)
H	Henry's law constant
H_{ex}	external holdup
ΔH_f	enthalpy change of formation [J/mol]
H_m	modified Henry's law constant
ΔH_R	reaction enthalpy [J/mol]
H_0	Hammett acidity function
HSAB	hard and soft acids and bases
h	hard
hcp	hexagonal close packing
ISS	ion scattering spectroscopy

K	equilibrium constant
K_i	adsorption equilibrium constant of component i
k	reaction rate constant
k_0	pre-exponential factor
$k_L a_L$	gas-liquid mass transfer coefficient
$k_S a_S$	liquid-solid mass transfer coefficient
k_{tot}	global mass transfer coefficient
L	liquid (subscript)
L	ligand
LEED	low-energy electron diffraction
LF	liquid flow [L/min]
M	metal
m	mass [kg]
$m_{cat.}$	mass of catalyst [kg]
n	number of moles [mol]
n	order of reaction
$\dot{n}$	flow rate [mol/s]
$\dot{n}_{A,0}$	feed flow rate of starting material A [mol/s]
Oxad	oxidative addition
P	total pressure [bar]
Ph	phenyl C_6H_5-
PPh_3	triphenylphosphine
p	pressure [bar]
p_i	partial pressure of component i [bar]
py	pyridine
R	ideal gas law constant [J mol^{-1} K^{-1}]
R	recycle ratio
R	alkyl
r	reaction rate [mol L^{-1} h^{-1}]
r_{eff}	effective reaction rate per unit mass of catalyst [mol kg^{-1} h^{-1}]
rel	relative (subscript)
r_d	deactivation rate
S	surface area [m^2/kg]
ΔS	entropy change [J mol^{-1} K^{-1}]
S_p	selectivity [mol/mol] or [%]
S	solid (subscript, too)
SCR	selective catalytic reduction
SIMS	secondary-ion mass spectroscopy
SLPC	supported liquid phase catalysts
SMSI	strong metal-support interaction
SSPC	supported solid phase catalysts
s	soft
s	sample standard deviation
s^2	experimental error variance

STY	space time yield [mol L^{-1} h^{-1}, kg L^{-1} h^{-1}]
T	temperature [K]
TEM	transmission electron microscopy
TON	turnover number [mol mol^{-1} s^{-1}]
t	time [s, h]
V	volume [m^3]
$\dot{V}$	volumetric flow-rate
V_R	reaction volume [m^3]
X	conversion [mol/mol] or [%]
$\bar{x}$	mean value of measurements
$\vec{x}$	positional vector (simplex method)
z	tube length [m]
δ	percentage d-band occupancy
ϵ	excitation energy of semiconductors [eV]
ϵ_P	void fraction of particle
η	catalyst effectiveness factor
θ_i	degree of coverage of the surface of component i
ν	stretching frequencies (IR) [cm^{-1}]
ν_i	stoichiometric coefficient
ρ	density [g/mL]
$\rho_{cat.}$	pellet density of the catalyst [g/mL]
τ	tortuosity
σ	interfacial tension
ϕ_0	work function [eV]
*	active centers on the catalyst surface

1 Introduction

1.1 The Phenomenon Catalysis

Catalysis is the key to chemical transformations. Most industrial syntheses and nearly all biological reactions require catalysts. Furthermore, catalysis is the most important technology in environmental protection, i. e., the prevention of emissions. A well-known example is the catalytic converter for automobiles.

Catalytic reactions were already used in antiquity, although the underlying principle of catalysis was not recognized at the time. For example, the fermentation of sugar to ethanol and the conversion of ethanol to acetic acid are catalyzed by enzymes (biocatalysts). However, the systematic scientific development of catalysis only began about 200 years ago, and its importance has grown up to the present day [2].

The term "catalysis" was introduced as early as 1836 by Berzelius in order to explain various decomposition and transformation reactions. He assumed that catalysts possess special powers that can influence the affinity of chemical substances.

A definition that is still valid today is due to Ostwald (1895): "a catalyst accelerates a chemical reaction without affecting the position of the equilibrium." Ostwald recognized catalysis as a ubiquitous phenomenon that was to be explained in terms of the laws of physical chemistry.

While it was formerly assumed that the catalyst remained unchanged in the course of the reaction, it is now known that the catalyst is involved in chemical bonding with the reactants during the catalytic process. Thus catalysis is a cyclic process: the reactants are bound to one form of the catalyst, and the products are released from another, regenerating the initial state.

In simple terms, the catalytic cycle can be described as shown in Figure 1-1 [T9]. The intermediate catalyst complexes are in most cases highly reactive and difficult to detect.

In theory, an ideal catalyst would not be consumed, but this is not the case in practice. Owing to competing reactions, the catalyst undergoes chemical changes, and its activity becomes lower (catalyst deactivation). Thus catalysts must be regenerated or eventually replaced.

Apart from accelerating reactions, catalysts have another important property: they can influence the selectivity of chemical reactions. This means that completely different products can be obtained from a given starting material by using different catalyst systems. Industrially, this targeted reaction control is often even more important than the catalytic activity [6].

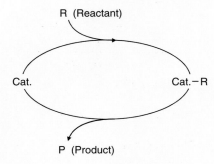

Fig. 1-1. Catalytic cycle

Catalysts can be gases, liquids, or solids. Most industrial catalysts are liquids or solids, whereby the latter react only via their surface. The importance of catalysis in the chemical industry is shown by the fact that 75% of all chemicals are produced with the aid of catalysts; in newly developed processes, the figure is over 90%. Numerous organic intermediate products, required for the production of plastics, synthetic fibers, pharmaceuticals, dyes, crop-protection agents, resins, and pigments, can only be produced by catalytic processes.

Most of the processes involved in crude-oil processing and petrochemistry, such as purification stages, refining, and chemical transformations, require catalysts. Environmental protection measures such as automobile exhaust control and purification of off-gases from power stations and industrial plant would be inconceivable without catalysts [5].

Catalysts have been successfully used in the chemical industry for more than 100 years, examples being the synthesis of sulfuric acid, the conversion of ammonia to nitric acid, and catalytic hydrogenation. Later developments include new highly selective multicomponent oxide and metallic catalysts, zeolites, and the introduction of homogeneous transition metal complexes in the chemical industry. This was supplemented by new high-performance techniques for probing catalysts and elucidating the mechanisms of heterogeneous and homogenous catalysis.

The brief historical survey given in Table 1-1 shows just how the closely the development of catalysis is linked to the history of industrial chemistry [4].

Table 1-1. History of the catalysis of industrial processes [4]

Catalytic reaction	Catalyst	Discoverer or company/year
Sulfuric acid (lead-chamber process)	NO_x	Désormes, Clement, 1806
Chlorine production by HCl oxidation	$CuSO_4$	Deacon, 1867
Sulfuric acid (contact process)	Pt, V_2O_5	Winkler, 1875; Knietsch, 1888 (BASF)
Nitric acid by NH_3-oxidation	Pt/Rh nets	Ostwald, 1906
Fat hardening	Ni	Normann, 1907
Ammonia synthesis from N_2, H_2	Fe	Mittasch, Haber, Bosch, 1908; Production, 1913 (BASF)
Hydrogenation of coal to hydrocarbons	Fe, Mo, Sn	Bergius, 1913; Pier, 1927
Oxidation of benzene, naphthalene to MSA or PSA	V_2O_5	Weiss, Downs, 1920
Methanol synthesis from CO/H_2	ZnO/Cr_2O_3	Mittasch, 1923
Hydrocarbons from CO/H_2 (motor fuels)	Fe, Co, Ni	Fischer, Tropsch, 1925
Oxidation of ethylene to ethylene oxide	Ag	Lefort, 1930
Alkylation of olefins with isobutane to gasoline	$AlCl_3$	Ipatieff, Pines, 1932
Cracking of hydrocarbons	Al_2O_3/SiO_2	Houdry, 1937
Hydroformylation of ethylene to propanal	Co	Roelen, 1938 (Ruhrchemie)
Cracking in a fluidized bed	Aluminosilicates	Lewis, Gilliland, 1939 (Standard Oil)
Ethylene polymerization, low-pressure	Ti compounds	Ziegler, Natta, 1954
Oxidation of ethylene to acetaldehyde	Pd/Cu chlorides	Hafner, Smidt (Wacker)
Ammoxidation of propene to acrylonitrile	Bi/Mo	Idol, 1959 (SOHIO process)
Olefin metathesis	Re, W, Mo	Banks, Bailey, 1964
Hydrogenation, isomerization, hydroformylation	Rh-, Ru complexes	Wilkinson, 1964
Methanol conversion to hydrocarbons	Zeolites	Mobil Chemical Co., 1975

1.2 Mode of Action of Catalysts

The suitability of a catalyst for an industrial process depends mainly on the following three properties:

- Activity
- Selectivity
- Stability (deactivation behavior)

The question which of these functions is the most important is generally difficult to answer because the demands made on the catalyst are different for each process. First, let us define the above terms [6, 7].

Activity

Activity is a measure of how fast one or more reactions proceed in the presence of the catalyst. Activity can be defined in terms of kinetics or from a more practically oriented viewpoint. In a formal kinetic treatment, it is appropriate to measure reaction rates in the temperature and concentration ranges that will be present in the reactor.

The reaction rate r is calculated as the rate of change of the amount of substance n_A of reactant A with time relative to the reaction volume or the mass of catalyst:

$$r = \frac{\text{converted amount of substance of a reactant}}{\text{volume or catalyst mass} \cdot \text{time}} \quad (\text{mol L}^{-1}\,\text{h}^{-1} \text{ or mol kg}^{-1}\,\text{h}^{-1}) \tag{1-1}$$

Kinetic activities are derived from the fundamental rate laws, for example, that for a simple irreversible reaction A → P:

$$\frac{dn_A}{dt} = kV\text{f}(c_A) \tag{1-2}$$

k = rate constant

$\text{f}(c_A)$ is a concentration term that can exhibit a first- or higher order dependence on adsorption equilibria (see Section 4.2).

The temperature dependence of rate constants is given by the Arrhenius equation:

$$k = k_0\, e^{-(E_a/RT)} \tag{1-3}$$

E_a = activation energy of the reaction
k_0 = pre-exponential factor
R = gas constant

As Equations 1-2 and 1-3 show, there are three possibilities for expressing catalyst activity, i.e., as:

- Reaction rate
- Rate constant k
- Activation energy E_a

Empirical rate equations are obtained by measuring reaction rates at various concentrations and temperatures. If, however, different catalysts are to be compared for a given reaction, the use of constant concentration and temperature conditions is often difficult because each catalyst requires it own optimal conditions. In this case it is appropriate to use the initial reaction rates r_0 obtained by extrapolation to the start of the reaction.

Another measure of catalyst activity is the turnover number TON, which originates from the field of enzymatic catalysis and is defined as the number of reactant molecules reacting per active center per second.

In the case of homogeneous catalysis, in which well-defined catalyst molecules are generally present in solution, the TON can be directly determined. For heterogeneous catalysts, this is generally difficult, because the activity depends on the size of the catalyst surface, which, however, does not have a uniform structure. For example, the activity of a supported metal catalyst is due to active metal atoms dispersed over the surface.

The number of active centers per unit mass or volume of catalyst can be determined indirectly by means of chemisorption experiments, but such measurements require great care, and the results are often not applicable to process conditions. Although the TON appears attractive due to its molecular simplicity, it should be used prudently in special cases.

In practice, readily determined measures of activity are often sufficient. For comparitive measurements, such as catalyst screening, determination of process parameters, optimization of catalyst production conditions, and deactivation studies, the following activity measures can be used:

- Conversion under constant reaction conditions
- Space velocity for a given, constant conversion
- Space–time yield
- Temperature required for a given conversion

Catalysts are often investigated in continuously operated test reactors, in which the conversions attained at constant space velocity are compared [6]

The space velocity is the volume flow rate $\dot{V}_0$, relative to the catalyst mass m_{cat}:

$$\text{Space velocity} = \frac{\dot{V}_0}{m_{cat}} \quad (\text{m}^3 \text{ kg}^{-1} \text{ s}^{-1}) \tag{1-4}$$

The conversion X_A is the ratio of the amount of reactant A that has reacted to the amount that was introduced into the reactor. For a batch reactor:

$$X_A = \frac{n_{A,0} - n_A}{n_{A,0}} \quad \text{(mol/mol or \%)} \tag{1-5}$$

If we replace the catalyst mass in Equation 1-4 with the catalyst volume, then we see that the space velocity is proportional to the reciprocal of the residence time.

Figure 1-2 compares two catalysts of differing activity with one another, and shows that for a given space velocity, catalyst A is better than catalyst B.

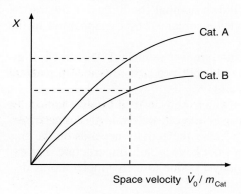

Fig. 1-2. Comparison of catalyst activities

Of course, such measurements must be made under constant conditions of starting material ratio, temperature, and pressure.

Often the performance of a reactor is given relative to the catalyst mass or volume, so that reactors of different size or construction can be compared with one another. This quantity is known as the space–time yield STY:

$$STY = \frac{\text{Desired product quantity}}{\text{Catalyst volume} \cdot \text{time}} \quad (\text{mol L}^{-1}\,\text{h}^{-1}) \tag{1-6}$$

Determination of the temperature required for a given conversion is another method of comparing catalysts. The best catalyst is the one that gives the desired conversion at a lower temperature. This method can not, however, be recommended since the kinetics are often different at higher temperature, making misinterpretations likely. This method is better suited to carrying out deactivation measurements on catalysts in pilot plants.

Selectivity

The selectivity S_p of a reaction is the fraction of the starting material that is converted to the desired product P. It is expressed by the ratio of the amount of desired product to the reacted quantity of a reaction partner A and therefore gives informa-

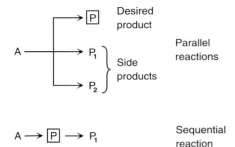

Scheme 1-1. Parallel and sequential reactions

tion about the course of the reaction. In addition to the desired reaction, parallel and sequential reactions can also occur (Scheme 1-1).

Since this quantity compares starting materials and products, the stoichiometric coefficients v_i of the reactants must be taken into account, which gives rise to the following equation [6]:

$$S_P = \frac{n_P/\nu_P}{(n_{A,0} - n_A)/|\nu_A|} = \frac{n_P |\nu_A|}{(n_{A,0} - n_A)\nu_P} \quad \text{(mol/mol or \%)} \quad (1\text{-}7)$$

In comparative selectivity studies, the reaction conditions of temperature and conversion or space velocity must, of course, be kept constant.

If the reaction is independent of the stoichiometry, then the selectivity $S_p = 1$. The selectivity is of great importance in industrial catalysis, as demonstrated by the example of synthesis gas chemistry, in which, depending on the catalyst used, completely different reaction products are obtained (Scheme 1-2) [2].

Selectivity problems are of particular relevance to oxidation reactions.

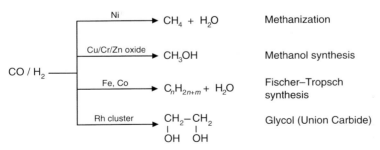

Scheme 1-2. Reactions of synthesis gas

Stability

The chemical, thermal, and mechanical stability of a catalyst determines its lifetime in industrial reactors. Catalyst stability is influenced by numerous factors, including decomposition, coking, and poisoning. Catalyst deactivation can be followed by measuring activity or selectivity as a function of time.

Catalysts that lose activity during a process can often be regenerated before they ultimately have to be replaced. The total catalyst lifetime is of crucial importance for the economics of a process.

Today the efficient use of raw materials and energy is of major importance, and it is preferable to optimize existing processes than to develop new ones. For various reasons, the target quantities should be given the following order of priority:

Selectivity > stability > activity

1.3 Classification of Catalysts

The numerous catalysts known today can be classified according to various criteria: structure, composition, area of application, or state of aggregation.

Here we shall classify the catalysts according to the state of aggregation in which they act. There are two large groups: heterogeneous catalysts (solid-state catalysts) and homogeneous catalysts (Scheme 1-3). There are also intermediate forms such as homogeneous catalysts attached to solids (supported catalysts), also known as immobilized catalysts [4]. The well-known biocatalysts (enzymes) also belong to this class.

In supported catalysts the catalytically active substance is applied to a support material that has a large surface area and is usually porous. By far the most impor-

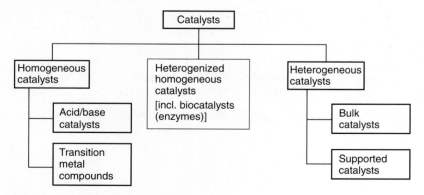

Scheme 1-3. Classification of catalysts

tant catalysts are the heterogeneous catalysts. The market share of homogeneous catalysts is estimated to be only ca. 10–15% [5, 6]. In the following, we shall briefly discuss the individual groups of catalysts.

Catalytic processes that take place in a uniform gas or liquid phase are classified as homogeneous catalysis. Homogeneous catalysts are generally well-defined chemical compounds or coordination complexes, which, together with the reactants, are molecularly dispersed in the reaction medium. Examples of homogeneous catalysts include mineral acids and transition metal compounds (e.g., rhodium carbonyl complexes in oxo synthesis).

Heterogeneous catalysis takes place between several phases. Generally the catalyst is a solid, and the reactants are gases or liquids. Examples of heterogeneous catalysts are Pt/Rh nets for the oxidation of ammonia to nitrous gases (Ostwald process), supported catalysts such as nickel on kieselguhr for fat hardening [1], and amorphous or crystalline aluminosilicates for cracking petroleum fractions.

Of increasing importance are the so-called biocatalysts (enzymes). Enzymes are protein molecules of colloidal size [e.g., poly(amino acids)]. Some of them act in dissolved form in cells, while others are chemically bound to to cell membranes or on surfaces. Enzymes can be classified somewhere between molecular homogeneous catalysts and macroscopic heterogeneous catalysts.

Enzymes are the driving force for biological reactions [4]. They exhibit remarkable activities and selectivities. For example, the enzyme catalase decomposes hydrogen peroxide 10^9 times faster than inorganic catalysts. The enzymes are organic molecules that almost always have a metal as the active center. Often the only difference to the industrial homogeneous catalysts is that the metal center is ligated by one or more proteins, resulting in a relatively high molecular mass.

Apart from high selectivity, the major advantage of enzymes is that they function under mild conditions, generally at room temperature in aqueous solution at pH values near 7. Their disadvantage is that they are sensitive, unstable molecules which are destroyed by extreme reaction conditions. They generally function well only at physiological pH values in very dilute solutions of the substrate.

Enzymes are expensive and difficult to obtain in pure form. Only recently have enzymes, often in immobilized form, been increasingly used for reactions of non-biological substances. With the increasing importance of biotechnological processes, enzymes will also grow in importance.

It would seem reasonable to treat homogeneous catalysis, heterogeneous catalysis, and enzymatic catalysis as separate disciplines. Enzymatic catalysis will not be treated further in this book; the interested reader is referred to the literature.

1.4 Comparison of Homogeneous and Heterogeneous Catalysis

Whereas for heterogeneous catalysts, phase boundaries are always present between the catalyst and the reactants, in homogeneous catalysis, catalyst, starting materials, and products are present in the same phase. Homogeneous catalysts have a higher degree of dispersion than heterogeneous catalysts since in theory each individual atom can be catalytically active. In heterogeneous catalysts only the surface atoms are active [3].

Due to their high degree of dispersion, homogeneous catalysts exhibit a higher activity per unit mass of metal than heterogeneous catalysts. The high mobility of the molecules in the reaction mixture results in more collisions with substrate molecules. The reactants can approach the catalytically active center from any direction, and a reaction at an active center does not block the neighboring centers. This allows the use of lower catalyst concentrations and milder reaction conditions.

The most prominent feature of homogeneous transition metal catalysts are the high selectivities that can be achieved. Homogeneously catalyzed reactions are controlled mainly by kinetics and less by material transport, because diffusion of the reactants to the catalyst can occur more readily. Due to the well-defined reaction site, the mechanism of homogeneous catalysis is relatively well understood. Mechanistic investigations can readily be carried out under reaction conditions by means of spectroscopic methods (Fig. 1-3). In contrast, processes occurring in heterogeneous catalysis are often obscure.

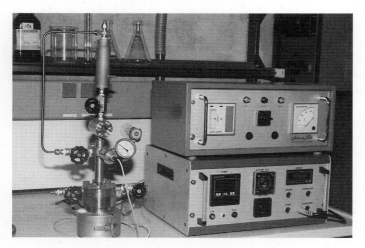

Fig. 1-3. Laboratory autoclave with dropping funnel, viewing window, and magnetic stirrer for the investigation of homogeneously catalyzed processes (high-pressure laboratory, FH Mannheim)

Owing to the thermal stability of organometallic complexes in the liquid phase, industrially realizable homogeneous catalysis is limited to temperatures below 200 °C. In this temperature range, homogeneous catalysts can readily be stabilized or modified by addition of ligands; considerable solvent effects also occur.

In industrial use, both types of catalyst are subject to deactivation as a result of chemical or physical processes. Table 1-2 summarizes the advantages and disadvantages of the two classes of catalyst.

The major disadvantage of homogeneous transition metal catalysts is the difficulty of separating the catalyst from the product. Heterogeneous catalysts are either automatically removed in the process (e.g., gas-phase reactions in fixed-bed reactors), or they can be separated by simple methods such as filtration or centrifugation. In the case of homogeneous catalysts, more complicated processes such as distillation, liquid–liquid extraction, and ion exchange must often be used [3].

Table 1-2. Comparison of homogeneous and heterogeneous catalysts

	Homogeneous	Heterogeneous
Effectivity		
Active centers	all metal atoms	only surface atoms
Concentration	low	high
Selectivity	high	lower
Diffusion problems	practically absent	present (mass-transfer-controlled reaction)
Reaction conditions	mild (50–200 °C)	severe (often >250 °C)
Applicability	limited	wide
Activity loss	irreversible reaction with products (cluster formation); poisoning	sintering of the metal crystallites; poisoning
Catalyst properties		
Structure/stoichiometry	defined	undefined
Modification possibilities	high	low
Thermal stability	low	high
Catalyst separation	sometimes laborious (chemical decomposition, distillation, extraction)	fixed-bed: unecessary suspension: filtration
Catalyst recycling	possible	unecessary (fixed-bed) or easy (suspension)
Cost of catalyst losses	high	low

The separability of homogeneous catalysts has been improved in the last few years by using organometallic complexes that are soluble in both organic and aqueous phases. These can readily be removed from the product stream at the reactor outlet by transferring them to the aqueous phase. This two-phase method has already been used successfully in large-scale industrial processes, for example:

- The Shell higher olefin process (SHOP), with nickel complex catalysts
- The Ruhrchemie/Rhône-Poulenc oxo synthesis with soluble rhodium catalysts (see Section 3.2)

There are of course also parallels between homogeneous and heterogeneous transition metal catalysts. Many reaction mechanisms of homogeneous and heterogeneous catalysts exhibit similarities with regard to the intermediates and the product distribution.

As shown in Table 1-3, the key reactions of homogeneous catalysis, such as hydride elimination and oxidative addition, correspond to dissociative chemisorption in heterogeneous catalysis (see Section 2.1).

The hope of increasing the separability of homogeneous catalysts by, for example, fixing them on solid supports has not yet been realized. The aim of many research projects is to maintain the high selectivity of homogeneous catalysts while at the same time exploiting the advantages of easier catalyst separation. The main problems are still catalyst "bleeding" and the relatively low stability and high sensitivity to poisoning of the heterogenized complexes.

An interesting intermediate between homogeneous and heterogeneous catalysts are the metal cluster catalysts. In many reactions that require several active centers of the catalyst, it is found that heterogeneous catalysts are active, while homogeneous catalysts give zero conversion. The reason is that crystallites on a metal surface exhibit several active centers, while conventional soluble catalysts generally contain only one metal center.

In contrast, metal clusters have several active centers or can form multi-electron systems. Metal clusters such as $Rh_6(CO)_{16}$, $Rh_4(CO)_{12}$, $Ir_4(CO)_{12}$, $Ru_3(CO)_{12}$, and more complex structures have been successfully tested in carbonylation reactions. Rhodium clusters catalyze the conversion of synthesis gas to ethylene glycol, albeit at very high pressures up to now.

With increasing size, the clusters become less soluble, and the precipitation of extremely small particles from solution is possible, that is, a transition from homogeneous to heterogeneous catalysis.

In conclusion, it can be stated that homogeneous and heterogeneous catalysts should be used to complement one another and not regarded as competitors, since each group has its special characteristics and properties.

1.4 Comparison of Homogeneous and Heterogeneous Catalysis 13

Table 1-3. Comparison of the key reactions of homogeneous and heterogeneous transition metal catalysis [10]

Homogeneous phase Oxad reactions	Heterogeneous phase dissociative chemisorption
$Ir(PPh_3)_3Cl + H_2 \rightleftharpoons H\text{-}Ir(PPh_3)_3Cl\text{-}H$	$H_2 + \text{-Pt-Pt-} \rightleftharpoons \text{-Pt(H)-Pt(H)-}$
$Pt(PPh_3)_2 + HC\equiv CR \rightleftharpoons Pt(H)(C\equiv CR)(PPh_3)_2$	$R\text{-}C\equiv CH + \text{-M-M-} \rightleftharpoons \text{-M(H)-M(C}\equiv\text{C-R)-}$
$Ph_2P\text{-}Ir(PPh_3)_2Cl\text{ (with phenyl ring, H)} \rightleftharpoons Ph_2P\text{-}Ir(PPh_3)_2Cl\text{ (ortho-metalated, H)}$	$C_6H_6 + \text{-Pt-Pt-} \rightleftharpoons \text{-Pt(H)-Pt(C}_6H_5)\text{-}$

Exercises for Chapter 1

Exercise 1.1

Classify the following reactions as homogeneous or heterogeneous catalysis and justify your answer:

a) The higher reaction rate for the oxidation of SO_2 with O_2 in the presence of NO.
b) The hydrogenation of liquid vegetable oil in the presence of a finely divided Ni catalyst.
c) The transformation of an aqueous solution of D-glucose into a mixture of the D and L forms, catalyzed by aqueous HCl.

Exercise 1.2

Compare homogeneous and heterogeneous catalysis according to the following criteria:

	Heterogeneous catalysts	Homogeneous catalysts
Active center		
Concentration		
Diffusion problems		
Modifiability		
Catalyst separation		

Exercise 1.3

Give four reasons why heterogeneous catalysts are preferred in industrial processes.

Exercise 1.4

a) Explain the difference between the activity and the selectivity of a catalyst.
b) Name three methods for measuring the activity of catalysts.

Exercise 1.5

Compare the key activation steps in the hydrogenation of alkenes with homogeneous and heterogeneous transition metal catalysts. What are the names of these steps?

	Homogeneous catalysis	Heterogeneous catalysis
Activation of H_2		
Activation of the olefin		

2 Homogeneous Catalysis with Transition Metal Catalysts

Most advances in industrial homogeneous catalysis are based on the development of organometallic catalysts. Thousands of organometallic complexes (i.e., compounds with metal–carbon bonds) have become known in the last few decades, and the rapid development of the organic chemistry of the transition metals has been driven by their potential applications as industrial catalysts [12].

The chemistry of organo transition metal catalysis is explained in terms of the reactivity of organic ligands bound to the metal center. The d orbitals of the transition metals allow ligands such as H (hydride), CO, and alkenes to be bound in such a way that they are activated towards further reactions.

The most important reactions in catalytic cycles are those involving ligands located in the coordination sphere of the same metal center. The molecular transformations generally require a loose coordination of the reactants to the central atom and facile release of the products from the coordination sphere. Both processes must proceed with an activation energy that is as low as possible, and thus extremely labile metal complexes are required. Such complexes have a vacant coordination site or at least one weakly bound ligand.

Reasons for the binding power of transition metals are that they can exist in various oxidation states and that they can exhibit a range of coordination numbers. The coordination complexes can be classified by dividing the ligands into two groups: ionic and neutral ligands [T11]. Ionic ligands include:

H^-, Cl^-, OH^-, $Alkyl^-$, $Aryl^-$, CH_3CO^-

and examples of neutral ligands are:

CO, alkenes, phosphines, phosphites, arsine, H_2O, amines

This distinction is useful for assigning oxidation states and in describing the course of reactions. However, it must be emphasized that this description is of a largely formal nature and sometimes does not describe the true bonding situation. Thus, although it is true that hydrogen ligands mostly react as H^- and alkyl groups as R^-, it is also possible that, for example, methyl groups react as CH_3^- or CH_3^+.

Rather than discussing the fundamentals of organometallic chemistry, this chapter is intended to give a survey of the most important types of reaction, a knowledge of which is sufficient for understanding the reaction cycles of homogeneous transition metal catalysis.

2.1 Key Reactions in Homogeneous Catalysis [9]

2.1.1 Coordination and Exchange of Ligands [18]

In many transition metal complexes, the coordination number is variable. Especially in solution or as the result of thermal dissociation, ligands can be released from the complex or undergo exchange, or free coordination sites can be occupied by solvent molecules. Therefore, most complexes do not react in their coordinatively saturated form, but via an intermediate of lower coordination number with which they are in equilibrium. For example, triphenylphosphine platinum complexes are involved in the following equilibrium reactions [T12]:

$$[Pt(PPh_3)_4] \underset{}{\overset{K_1}{\rightleftharpoons}} [Pt(PPh_3)_3] + PPh_3 \qquad K_1^{300\,K} = 1 \text{ mol/L} \qquad (2\text{-}1)$$

$$[Pt(PPh_3)_3] \underset{}{\overset{K_2}{\rightleftharpoons}} [Pt(PPh_3)_2] + PPh_3 \qquad K_2^{300\,K} \cong 10^{-6} \text{ mol/L} \qquad (2\text{-}2)$$

In aromatic solvents, the first equilibrium constant K_1 indicates rapid dissociation, but the second equilibrium constant is very small. However, the extremely high reactivity of $[Pt(PPh_3)_2]$ compensates for this concentration effect, and complete reaction occurs with π-acidic molecules such as CO and NO:

$$[Pt(PPh_3)_2] \begin{array}{c} \overset{2\,CO}{\longrightarrow} [Pt(CO)_2(PPh_3)_2] \\ \underset{2\,NO}{\longrightarrow} [Pt(NO)_2(PPh_3)_2] \end{array} \qquad (2\text{-}3)$$

The rapid dissociation of many complexes is explained in terms of steric hindrance of the ligands. With increasing space requirements of the phosphine or phosphite ligands, the rate of dissociation increases. A semi-quantitive measure for steric demand is the cone angle of the ligand (Table 2-1), as introduced by Tolman [20].

Accordingly, the sterically most demanding ligands should exhibit the fastest dissociation. This is demonstrated by the dissociation constants for complexes of nickel. For the reaction

$$NiL_4 \overset{K}{\rightleftharpoons} NiL_3 + L \qquad (2\text{-}4)$$

the following sequence, which correlates with the cone angles listed in Table 2-1, was found:

Table 2-1. Typical cone angles for trivalent phosphorus ligands [20]

Ligand	Cone angle [°]
PH_3	87
$P(OMe)_3$	107
$P(OEt)_3$	109
PMe_3	118
$P(OPh)_3$	121
$P(O\text{-}^iPr)_3$	130
PEt_3	132
PMe_2Ph	136
PPh_3	145
$P(^iPr)_3$	160
$P(cyclohexyl)_3$	170
$P(^tBu)_3$	182

$$L = \frac{P(OEt)_3 < PMe_3 < P(O-{}^iPr)_3 < PEt_3 < PMe_2Ph \ll PPh_3}{K\,(\text{Eq. 2-4})\,[\text{mol}]} \longrightarrow \text{completely dissociated}$$

However, care should be taken before making general statements, since the cone angles refer to a constant metal–phosphorus bond length and therefore do not reflect the true space filling in the coordinated state. Even complexes cotaining voluminous ligands can undergo addition of one or two small molecules:

$$[RhCl\{P(^tBu)_3\}_2] \xrightarrow[CO]{H_2} \begin{array}{c} [RhH_2Cl\{P(^tBu)_3\}_2] \\ [RhCl(CO)\{P(^tBu)_3\}_2] \end{array} \qquad (2\text{-}5)$$

For ligand dissociation/association processes, Tolman introduced the 16/18-electron rule [19] (see Section 2.2.1). For each covalently bonded ligand, two electrons are added to the number of d electrons of the central transition metal atom (corresponding to its formal oxidation state) to give a total valence electron count. An example for a complex involved in a thermal dissociative equilibrium is the well-known Wilkinson's catalyst:

$$HRh(CO)(PPh_3)_3 \xrightleftharpoons{-PPh_3} HRh(CO)(PPh_3)_2 \xrightleftharpoons{-PPh_3} HRh(CO)(PPh_3) \qquad (2\text{-}6)$$

The active form of the catalyst is generated by loss of PPh_3 ligands in solution (Eq. 2-6). An important step in the catalytic reactions of alkenes is the complexation of the substrate at the transition metal center to give a so-called π complex

[18]. The differing tendency of metals to bind alkenes is illustrated by the following trend (given isostructural complexes):

$$M(C_2H_4)(PPh_3)_2 \underset{K}{\overset{K}{\rightleftharpoons}} M(PPh_3)_2 + C_2H_4$$

$$M = Pd > Pt > Ni \qquad (2\text{-}7)$$

The tendency of ethylene complexes to dissociate (Eq. 2-7) can be explained in terms of the strength of the backbonding from the metal to the alkene $M \rightarrow C_2H_4$ (M = Ni > Pt > Pd; Co > Ir > Rh; Fe > Os > Ru). Care must be taken, however, in predicting the coordination equilibria of labile metal olefin or similar complexes since steric and electronic factors also play a role.

The coordination of certain ligands to a transition metal center can be facilitated by exploiting the *trans* effect. For example, the reaction

$$[PtCl_4]^{2-} + C_2H_4 \longrightarrow [PtCl_3(C_2H_4)]^- + Cl^- \qquad (2\text{-}8)$$

is slow, but can be accelerated by adding $SnCl_2$. This leads to formation of a $SnCl_3^-$ group, whose strong *trans* effect labilizes the chloro ligand in the *trans* position (Eq. 2-9).

$$\begin{bmatrix} Cl & Cl \\ & Pt \\ Cl_3Sn & Cl \end{bmatrix}^{2-} + C_2H_4 \longrightarrow [Pt(SnCl_3)Cl_2(C_2H_4)]^- + Cl^- \qquad (2\text{-}9)$$

The trichlorotin(II) ion $SnCl_3^-$ can replace the ligands Cl^-, CO, and PF_3 in nucleophilic ligand-exchange reactions (e.g., Eq. 2-10).

$$PtCl_4^{2-} + 2\ SnCl_3^- \longrightarrow [PtCl_2(SnCl_3)_2]^{2-} \xrightarrow{3\ SnCl_3^-} [Pt(SnCl_3)_5]^{3-} \qquad (2\text{-}10)$$

Ligand-substitution reactions, particularly those involving readily accessible square-planar Pd^{II} or Pt^{II} complexes are often used as model reactions for ranking ligands in order of their nucleophilicity (Eq. 2-11).

$$[PtX_4]^{2-} + Y^- \longrightarrow [PtX_3Y]^{2-} + X^- \qquad (2\text{-}11)$$

The reactions, which proceed by an S_N2 mechanism, give the following series [14] for the nucleophilicity of the incoming ligand Y:

$$F^- \sim H_2O \sim OH^- < Cl^- < Br^- \sim NH_3 \sim C_2H_4 < py < NO_2^- < N_3^- < I^- \sim SCN^- \sim R_3P$$

The Pt^{II} complex of Equation 2-11 has a soft, electrophilic center. Therefore, according to the HSAB (hard and soft acids and bases) concept, fast substitution reactions should occur with soft reagents such as phosphines, thiosulfate, iodide, and olefins. Ligand-exchange processes can often be explained in terms of the higher stability of the product complex:

$$[Co(NH_3)_5I]^{2+} + H_2O \rightleftharpoons [Co(NH_3)_5(H_2O)]^{3+} + I^- \qquad (2\text{-}12)$$
$$\mathbf{A} \mathbf{B}$$

The HSAB concept is helpful here, too: each soft or hard fragment strives for stabilization on a corresponding center (symbiotic effect). Complex **A** exhibits a hard/soft dissymmetry (NH_3 is hard, I^- soft), whereas in complex **B** the hard Co^{3+} center is stabilized exclusively by hard ligands.

The final example of ligand-exchange processes to be treated in this chapter is the heterolytic addition of reagents [T11]. Here a substrate XY undergoes addition to the metal center without changing the formal oxidation state or coordination number of the metal center. The molecular fragments X or Y are bound to the metal center as shown schematically in Equation 2-13.

$$M^xL_y + XY \longrightarrow M^xL_{y-1} + X + Y^+ + L^- \qquad (2\text{-}13)$$

Often, one anionic ligand is replaced by another, as in the addition of hydrogen to ruthenium(II) complexes:

$$[Ru^{II}Cl_2(PPh_3)_3] + H_2 \rightleftharpoons [Ru^{II}ClH(PPh_3)_3] + H^+ + Cl^- \qquad (2\text{-}14)$$

The activation of molecular hydrogen by Pt^{II}, Ru^{III}, and Pd/Sn catalyst systems can be explained analogously (Eq. 2-15).

$$[Pd(SnCl_3)_2(PPh_3)_2] \underset{}{\overset{H_2}{\rightleftharpoons}} [HPd(SnCl_3)(PPh_3)_2] + H^+ + SnCl_3^- \qquad (2\text{-}15)$$

In each case, hydrido metal compounds are formed as catalytically active complexes. Finally, it should be mentioned that in practice heterolytic addition can often not be distinguished from oxidative addition followed by reductive elimination, which is discussed later in this book.

2.1.2 Complex Formation [7]

An important step in the catalytic reactions of alkenes is the complexation of the substrate at the transition metal center. Differences in ability of olefins to coordinate can influence the selectivity of a catalytic process to such an extent that, for example, in a positionally isomeric olefins, the terminal olefins react preferentially to give the desired product.

In alkene complexes, the transition metal can have oxidation state 0 or higher. The olefin ligands are bound to the transition metal through one or more double bonds, the exact number depending on the number of free sites in the electron shell of the metal atom. Generally sufficient olefins or other Lewis bases are added to give the transition metal the electron configuration of the next higher noble gas, for example [T1]:

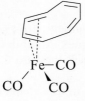

Cyclooctatetraene irontricarbonyl.
Formal Fe charge: 0.
Number of π electrons involved: 4

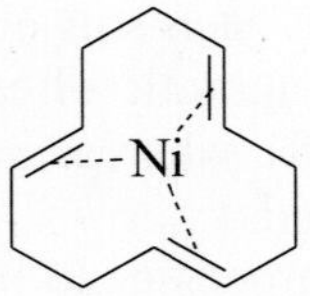

1,5,9-Cyclododecatriene nickel.
Formal Ni charge: 0.
Number of π electrons involved: 6

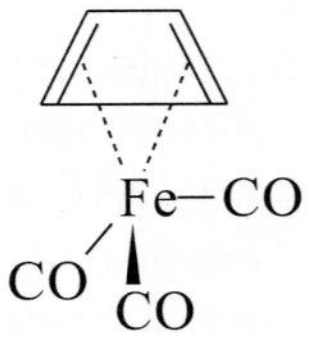

Cyclobutadiene irontricarbonyl.
Formal Fe charge: 0.
Number of π electrons involved: 4

In olefin–metal bonding, a distinction is made between σ and π bonding contribitions. The π bonding contribution for several metals increases as follows:

$$Al^{3+} \ll Ti^{4+} < Pt^{2+} < Ni^0$$
$$\phantom{Al^{3+} \ll\ } d^0 \phantom{\ <\ } d^8 \phantom{\ <\ } d^{10}$$

$$Pt^{II} < Rh^{I} < Fe^0 < Ni^0$$

$$Ag^{I} < Pd^{II} \ll Rh^{II} \sim Pt^{II} < Rh^{I}$$

———————————————————→
π-Bonding contribution (softness), stability of the metal olefin complex

Metal–olefin backbonding is particularly strong for soft metals that are rich in d electrons, but negligible at low d electron densities. For silver and palladium complexes, with their dominant σ contributions, the metal–olefin bond strength can be increased by donor substituents on the olefin, while in the case of the soft platinum complexes, it is increased by electron-withdrawing groups on the olefin. For a given metal ion, the σ-acceptor property becomes stronger with increasing positive charge. Thus Rh^{II} (d^7) is a stronger σ acceptor than Rh^{I} (d^8).

The coordination ability of olefins can also be compared. The following series, obtained for a nickel(0) complex, illustrates the importance of electronic effects in the olefin:

Ethene > Propene > 1-Butene ≈ 1-Hexene
$NC–CH=CH–CN > CH_2=CH–CN > CH_2=CH–COOCH_3 > CH_2=CH–CO–CH_3$
$> CH_2=CH–C_6H_5 > $ 1-Hexene $> CH_2=CH–O–(CH_2)_3–CH_3$

The strength of the nickel–olefin bond is increased by electron-withdrawing substituents such as cyano and carboxyl groups, and decreased by electron-donating groups. Donor ability increase in the series

Methyl < Ethyl < Alkoxyl

This behavior shows that for soft metal centers like Ni^0 (d^{10}), backbonding of electrons from filled d orbitals of the metal into empty olefin π^* orbitals (i.e., π bonding) dominates.

However, even relatively hard metal centers such as Ti^{III}, V^{III}, V^{II}, and Cr^{III} form unstable olefin complexes that are important intermediates in catalytic reactions.

With their delocalized π-electron system, allyl ligands can bond to metals in a manner similar to olefins. Allyl complexes have been detected as intermediates in catalytic processes involving propene or higher olefins and dienes. Examples include the cyclooligomerization of butadiene and the codimerization of butadiene with ethylene.

The abstraction of a hydrogen atom from an alkyl group next to a double bond (1,3 hydride shift) leads to formation of hydrido metal π-allyl complexes via intermediate σ-allyl compounds:

$$L_nM + \underset{}{\overset{H}{\underset{H}{R-C}}\underset{CH_2}{\overset{}{C}}H} \rightleftharpoons \underset{\sigma\text{-Allyl complex}}{\overset{R}{\underset{L_nM-CH_2}{\overset{CH}{\underset{}{\|}}\underset{}{CH}}}} \underset{Donor}{\overset{Acceptor}{\rightleftharpoons}} \underset{\pi\text{-Allyl complex}}{\overset{R-CH}{\underset{H\ CH_2}{L_nM-CH}}} \quad (2\text{-}16)$$

This reaction occurs mainly in metal complexes of low oxidation state. Typical examples of this class of compounds are $[Mn(\eta^3\text{-}C_3H_5)(CO)_4]$ and the dimer $[\{Pd(\eta^3\text{-}C_3H_5)Cl\}_2]$ with the structure:

$$\underset{CH_2}{\overset{CH_2}{HC-Pd}}\overset{Cl}{\underset{Cl}{<>}}\underset{CH_2}{\overset{CH_2}{Pd-CH}}$$

The equilibrium between σ- and π-allyl complexes can be influenced by the ligands. Thus strongly basic alkyl phosphine ligands favor the σ structure, as has been shown for allyl metal halide complexes of Pt^{II} and Ni^{II}. Soft π-acceptor ligands such as CO favor the formation of π-allyl complexes.

2.1.3 Acid–Base Reactions

According to the general acid–base concepts of Brønsted and Lewis, metal cations are generally regarded as acids. Therefore, transition metal cations or coordinatively unsaturated compounds can undergo addition of neutral or anionic nucleophiles to give cationic (Eq. 2-17), anionic (Eq. 2-18), and π-acceptor complexes (Eq. 2-19).

$$Cu^{2+} + 4\,NH_3 \longrightarrow [Cu(NH_3)_4]^{2+} \tag{2-17}$$

$$[PdCl_4]^{2-} + Cl^- \rightleftharpoons [PdCl_5]^{3-} \tag{2-18}$$

$$Ag(H_2O)_n^+ + C_2H_4 \rightleftharpoons Ag(H_2O)_{n-1}^+ \cdot C_2H_4 + H_2O \tag{2-19}$$

Another example of Lewis acid behavior is shown in Equation 2-20, in which an iridium complex takes up a CO ligand to form a dicarbonyl complex.

$$\textit{trans-}[IrCl(CO)(PPh_3)_2] + CO \rightleftharpoons [IrCl(CO)_2(PPh_3)_2] \tag{2-20}$$

In the reverse of dissociation, 16-electron species can add a ligand to give 18-electron complexes [19]:

$$\underset{16e}{[(acac)Rh(C_2H_4)_2]} + C_2H_4 \longrightarrow \underset{18e}{[(acac)Rh(C_2H_4)_3]} \tag{2-21}$$

The Brønsted theory states that the acid/base character of a compound depends on its reaction partner and is therefore not an absolute. An indication that transition metal compounds can act as bases is provided by the long-known protonation reactions of transition metal complexes, generally of low oxidation state. An example is cobalt carbonyl hydride, the true catalyst in many carbonylation reactions:

$$[Co(CO)_4]^- + H^+ \longrightarrow [HCo(CO)_4] \tag{2-22}$$

Metal basicity is also exhibited by phosphine and phosphite complexes of nickel(0), which can be protonated by acids of various strengths:

$$[Ni\{P(OEt)_3\}_4] + H^+ \longrightarrow [HNi\{P(OEt)_3\}_4]^+ \tag{2-23}$$

The hydride formation constant K for the general reaction of Equation 2-24

$$[NiL_4] + H^+ \overset{K}{\rightleftharpoons} [HNiL_4]^+ \tag{2-24}$$

can be strongly influenced by the donor character of the phosphine ligand L:

L = Ph$_2$P–CH$_2$–CH$_2$–PPh$_2$
PPh(OEt)$_2$
P(OEt)$_3$
P(OMe)$_3$
P(OCH$_2$–CH$_2$Cl)$_3$
P(OCH$_2$–CCl$_3$)$_3$

K (Eq. 2-24) ↑

With very good σ donors like diphosphines, nickel(0) becomes a strong metal base, and the corresponding hydride is highly stable. Phosphine ligands that remove electron density from the metal center lower the complex-formation constant. Thus trialkylphosphine ligands, which primarily act as σ donors, increase the electron density at nickel atom and give rise to strong metal bases.

For example, [Ni(PEt$_3$)$_4$] can be protonated with weak acids such as ethanol. An intermediate basicity is obtained with triarylphosphines and -phosphites, and protonation of the corresponding nickel complexes requires strong mineral acids. In contrast, PF$_3$ complexes exhibit negligible basicity because PF$_3$ is a strong electron acceptor, like CO.

Some transition metal hydrides are also strong bases; [π-Cp$_2$ReH] (Eq. 2-25) has a basicity comparable to that of ammonia.

$$[(\pi\text{-Cp})_2\text{ReH}] + \text{H}^+ \rightleftharpoons [(\pi\text{-Cp})_2\text{ReH}_2]^+ \qquad (2\text{-}25)$$

Many neutral carbonyl complexes can also be protonated; examples are given in Equations 2-26 and 2-27.

$$[\text{Os}_3(\text{CO})_{12}] + \text{H}^+ \rightleftharpoons [\text{HOs}_3(\text{CO})_{12}]^+ \qquad (2\text{-}26)$$

$$[\text{Ru(CO)}_3(\text{PPh}_3)_2] + \text{H}^+ \rightleftharpoons [\text{HRu(CO)}_3(\text{PPh}_3)_2]^+ \qquad (2\text{-}27)$$

Shriver has presented extensive data on transition metal basicity and described trends acoording to the position of the metal in the periodic table [15]. On the basis of IR spectroscopic data, the following rules can be drawn up:

1) Low oxidation states, especially negative ones or metal(0) complexes, increase the metal basicity. With increasing oxidation state, metals become more acidic.
2) Transition metal basicity increases from right to left in a period, and from top to bottom in a group; for example:

[Mn(CO)$_5$]$^-$ > [HFe(CO)$_4$]$^-$ > [Co(CO)$_4$]$^-$

[Re(CO)$_5$]$^-$ ≫ [Mn(CO)$_5$]$^-$

3) Electron-donor ligands such as phosphines increase the metal basicity:

[Fe(CO)$_4$PPh$_3$] > [Fe(CO)$_5$]

Another possibility for classifying transition metal basicity is complex formation with various Lewis acids. Numerous stable adducts can be regarded as the result of acid–base reactions of transition metal complexes (Eqs. 2-28 and 2-29).

$$[(\pi\text{-Cp})_2WH_2] + BF_3 \longrightarrow [(\pi\text{-Cp})_2WH_2\text{-}BF_3] \qquad (2\text{-}28)$$

$$[(\pi\text{-Cp})Co(CO)_2] + HgCl_2 \rightleftharpoons [(\pi\text{-Cp})Co(CO)_2\text{-}HgCl_2] \qquad (2\text{-}29)$$

The tendency of σ donors to increase basicity is also observed in complex formation. Thus the iron complexes $[Fe(CO)_3(EPh_3)_2]$ (E = P, As, Sb) form stable 1:1 adducts with $HgCl_2$ and $HgBr_2$. Adducts with the unsubstituted $[Fe(CO)_5]$ are less stable.

The oxidative addition reactions treated in the next section can in principle be interpreted as acid–base reactions. In the oxidative addition of hydrogen to a square-planar d^8 iridium complex (Eq. 2-30), the transition metal complex acts as an electron-providing metal base, and the substance undergoing addition can be regarded as an acid [10]:

$$[Ir^I Cl(CO)(PPh_3)_2] + H_2 \rightleftharpoons [Ir^{III} H_2 Cl(CO)(PPh_3)_2] \qquad (2\text{-}30)$$

Reducing agent   Oxidizing agent
(metal base)   (acid)

2.1.4 Redox Reactions: Oxidative Addition and Reductive Elimination

Coordinatively unsaturated transition metal complexes can in general add neutral or anionic nucleophiles. Oxidative addition to coordinatively unsaturated transition metal compounds has opened up undreamt of synthetic possibilities [18]. This reaction and its reverse — reductive elimination — are formally described by the following equilibrium:

$$L_x M^n + X\text{-}Y \underset{\text{Red. elim.}}{\overset{\text{Oxid. add.}}{\rightleftharpoons}} L_x M^{n+2} XY \qquad (2\text{-}31)$$

In general, the bonds of small covalent molecules XY (H–X, C–X, H–H, C–H, C–C, etc.) add to a low oxidation state transition metal, whose oxidation state then increases by two units. The reaction is mainly observed with complexes of d^8 and d^{10} transition metals (e. g., Fe^0, Ru^0, Os^0, Rh^I, Ir^I, Ni^0, Pd^0, Pt^0, Pd^{II}, Pt^{II}). The reaction can take two possible courses:

1) The molecules being added split into two η^1 ligands, which are both formally anionically bound to the metal center. One of the most thoroughly investigated

compounds is a square-planar iridium complex whose central atom gives up two electrons and is oxidized to Ir^{III} (Eq. 2-32).

$$\text{trans-}[Ir^{I}Cl(CO)(PPh_3)_2] + HCl \rightleftharpoons [Ir^{III}HCl_2(CO)(PPh_3)_2] \quad (2\text{-}32)$$
$$d^8 \qquad\qquad\qquad\qquad d^6$$

2) The molecules being added contain multiple bonds and are bound as η^2 ligands, without bond cleavage. The resulting complexes contain three-membered rings, as shown in Equations 2-33 and 2-34.

$$\begin{array}{c}\text{Cl}\diagdown\quad\diagup\text{PPh}_3\\ \text{Ir}\\ \text{Ph}_3\text{P}\diagup\quad\diagdown\text{CO}\end{array} + O_2 \rightleftharpoons \begin{array}{c}\text{PPh}_3\\ \text{Cl}\diagdown\ |\ \diagup\text{O}\\ \text{Ir}\\ \text{OC}\diagup\ |\ \diagdown\text{O}\\ \text{PPh}_3\end{array} \qquad (2\text{-}33)$$

$$Pt(PPh_3)_4 + (CF_3)_2C{=}O \longrightarrow \begin{array}{c}\text{CF}_3\\ \text{Ph}_3\text{P}\diagdown\ |\ \text{C}\diagup\text{CF}_3\\ \text{Pt}\\ \text{Ph}_3\text{P}\diagup\ \diagdown\text{O}\end{array} + 2\ PPh_3 \qquad (2\text{-}34)$$

The most important molecules for oxidative addition reactions are listed in Table 2-2.

Some illustrative examples of the possible types of reaction follow [17]:

$$[Pt^{II}Cl(SnCl_3)(PPh_3)_2] + H_2 \longrightarrow [Pt^{IV}ClH_2(SnCl_3)(PPh_3)_2] \qquad (2\text{-}35)$$

Equation 2-35 describes the addition and simultaneous activation of molecular hydrogen, an important step in homogeneous hydrogenation reactions.

$$[IrCl(CO)(PPh_3)_2] + R_3SiH \longrightarrow [IrHCl(SiR_3)(CO)(PPh_3)_2] \qquad (2\text{-}36)$$

Equation 2-36 can be regarded as a model reaction for the first step of hydrosilylation.

Anionic Rh^I complexes readily undergo addition of alkyl halides (Eq. 2-37).

$$[Rh^{I}(CO)_2I_2]^- + CH_3I \longrightarrow [CH_3Rh^{III}(CO)_2I_3]^- \qquad (2\text{-}37)$$

The formation of η^3-allyl complexes can also be regarded as an oxidative addition reaction. Proton abstraction from an olefin leads to the formally anionic allyl group (Eq. 2-38).

Table 2-2. Oxidative addition reactions on transition metal complexes; classification of the adding compounds

Bond cleavage (Addends dissociate)	No bond cleavage (Addends stay associated)
H_2	O_2
X_2	SO_2
HX (X = Hal, CN, RCOO, ClO_4)	CS_2
H_2S	$CF_2=CF_2$
C_6H_5SH	$(NC)_2C=C(CN)_2$
RX	$R-C\equiv C-R'$
RCOX	$(CF_3)_2CO$
RSO_2X	RNCO
R_3SnX	$R_2C=C=O$
R_3SiX	
HgX_2	
CH_3HgX	
$SiCl_4$	
C_6H_6	

R = alkyl, aryl, CF_3 etc.
X = Hal

$$\begin{array}{c} CH_3 \\ CH \\ \parallel\!\!-Ni^0-PF_3 \\ CH_2 \end{array} \rightleftharpoons \begin{array}{c} CH_2 \\ CH-Ni^{II} \\ CH_2 \end{array}\!\!\!\begin{array}{c} H \\ \\ PF_3 \end{array} \qquad (2\text{-}38)$$

Nickel, palladium, and platinum d^{10} complexes preferentially add polar reagents (acids; alkyl, acyl, and metal halides), whereby a ligand must dissociate to give a free coordination site (Eq. 2-39).

In Equation 2-40, addition of alkyl halide occurs first and is followed by dissociation of a phosphine ligand.

$$[Ni(CO)_4] + HCl \longrightarrow [Ni^{II}HCl(CO)_2] + 2\,CO \qquad (2\text{-}39)$$

$$RCl + [Pt(PEt_3)_3] \longrightarrow [RPtCl(PEt_3)_3] \longrightarrow R\!-\!\underset{\underset{PEt_3}{|}}{\overset{\overset{PEt_3}{|}}{Pt}}\!-\!Cl + PEt_3 \qquad (2\text{-}40)$$

With Brønsted acids the reaction can proceed via an ionic intermediate (Eq. 2-41).

$$[Pt^0(PPh_3)_3] \underset{}{\overset{HX}{\rightleftharpoons}} [Pt^{II}H(PPh_3)_3]^+X^- \underset{}{\overset{-PPh_3}{\rightleftharpoons}} [Pt^{II}HX(PPh_3)_2] \qquad (2\text{-}41)$$

Other oxidative addition reactions that involve simultaneous ligand dissociation can be explained by applying the 18-electron rule [19]. These usually involve coordinatively saturated 18-electron complexes (d electrons + electron lone pairs of the ligands), which must first lose a ligand to provide a vacant coordination site for oxidative addition (Eqs. 2-42 and 2-43).

$$[Ru^0(CO)_3(PPh_3)_2] + I_2 \longrightarrow [Ru^{II}I_2(CO)_2(PPh_3)_2] + CO \quad (2\text{-}42)$$
$$d^8 \quad 6e \quad 4e$$

$$[Mo^0(CO)_4 bipy] + HgCl_2 \longrightarrow [Mo^{II}Cl(HgCl)(CO)_3 bipy] + CO \quad (2\text{-}43)$$
$$d^6 \quad 8e \quad 4e$$

The mechanisms of oxidative addition reactions, which in some cases are complicated, will not be discussed further here. What is of interest, however, is the general reactivity of the transition metals. For the metals of group VIII, the trend shown in Scheme 2-1 was found for oxidative addition reactions of the type $d^8 \rightarrow d^6$, given the same ligands.

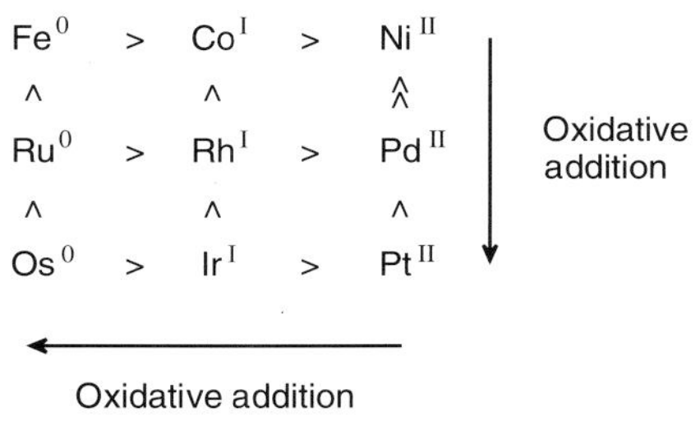

```
Fe⁰    >   Co^I   >   Ni^II
 ∧          ∧          ∧
Ru⁰    >   Rh^I   >   Pd^II       Oxidative
 ∧          ∧          ∧          addition
Os⁰    >   Ir^I   >   Pt^II
                                  ↓

←──────────────────────────
Oxidative addition
```

Scheme 2-1. Tendency to undergo oxidative addition for the metals of groups 8–10

The tendency to undergo oxidative addition increases from top to bottom in a group and from right to left in a period, as does the metal basicity. This is shown by numerous empirical orders of reactivity [10]:

$[Ir^I(PPh_2Me)_2(CO)Cl] > [Ir^I(PPh_3)_2(CO)Cl] > [Rh^I(PPh_3)_2(CO)Cl];$

$[Ir^I(PPh_3)_2(CO)Cl] > [Pt^{II}(PPh_3)_2(CO)Cl]^+;$

$Ir^I > Pt^{II} \gg Au^{III}$

Ligand effects are of major importance in oxidative addition reactions. Increasing donor character of a ligand increases the electron density at the metal center and favors oxidative addition. This means that electron-releasing (basic) ligands make the metal base stronger, while electron-withdrawing ligands weaken it. Some examples for ligand influences are given in the following:

PEt$_3$ > PPh$_3$;
PPhEt$_2$ > PPh$_2$Me > PPh$_3$;
PPhMe$_2$ > PPh$_3$ > CO;
I > Br > Cl

←───────────────────────────
σ-donor strength, oxidative addition

Alkylphosphines, which are good σ donors, facilitate oxidative addition, while π-acceptor ligands make it more difficult. However, steric effects must also be considered. For instance, a low reaction rate is observed for the strongly basic, bulky ligand tri-*tert*-butylphosphine.

For the square-planar iridium complex [Ir(CO)(PPh$_3$)$_2$X], the ligand effects shown in Equation 2-44 were found.

$$[Ir(CO)(PPh_3)_2X] \begin{array}{c} \xrightarrow{+H_2} [Ir(CO)(H)_2(PPh_3)_2X] \quad X = I > Br > Cl; \\ F > Br > Cl \\ \xrightarrow{+O_2} [Ir(CO)(O_2)(PPh_3)_2X] \quad X = I > Br, Cl \end{array} \quad (2\text{-}44)$$

Although the fluoro ligand lowers the σ basicity, it is also a good π donor that increases the π basicity of the metal, and it is the latter effect that predominates in the oxidative addition of hydrogen (Eq. 2-44).

The complex [Ir(CO)(PPh$_3$)$_2$Cl] reacts with hydrogen at room temperature to give a dihydride complex, but the analogous rhodium complex [Rh(CO)(PPh$_3$)$_2$Cl] does not; only the chloro complex [Rh(PPh$_3$)$_3$Cl] forms a hydrogen adduct. The comparison once again demonstrates the effect of the metal basicity (Ir > Rh), but also the influence of the ligands: donor ligands (PPh$_3$) increase the reactivity, and the stronger π acid CO lowers it. The even stronger π acid N$_2$ behaves in a similar manner compared to CO: the dinitrogen complex [IrCl(PPh$_3$)$_2$N$_2$] does not undergo addition of hydrogen. If σ-donor and π-acceptor ligands are approximately in balance, as in the complex [Ni(CO)$_2$(PPh$_3$)$_2$], then the compound is relatively stable and unreactive towards oxidative addition. Dissociation of ligands is also more difficult.

As would be expected, reductive elimination, the reverse of oxidative addition, is favored by ligands that lower the electron density at the metal center. The last step of a catalytic cycle is often an irreversible reductive elimination in which the product is released. Equation 2-45 shows the formation of an alkane from a alkyl hydride complex.

$$[Rh^{III}ClH(C_2H_5)(PPh_3)_3] \longrightarrow [Rh^{I}Cl(PPh_3)_3] + C_2H_6 \qquad (2\text{-}45)$$

In the rhodium-catalyzed carbonylation of methanol via methyl iodide, acetyl iodide is formed by reductive elimination from an anionic rhodium^{III} acyl complex [T14]:

$$\left[\begin{array}{c} \text{I} \\ \text{I} - \overset{|}{\underset{|}{\text{Rh}^{\text{III}}}} - \overset{\text{O}}{\overset{\|}{\text{C}}} - \text{CH}_3 \\ \text{OC} - \overset{|}{\underset{|}{\text{I}}} - \text{CO} \end{array}\right]^{-} \longrightarrow [\text{Rh}^{\text{I}}\text{I}_2(\text{CO})_2]^{-} + \text{CH}_3\text{C}\overset{\diagup\text{O}}{\diagdown\text{I}} \qquad (2\text{-}46)$$

In the same manner, aldehydes are formed as the final products of cobalt-catalyzed hydroformylation:

$$\text{Co}^{\text{III}}(\text{H})_2(\overset{\text{O}}{\overset{\|}{\text{C}}}-\text{R})(\text{CO})_2\text{L} \longrightarrow \text{Co}^{\text{I}}\text{H}(\text{CO})_2\text{L} + \text{R}-\text{C}\overset{\diagup\text{O}}{\diagdown\text{H}} \qquad (2\text{-}47)$$

Reductive elimination is generally not the rate-determining step in a catalytic process.

Besides oxidative addition, there is also another type of homolytic addition in transition metal chemistry [11]. By definition, in this type of reaction, a substrate XY adds to two metal centers in such a way that the formal oxidation state of each metal increases or decreases by one unit (Eq. 2-48).

$$2 \, M^n L_x + X\text{–}Y \longrightarrow 2 \, M^{n \pm 1}(X)(Y)L_x \qquad (2\text{-}48)$$

An industrially important example is the activation of $[\text{Co}_2(\text{CO})_8]$ with hydrogen (Eq. 2-49), the resulting complex being the active catalyst in carbonylation reactions.

$$[\text{Co}_2^0(\text{CO})_8] + \text{H}_2 \rightleftharpoons [2 \, \text{HCo}^{-\text{I}}(\text{CO})_4] \qquad (2\text{-}49)$$

In this case, the metal is assigned a formal negative oxidation state since the product behaves as a strong acid and should therefore be regarded as a hydro compound. As with oxidative addition, electron-donating ligands such as trialkylphosphines increase the rate of reaction. For hydrogen addition:

$$\text{Co}_2(\text{CO})_6(\text{PBu}_3)_2 > \text{Co}_2(\text{CO})_8$$

When hydrogen is passed into an aqueous cobalt cyanide solution, hydridopentacyanocobalt ions are formed (Eq. 2-50) and can be used for the reduction of organic and inorganic substrates:

$$2 \, [\text{Co}^{\text{II}}(\text{CN})_5]^{3-} + \text{H}_2 \rightleftharpoons 2 \, [\text{Co}^{\text{III}}\text{H}(\text{CN})_5]^{3-} \qquad (2\text{-}50)$$

Another example is the addition of hydrogen halides to metal–metal bonds, as in, for example, $[\text{Mo}_2^{\text{II}}X_8]^{4-}$ (Eq. 2-51; X = Cl, Br). This type of reaction could be of interest for catalysis with clusters.

$$[\text{Mo}_2^{\text{II}}X_8]^{4-} + \text{HX} \longrightarrow [\text{Mo}_2^{\text{III}}(\text{H})X_8]^{3-} + X^- \qquad (2\text{-}51)$$

Oxidative coupling, as defined by Tolman, is a reaction in which the oxidation state of the metal increases by two units and the coordination number remains unchanged [19]. Hence it is a special case of oxidative addition. Many C–C coupling reactions proceed according to this scheme, in which an unsaturated ligand accepts two electrons from the transition metal. The resulting dicarbanion is bound to the metal center in a chelating fashion (Eqs. 2-52 and 2-53).

$$[(C_2H_4)Ni^0(PPh_3)_2] + 2\ CF_2=CFH \longrightarrow \underset{d^8}{(Ph_3P)_2Ni^{II}(CF_2-CFH-CFH-CF_2)} + C_2H_4 \quad (2\text{-}52)$$

$$\underset{d^8}{(CH_2=CH-X)_2Fe(CO)_3} \rightleftharpoons \underset{d^6}{[\text{cyclo-}(CHX-CH_2-CH_2-CHX)]Fe(CO)_3} \quad X = -COOCH_3 \quad (2\text{-}53)$$

In the above two examples, oxidative coupling of two olefin molecules occurs. It is likely that the catalysis of numerous cyclooligomerization reactions of unsaturated hydrocarbons proceeds in this manner, as shown for the example of butadiene in Equation 2-54.

$$Ni^0(CDT)PR_3 \xrightarrow{-CDT} \underset{\mathbf{A}}{[R_3P-Ni^0(\text{butadiene})_2]} \rightleftharpoons \underset{\mathbf{B}}{[R_3P-Ni^{II}(\text{C}_8\text{-chain})]} \longrightarrow \text{1,5-COD} + \text{4-vinylcyclohexene} \quad (2\text{-}54)$$

First, a ligand displacement reaction with butadiene gives a nickel(0) π complex **A**, which undergoes oxidative coupling to give the metal-containing ring **B**, a π-allyl σ-alkyl complex. Finally, reductive elimination gives the main products 1,5-cyclooctadiene and 4-vinylcyclohexene.

2.1.5 Insertion and Elimination Reactions

Insertion reactions play an important role in the catalysis of C–C and C–H coupling [1]. Insertion of CO and olefins into metal–alkyl and metal–hydride bonds are of major importance in industrial chemistry. Insertion reactions take place according to the following scheme:

$$L_xM^n-X + YZ \longrightarrow L_xM^n-(YZ)-X \qquad (2\text{-}55)$$

X = H, C, N, O, Cl, metal
YZ = CO, olefin, diene, alkyne, nitrile, etc.

Initially, a molecule XY is inserted into an M–X bond without changing the formal oxidation state of the metal M. A simple example is the insertion of an olefin into a Pt–H bond to give an alkyl complex (Eq. 2-56).

$$\begin{array}{c}\text{PEt}_3\\|\\\text{Cl}-\text{Pt}-\text{H}\\|\\\text{PEt}_3\end{array} \underset{\pi \text{ complex formation}}{\overset{C_2H_4}{\rightleftharpoons}} \begin{array}{c}CH_2{=}CH_2\\|\\\text{PEt}_3\\|\\\text{Cl}-\text{Pt}-\text{H}\\|\\\text{PEt}_3\end{array} \underset{cis \text{ insertion}}{\rightleftharpoons} \begin{array}{c}\text{PEt}_3\\|\\\text{Cl}-\text{Pt}-CH_2CH_3\\|\\\text{PEt}_3\end{array}$$

$$(2\text{-}56)$$

Formally speaking, the above insertion reaction is a nucleophilic attack of a base (hydride ion) on a positively polarized olefin (coordination of the olefin lowers its electron density and thus facilitates nucleophilic attack).

Olefin insertion is particularly facile in the case of the complexes $[PtH(SnCl_3)(PR_3)_2]$. The soft π-acceptor ligand $[SnCl_3]^-$ stabilizes the metal–hydride bond (symbiosis of soft ligands) and hence catalyzes the insertion reaction as a preliminary step of hydrogenation. Pt/Sn systems are known to be good hydrogenation catalysts.

Equation 2-57 describes the insertion of acetylene into a Pt–H bond to give a vinylplatinum complex.

$$[HPtCl(PEt_3)_2] + RC{\equiv}CR \longrightarrow \begin{array}{c}R\\ \diagdown\\ C-PtCl(PEt_3)_2\\ \diagup\!\!\diagup\\ C\\ \diagup\ \ \diagdown\\ H\ \ \ \ R\end{array} \qquad (2\text{-}57)$$

An important step in industrial carbonylation reactions is the insertion of CO into metal–carbon bonds (Eq. 2-58) [3, 4], which was described as early as 1957.

$$[R\text{-}Mn(CO)_5] + CO \longrightarrow [R\text{-}CO\text{-}Mn(CO)_5] \qquad (2\text{-}58)$$

Formally, CO inserts into the polarized metal–carbon bond to give an acyl metal complex. However, it has been shown that in fact an alkyl group migration to a CO group coordinated in the *cis* position occurs. This migration probably occurs via a three-center transition state (Eq. 2-59).

$$\overset{R}{\underset{|}{M}}-C\equiv O \longrightarrow \left[\overset{R}{M\cdots C\equiv O}\right]^{\ddagger} \longrightarrow M-C\overset{R}{\underset{\diagdown O}{\diagup}} \qquad (2\text{-}59)$$

For the carbonylation of manganese complexes of the type $[RMn(CO)_5]$, the following influence of the substituents has been found:

$$R = \underset{\xleftarrow{\text{CO insertion in [R-Mn(CO)}_5\text{], reactivity}}}{n\text{-}C_3H_7 > Et > Ph > Me \gg CH_2Ph, CF_3, Cl}$$

The trend can be explained as follows: the electron-releasing alkyl groups cause a stronger polarization of the metal–carbon bond, but more electronegative electron-withdrawing ligands lower the reaction rate. This σ effect has been confirmed by model calculations [1].

If the stability and reactivity of the metal complexes in a triad of the periodic table are compared, two counteropposing trends become apparent [3]:

4d ↓ 5d 6d	Stability of M–C bond, polarizability, softness of metal	Reactivity in CO insertion reactions ↑

The harder metals at the top of the groups are more reactive towards carbonyl insertion. Thus iridium carbonyl complexes are less reactive than the rhodium and cobalt homologues. The following also applies:

$Pd^{II} > Pt^{II}$; $Mn > Re$; $Cr > Mo > W$

The influence of nucleophilic ligands on the CO insertion reaction has been investigated for molybdenum complexes (Eq. 2-60).

$$[CH_3\text{-}Mo(\pi\text{-}Cp)(CO)_3] + L \longrightarrow [CH_3\text{-}CO\text{-}Mo(\pi\text{-}Cp)(CO)_2L] \qquad (2\text{-}60)$$

In the nonpolar solvent toluene, the reaction rate decreases in the sequence:

$L = P(^nBu)_3 > P(O\text{-}^nBu)_3 > PPh_3 > P(OPh)_3 > AsH_3$

As expected, the alkylphosphines of higher σ basicity activate the CO insertion reaction, as do polar solvents such as ether, which can increase the reaction rate by a factor of 10^3 to 10^4. Examples of very fast insertions are the reactions of carbenes with M–H, M–C, and M–Cl bonds (Eqs. 2-61 and 2-62).

$$[HMn(CO)_5] + :CH_2 \longrightarrow [CH_3Mn(CO)_5] \quad (2\text{-}61)$$
(from CH_2N_2)

$$[IrCl(CO)(PPh_3)_2] + :CH_2 \longrightarrow [Ir(CH_2Cl)(CO)(PPh_3)_2] \quad (2\text{-}62)$$
(from CH_2N_2)

The insertion reactions discussed below can be explained well by using the HSAB concept [7].

Carboxylation reactions with the hard Lewis acid CO_2 are of potential interest for future industrial syntheses. Understandably, the hardest alkyl metal compounds are required to facilitate reactions of the type:

$$\overset{\delta+ \;\; \delta-}{M-R} + \overset{\delta- \;\; \delta+}{O=C=O} \longrightarrow M-O-\underset{R}{\overset{O}{C}} \quad (2\text{-}63)$$

This is shown by the following reactions:

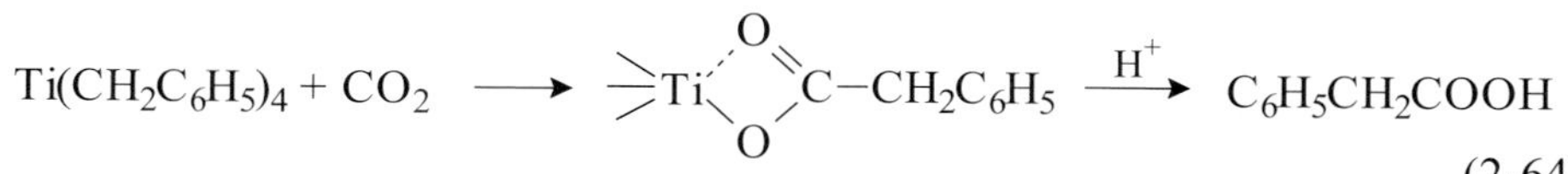

$$Ti(CH_2C_6H_5)_4 + CO_2 \longrightarrow Ti\underset{O}{\overset{O}{\diagup\!\!\!\diagdown}}C-CH_2C_6H_5 \xrightarrow{H^+} C_6H_5CH_2COOH \quad (2\text{-}64)$$

$$[W(NMe_2)_6] + 3\,CO_2 \longrightarrow [W(NMe_2)_3(O_2CNMe_2)_3] \quad (2\text{-}65)$$

The benzyl complex of titanium is a very hard starting material (Eq. 2-64), as is the tungsten dialkylamide (Eq. 2-65).

Elimination reactions can proceed as the direct reverse of insertion reactions. Thus the elimination of CO from acyl complexes (Eq. 2-66) and of CO_2 from carboxylates (Eq. 2-67) can result in the formation of metal–aryl bonds. Such eliminations occur under the influence of heat and light.

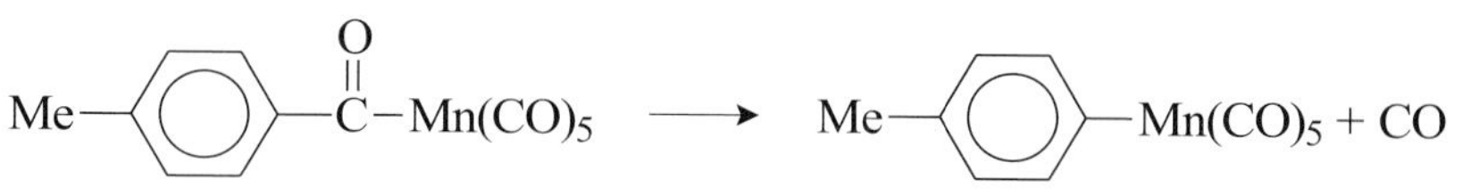

$$Me\text{–}\langle C_6H_4\rangle\text{–}\overset{O}{\underset{\|}{C}}\text{–}Mn(CO)_5 \longrightarrow Me\text{–}\langle C_6H_4\rangle\text{–}Mn(CO)_5 + CO \quad (2\text{-}66)$$

$$[Ni(bipy)(COOPh)_2] \longrightarrow [Ni(bipy)Ph_2] + 2\,CO_2 \quad (2\text{-}67)$$

Decomposition reactions can proceed by another mechanism, namely, β elimination. In particular, β-hydride elimination is an important mechanism for the decomposition of σ-organyl complexes (Eq. 2-68).

$$L_xM-X-CH_2-R \rightleftharpoons \underset{\underset{R}{H\text{-----}CH}}{L_xM\text{-----}X} \rightleftharpoons L_xM-H + R-CH=X \quad (2\text{-}68)$$
$$\alpha\beta$$

The products of this intramolecular rearrangement are a metal hydride complex and a stable unsaturated compound. Formally, it can be regarded as a competive reaction between the metal center and the the unsaturated ligand fragment or the soft base H^-. Generally, such elimination reactions are favored by high reaction temperatures and low oxidation states of the transition metal. The HSAB concept predicts that β-hydride elimination is favored when the metal center is made softer and the unsaturated product harder. For example, acetaldehyde is more readily eliminated than ethylene (Eq. 2-69).

$$\begin{array}{c}
\text{Ph}_3\text{P}\diagdown\phantom{\text{Pt}}\diagup\text{O-CH}_2\text{CH}_3 \\
\phantom{\text{Ph}_3\text{P}}\text{Pt} \\
\text{Ph}_3\text{P}\diagup\phantom{\text{Pt}}\diagdown\text{Cl}
\end{array}
\longrightarrow
\begin{bmatrix}
\text{Ph}_3\text{P} & \text{O} & & \text{H} \\
| & | & & \diagup \\
\text{Cl}-\text{Pt} & \cdots & \text{C} & \\
| & & & \diagdown \\
\text{Ph}_3\text{P} & \text{H} & & \text{CH}_3
\end{bmatrix}$$

$$\longrightarrow \begin{array}{c}
\text{Ph}_3\text{P}\diagdown\phantom{\text{Pt}}\diagup\text{H} \\
\phantom{\text{Ph}_3\text{P}}\text{Pt} \\
\text{Cl}\diagup\phantom{\text{Pt}}\diagdown\text{PPh}_3
\end{array} + \text{CH}_3\text{C}\begin{array}{c}\diagup\text{O}\\\diagdown\text{H}\end{array} \quad (2\text{-}69)$$

Alkoxy complexes of transition metals are generally less stable because of the presence of a hard(OR)/soft(M) dissymmetry. The elimination of alkene from the more stable alkyl metal complexes generally requires drastic conditions (Eq. 2-70).

$$\textit{trans-}[(\text{PEt}_3)_2\text{PtCl}(\text{C}_2\text{H}_5)] \xrightleftharpoons[95°\text{C, 40 bar}]{180°} \textit{trans-}[(\text{PEt}_3)_2\text{PtHCl}] + \text{CH}_2=\text{CH}_2 \quad (2\text{-}70)$$

This can be explained in terms of the softness of the R^- group, which stabilizes the complex.

β-Hydride elimination is favored by a free coordination site at the metal center, as exemplified by the complex [^{n}Bu$_2$Pt(PPh$_3$)$_2$], thermal decomposition of which is inhibited by the presence of an excess of triphenylphosphine. This shows that dissociation of a PPh$_3$ ligand is required for the elimination reaction to occur.

Understandably, metal complexes containing alkyl groups that have no hydrogen atoms in the β position, such as methyl, benzyl, and neopentyl, are more stable than other alkyl derivatives. The decomposition of metal alkyls — the reverse of olefin insertion — is of importance in the transition metal catalyzed isomerization of olefins and as a chain-termination reaction in olefin polymerization. The α elimination reaction should also be mentioned here. It is mainly of importance in W and Mo complexes [T11]. Extraction of an α hydrogen atom from methyl compounds leads to intermediate alkylidene species:

$$W-CH_3 \rightleftharpoons \overset{H}{\underset{}{W}}=CH_2 \qquad (2\text{-}71)$$

The decomposition of methyltungsten compounds with formation of methane is believed to involve such hydrido carbene intermediates:

$$Cl_4W\begin{array}{c}CH_3\\CH_3\end{array} \rightleftharpoons Cl_4\overset{H}{\underset{CH_3}{W}}=CH_2 \longrightarrow Cl_4W=CH_2 + CH_4 \qquad (2\text{-}72)$$

In ethyltungsten complexes, for which β elimination of alkene would be expected, the α elimination according to Equation 2-73 is favored.

$$Cl_5W-\overset{H}{\underset{H}{C}}-CH_3 \longrightarrow Cl_5W=C\begin{array}{c}H\\CH_3\end{array} \qquad (2\text{-}73)$$

Metal carbene complexes are discussed as intermediates in metathesis reactions (olefin disproportionation).

2.1.6 Reactions at Coordinated Ligands

Nucleophilic attack on coordinated ligands is a widely encountered type of reaction. For example, carbonyl complexes are readily attacked by various nucleophiles, including OH^-, OR^-, NR_3, NR_2^-, H^-, and CH_3^-. A well-known example is the base reaction of carbonyl complexes (Eq. 2-74).

$$(CO)_4Fe-\overset{\delta+}{C}\equiv\overset{\delta+}{O}| + \overset{\delta-}{OH^-} \longrightarrow \left[(CO)_4-Fe-C\begin{array}{c}O\\\overline{O}-H\end{array}\right]$$

$$\xrightarrow{-CO_2} (CO)_4Fe^{2-} + H^+ \qquad (2\text{-}74)$$

The carbonyl carbon atom of carbonyl complexes is an electrophilic center that according to the HSAB concept can be regarded as a hard acid (similar to H^+). The attack of the hard base OH^- initially gives a hydroxycarbonyl species, which, however, is unstable and loses CO_2, forming a cabonyl metallate anion. The effectiveness of nucleophiles with respect to the carbonyl carbon atom decreases in the following sequence [T12]:

$$EtO^- > PhO^- > OH^- > AcO^- > N_3^- > F^- > H_2O > Br^- \sim I^-$$

Thus the hard oxygen bases react more readily with metal carbonyls than the softer bases. Alkoxide ions attack coordinated carbon monoxide to form alkoxy carbonyl complexes. This reaction (Equation 2-75) has been observed for many complexes of the metals Mn, Re, Fe, Ru, Os, Co, Rh, Ir, Pd, Pt, and Hg.

$$[Ir(CO)_3(PPh_3)_2]^+ \underset{H^+}{\overset{CH_3O^-}{\rightleftharpoons}} Ir\left(C\overset{O}{\underset{OCH_3}{\diagdown}}\right)(CO)_2(PPh_3)_2 \qquad (2\text{-}75)$$

As a final example of ligand reactions of carbonyls, the rhodium-catalyzed CO conversion reaction will be mentioned. Anionic rhodium complexes such as $[Rh(CO)_2I_2]^-$ undergo nucleophilic attack by water with formation of CO_2 (Eq. 2-76).

$$Rh^{III}-\overset{\delta+}{C}\equiv O + \overset{\frown}{O}H_2 \longrightarrow Rh^I + CO_2 + 2\,H^+ \qquad (2\text{-}76)$$

The resulting rhodium^I carbonyl complex can be oxidized back to rhodium^{III} by protons (Eq. 2-77); the final products are CO_2 and H_2.

$$Rh^I(CO) + 2\,H^+ \longrightarrow Rh^{III}(CO) + H_2 \qquad (2\text{-}77)$$

Electrophilic attack on a ligand is often observed for complexes of olefins and aromatic compounds. The electrophilic or nucleophilic behavior of these π ligands can be predicted on the basis of the σ/π bonding model. The olefin reacts not only as a σ donor but also as a π acceptor.

When π backbonding from the metal to the olefin predominates, electrons flow from the metal to the olefin, which then exhibits carbanion behavior. In this case, electrophilic attack readily occurs [21]. Low metal oxidation states (0, +1) and anionic complexes favor electrophilic attack on coordinated ligands. Of course, substituent effects also play a role: electron-withdrawing groups can inhibit electrophilic attack.

The reactivity of ligands towards nucleophiles increases for higher oxidation states of the metal (+2, +3) and for cationic complexes. The following examples illustrate this rule:

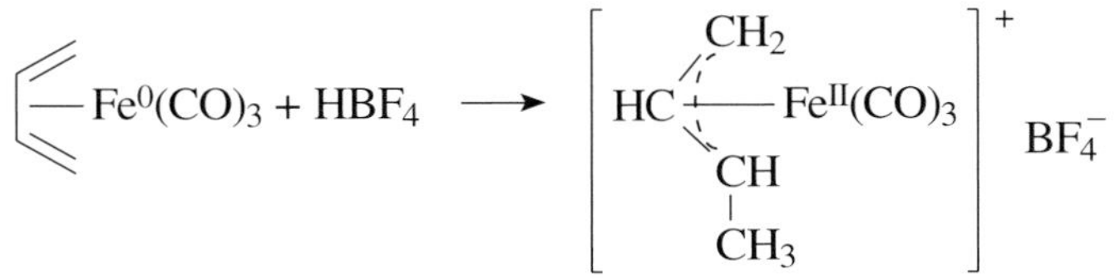

$$\text{(butadiene)}-Fe^0(CO)_3 + HBF_4 \longrightarrow \left[HC \overset{CH_2}{\underset{\underset{CH_3}{CH}}{\Big|}} Fe^{II}(CO)_3 \right]^+ BF_4^- \quad (2\text{-}78)$$

In Equation 2-78 a proton attacks a diene ligand to give an π-allyl complex.

In the following reaction hydride ions are removed from a diene complex to give an arene complex (Eq. 2-79)

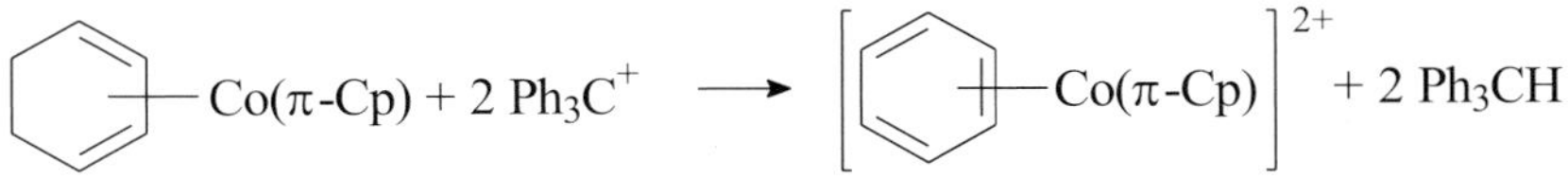

$$\text{(cyclohexadiene)}-Co(\pi\text{-}Cp) + 2\,Ph_3C^+ \longrightarrow \left[\text{(benzene)}-Co(\pi\text{-}Cp) \right]^{2+} + 2\,Ph_3CH \quad (2\text{-}79)$$

The final example (Eq. 2-80) shows that alkyl complexes can undergo irreversible cleavage of alkane on reaction with acids.

$$(\pi\text{-}Cp)(CO)_2Fe\text{-}CH_2CH_3 + HCl \longrightarrow (\pi\text{-}Cp)(CO)_2FeCl + CH_3\text{-}CH_3 \quad (2\text{-}80)$$

The dual activation of ligands is also of interest for catalytic reactions. Carbon monoxide is classified according to the HSAB concept as a very soft Lewis base. Thus activation occurs by coordination of the C atom to soft transition metals. The CO ligand can, however, react as a hard Lewis base via the oxygen atom. Sufficiently hard Lewis acids A can therefore coordinate to the oxygen atom and further weaken the C–O bond [16]:

M–CO–A
s   h

Hard Lewis acids (A = $AlCl_3$, AlR_3, BCl_3) preferentially attack bridging CO ligands:

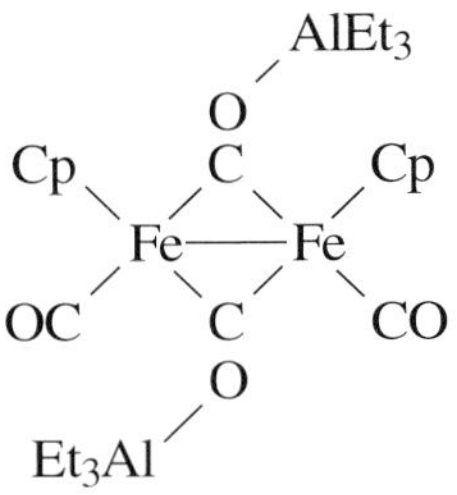

$$\begin{array}{c}
\text{AlEt}_3 \\
\diagup \\
\text{Cp} \quad \overset{O}{\underset{}{C}} \quad \text{Cp} \\
\diagdown \diagup \diagdown \diagup \\
\text{Fe}\text{—}\text{Fe} \\
\diagup \diagdown \diagup \diagdown \\
\text{OC} \quad \underset{O}{C} \quad \text{CO} \\
\diagup \\
\text{Et}_3\text{Al}
\end{array}$$

But examples are also known for the coordination of Lewis acids to terminal CO ligands:

$$\begin{array}{c} \text{N} \quad\; \text{PPh}_3 \\ \diagdown \;\;|\;\; \diagup \text{CO}-\text{AlR}_3 \\ \text{Mo} \\ \diagup \;\;|\;\; \diagdown \text{CO}-\text{AlR}_3 \\ \text{N} \quad\; \text{PPh}_3 \end{array}$$

Bifunctional activation of CO leads to carbene-like resonance structures of the type shown in Equation 2-81.

$$L_nM-C\equiv O| + AlX_3 \longrightarrow L_nM=C=O{\diagdown \atop AlX_3} \qquad (2\text{-}81)$$

The attack of the electrophile is particularly facile in the case of anionic and other electron-rich complexes. The dual activation of CO ligands weakens the C–O bond and lowers the C–O stretching frequency in the IR spectrum.

It has been found that the presence of Lewis acids or protons can accelerate carbonyl insertion reactions, providing another possibility of modifying catalysts. Mixtures of transition metal carbonyls and Lewis acids could in future be of interest as catalysts for CO hydrogenation, for example, in Fischer–Tropsch reactions (Eq. 2-82) [T11].

$$\overset{\text{H}}{\underset{|}{\text{M}}} + |\text{C}\equiv \text{O}| + \text{A} \longrightarrow \overset{\text{H}}{\underset{|}{\text{M}-\text{C}}}=\bar{\text{O}}| \rightarrow \text{A} \qquad (2\text{-}82)$$

The hard electron acceptor A lowers the electron density in the CO moiety, facilitating attack of the hydride on the carbonyl carbon atom.

In dinitrogen complexes, polarization of the N_2 ligand occurs with an electron-rich metal center on one side and a strongly polarizing, hard cation on the other [13]. As an example, the following resonance structure can be given for the cobalt complex $[\{KCo(N_2)(PMe_3)_3\}_6]$:

$$L_3Co^-=N^+=\underline{\bar{N}}^- \leftarrow K^+$$

This ligand polarization favors electrophilic attack on the terminal nitrogen atom. Reactions that activate dinitrogen are of interest as the basis for the fixation of nitrogen as ammonia.

Exercises for Section 2.1

Exercise 2.1

What is the oxidation state of the transition metal in the following complexes?

a) $[V(CO)_6]^-$
b) $[Mn(NO)_3CO]$
c) $[Pt(SnCl_3)_5]^{3-}$
d) $[RhCl(H_2O)_5]^{2+}$
e) $[(\pi\text{-}C_5H_5)_2Co]^+$
f) $[H_2Fe(CO)_4]$
g) $[Ni_4(CO)_9]^{2-}$
h) $[Fe(CO)_3(SbCl_3)_2]$
i) $O_2[PtF_6]$
j) $[HRh(CO)(PPh_3)_3]$

Exercise 2.2

What type of reaction is occurring in the following:

a) $trans\text{-}[PtCl_2(PEt_3)_2] + HCl \longrightarrow [PtCl_3H(PEt_3)_2]$
b) $[W(CH_3)_6] \longrightarrow 3\,CH_4 + \text{„}W(CH_2)_3\text{"}$
c) $[Co(H)_2\{P(OMe)_3\}_4]^+ \rightleftharpoons [Co\{P(OMe)_3\}_4]^+ + H_2$
d) $[(\pi\text{-}C_5H_5)W(CO)_3]Na + CH_3I \longrightarrow [(\pi\text{-}C_5H_5)W(CO)_3Me] + NaI$
e) $[IrCl(CO)(PPh_3)_2] + Me_3O^+BF_4^- \longrightarrow [IrMeCl(CO)(PPh_3)_2]^+BF_4^- + Me_2O$
f) $[(\pi\text{-}C_5H_5)Mn(CO)_3] + C_2F_4 \longrightarrow [(\pi\text{-}C_5H_5)Mn(CO)_2C_2F_4] + CO$

g) 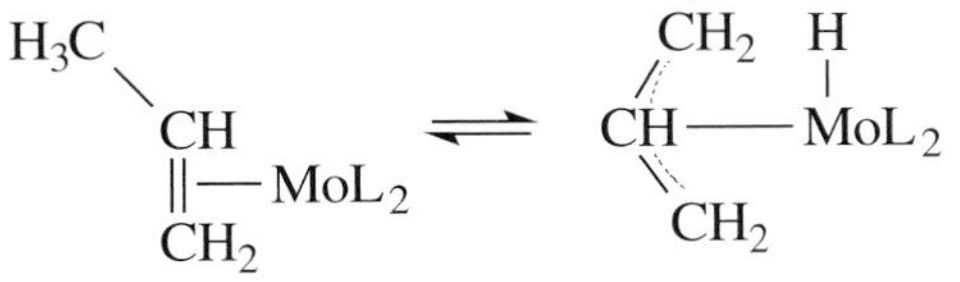
$$\begin{array}{c} H_3C \\ \diagdown \\ CH \\ || \text{---} MoL_2 \\ CH_2 \end{array} \rightleftharpoons \begin{array}{c} CH_2 \quad H \\ \diagup \quad | \\ CH \text{-----} MoL_2 \\ \diagdown \\ CH_2 \end{array}$$

h) $[(\pi\text{-}C_5H_5)_2ReH] + BF_3 \rightleftharpoons [(\pi\text{-}C_5H_5)_2ReHBF_3]$

Exercise 2.3

Classify the following reactions by means of the oxidation states:

a) $CoCO_3 + 2\,H_2 + 8\,CO \longrightarrow [Co_2(CO)_8] + 2\,CO_2 + 2\,H_2O$
b) $2\,[Fe(CO)_2(NO)_2] + I_2 \longrightarrow [FeI(NO)_2]_2 + 4\,CO$
c) $[Pt(PPh_3)_3] + CH_3I \longrightarrow [CH_3PtI(PPh_3)_2] + PPh_3$
d) $[Mn(CO)_5Cl] + AlCl_3 + CO \longrightarrow [Mn(CO)_6]^+[AlCl_4]^-$
e) $[PtCl_2(PR_3)_2] + 2\,N_2H_4 \longrightarrow [PtHCl(PR_3)_2] + N_2 + NH_3 + NH_4Cl$

Exercise 2.4

Interpret the following ligand-exchange reactions and explain how they differ from one another:

a) $[W(CO)_6] + Si_2Br_6 \longrightarrow [W(CO)_5SiBr_2] + SiBr_4 + CO$
b) $[Pt(PPh_3)_4] + Si_2Cl_6 \longrightarrow [Pt(PPh_3)_2(SiCl_3)_2] + 2\ PPh_3$
c) $[Fe(CO)_5] + PEt_3 \longrightarrow [(PEt_3)Fe(CO)_4] + CO$

Exercise 2.5

Rhodium complexes react with ethylene according to Equations (a) and (b). Comment on the two reactions.

a) $[Rh(NH_3)_5H]^{2+} + CH_2=CH_2 \longrightarrow [Rh(NH_3)_5C_2H_5]^{2+}$
b) $[RhCl(PPh_3)_3] + CH_2=CH_2 \rightleftharpoons [RhCl(C_2H_4)(PPh_3)_2] + PPh_3$

Exercise 2.6

Complete the following equations and name the type of reaction involved in each case.

a) $[IrCl(CO)(PR_3)_2] + SnCl_4 \longrightarrow$
b) $[(\pi\text{-}C_5H_5)_2(CO)_3WH] + CH_2N_2 \longrightarrow$
c) $\left[(\pi\text{-}C_5H_5)(CO)_2Fe \cdots \parallel \begin{array}{c} CH_3 \\ / \\ CH \\ CH_2 \end{array} \right]^{+} + BH_4^{-} \longrightarrow$

d) $[RuCl_2(PPh_3)_3] + H_2 + Et_3N \longrightarrow$

Exercise 2.7

Define the term oxidative addition. What conditions are required for reactions of this type? What is the reverse reaction called?

Exercise 2.8

Define the term "insertion reaction" and give an example.

Exercise 2.9

What reaction occurs when the platinum hydride complex $[PtH(SnCl_3)(CO)(PPh_3)]$ is treated with hydrogen under pressure in an autoclave?

Exercise 2.10.

Transition metal complexes can readily catalyze olefin isomerization. This can occur without cocatalysts via π-allyl complexes. Formulate such a reaction between the coordinatively unsaturated complex $\square$-ML_m and the olefin $RCH_2CH=CH_2$.

Exercise 2.11

a) Many d^8 transition metal complexes react with molecular hydrogen under nonpolar conditions. This is surprising given the high intramolecular bond energy of H–H (ca. 450 kJ/mol). Give an explanation.
b) Explain the following H_2 activation reaction:

$$[RuCl_2(PPh_3)_3] + H_2 + Et_3N \longrightarrow [RuHCl(PPh_3)_3] + Et_3NH^+Cl^-$$

Exercise 2.12.

α-Olefins readily undergo addition to Pd complexes. Which reactions can subsequently occur?

$$\text{R–CH}_2\text{–CH}{=}\text{CH}_2 \longrightarrow$$
$$\quad\quad\quad\; |$$
$$\quad\quad\quad L_n Pd$$

Exercise 2.13.

Why is no 2-butene formed in the nickel-catalyzed dimerization of ethylene?

Exercise 2.14

Addition of PPh_3 to Wilkinson's catalyst $[RhCl(PPh_3)_3]$ lowers the turnover rate in the hydrogenation of propene. Give a plausible mechanistic explanation for this observation.

2.2 Catalyst Concepts in Homogeneous Catalysis

A catalytic process can be depicted as a reaction cycle in which substrates are converted to products with regeneration of the catalytically active species. At the end of the process, the catalyst is present in its original form. The cyclic depiction of catalytic processes is particularly clear and is also helpful in developing new processes.

2.2.1 The 16/18-Electron Rule

As we have already seen, transition metal catalyzed reactions proceed stepwise according to fixed rules regarding the oxidation state and coordination number of the metal center.

Particularly useful is the 16/18-electron rule proposed by Tolman [19], which has been successfully employed to specify preferred reaction paths in homogeneous catalysis.

The rule is based on the observation that the well-characterized diamagnetic complexes of the transition metals in particular have 16 or 18 valence electrons. All ligands bound covalently to the metal center contribute two electrons to the valence shell, and the metal atom provides all the d electrons, corresponding to its formal oxidation state.

Examples:

$[Rh^I Cl(PPh_3)_3]$ has $8 + (4 \times 2) = 16$ valence electrons
  8 e
$[CH_3 Mn^I (CO)_5]$ has $6 + (6 \times 2) = 18$ valence electrons
    6 e

Tolman specified the following rules for organometallic complexes and their reactions:

1) Under normal conditions, diamagnetic organometallic complexes of the transition metals exist in measurable concentrations only as 16- or 18-electron complexes.
2) Organometallic reactions, including catalytic processes, proceed by elemental steps involving intermediates with 16 or 18 valence electrons.

The second rule can be depicted schematically for the key reactions of homogeneous catalysis as shown in Scheme 2-2.

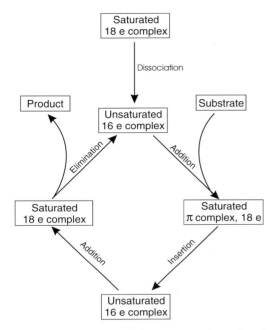

Scheme 2-2. Course of a homogeneously catalyzed reaction according to the 16/18-electron rule

2.2.2 Catalytic Cycles

With a knowledge of the key reactions of homogeneous catalysis and the 16/18-electron rule, homogeneously catalyzed processes can be depicted as cyclic processes. This way of describing catalytic mechanisms was also introduced by Tolman.

We will now discuss the industrially important hydroformylation of a terminal alkene in terms of a cyclic process (Scheme 2-3) [T11].

The catalyst precursor is the 18-electron hydrido cobalt tetracarbonyl complex **A**, which dissociates a CO ligand to give the 16-electron active catalyst **B**. The next step is the coordination of alkene to give the 18-electron π complex **C**. This is followed by rapid insertion of the alkene into the metal–hydrogen bond by hydride migration to form the cobalt^I alkyl complex **D**. The next step is addition of CO from the gas phase to afford the 18-electron tetracarbonyl complex **E**, which undergoes CO insertion to give the 16-electron acyl complex **F**. This is followed by oxidative addition of H_2 to the Co^I acyl complex to form the 18-electron Co^{III} dihydrido complex **G**.

The final, rate-determining step of the catalytic cycle is the hydrogenolysis of the acyl complex to aldehyde, which is reductively eliminated from the complex,

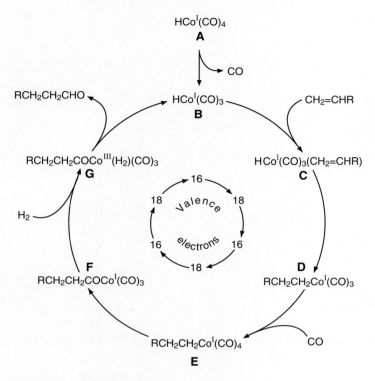

Scheme 2-3. Cobalt-catalyzed hydroformylation of a terminal alkene in terms of the 16/18-electron rule

re-forming the active catalyst **B**, which can then start a new cycle. Thus the cycle consists of a series of 16/18-electron processes, as shown in the inner circle of Scheme 2-3.

As we have seen in this example of an industrial reaction, the cobalt passes through a series of intermediates, each of which promotes a particular step of the total reaction. Thus there is not a single catalyst, but various catalyst species that take part in the entire process. This is typical of homogeneous catalysis. Generally, the complex that is introduced into the system is referred to as the catalyst, although strictly speaking this is incorrect.

In contrast to heterogeneous catalysts, the compounds used in homogeneous transition metal catalysis have well-defined structures, as can be shown directly by analytical methods. Often, however, it is difficult to identify the species that is truly catalytically active because of the numerous closely interrelated reactions, which can often not be independently investigated. However, detailed knowledge of the reaction mechanism of homogeneous catalysis is a prerequisite for making optimal use of the reactions.

2.2.3 Hard and Soft Catalysis

As we have already seen, catalytic processes generally consist of complicated series of reactions, whereby the activation of individual steps can place different demands on the catalyst. Ugo has classified the homogeneous catalysis of organic reactions on the basis of the HSAB concept [21].

If the first step of a reaction cycle is regarded as an acid–base reaction between the catalyst and the organic substrate, then a distinction can be made between "hard" and "soft" catalysis, providing a simple basis for understanding transition metal catalyzed processes (Scheme 2-4).

Homogeneous catalysis

Hard catalysis

- With H^+ or transition metal ions in high oxidation states, e.g. Mo^{6+}, VO^{2+}, $FeCl_3$, $TiCl_4$, Zn^{2+}

- Acid-base catalysis: generation of electrophilic and nucleophilic centers

- Examples: Friedel-Crafts reactions, oxidation processes, epoxidation, ester hydrolysis

Soft catalysis

- With transition metal complexes in low oxidation states, e.g. Co^-, Rh^+, Ni^0, Fe^0, Cu^+, Ir^+

- Good electron exchange between metal and substrate (covalent interaction)

- Soft substrates (olefins, dienes, aromatics)

- Soft ligands and reagents (H_2, CO, CN^-, PR_3, $SnCl_3^-$ etc.)

- Soft solvents (benzene, acetone, Me_2SO)

- Examples: carbonylation, hydrogenation, olefin oligomerization

Scheme 2-4. Hard and soft catalysis with transition metal compounds

Petrochemical catalytic reactions are predominantly soft; hard catalysis with transition metal ions is less important.

A second possibility for classifying homogeneous catalysis with transition metal complexes is the redox mechanism of such reactions [T16]. During the sequence of reactions, the transition metal formally changes its oxidation state by two units. The catalytic cycle begins, for example, with a coordinatively unsaturated soft metal complex, passes through an oxidation state two units higher as the result of an oxidative addition reaction with the reagent, and re-forms the starting complex by reductive elimination of the product.

In the intermediate of higher oxidation state, the metal is harder, and a hard–soft dissymmetry of the ligands favors the irreversible elimination of the product. In this way the key role of oxidative addition and its reverse reaction in the homogeneous catalysis of C–C and C–H coupling reactions can be understood: in the simplest case, the interplay of oxidative addition and reductive elimination can be represented by a redox reaction (Eq. 2-83).

$$\underset{s}{\boxed{M^n}} \xrightarrow{+S} \underset{\substack{\pi\,(\sigma) \\ \text{complex} \\ \text{formation}}}{M^n\text{—}S} \xrightarrow{+R} \underset{\substack{\text{Oxidative} \\ \text{addition}}}{} \underset{\substack{\text{Oxad product,} \\ h\text{-}s\text{ dissym-} \\ \text{metry}}}{R-M^{n+2}-S} \xrightarrow{} \underset{\substack{\text{Red.} \\ \text{elim.}}}{} \underset{s}{\boxed{M^n}} + R-S \qquad (2\text{-}83)$$

S = substrate, R = reagent

2.2.3.1 Hard Catalysis with Transition Metal Compounds

An example of hard catalysis is the oxidation of aldehydes with Co^{III} or Mn^{III} salts (Eq. 2-84).

$$CH_3CHO + O_2 \xrightarrow{Mn^{3+},\, Co^{3+}} CH_3COOH \qquad (2\text{-}84)$$

Initially, acetaldehyde is oxidized to peracetic acid via hard acetyl radical intermediates (Eq. 2-85). The peracetic acid then oxidizes acetaldehyde to acetic acid.

$$\underset{h\quad h}{CH_3CHO + M^{3+}} \xrightarrow{-H^+,\, -M^{2+}} \underset{h}{CH_3\overset{O}{\overset{\|}{C}}\!\cdot} \xrightarrow{+O_2} CH_3-\overset{O}{\overset{\|}{C}}\diagdown_{OO\cdot}$$

$$\xrightarrow{+CH_3CHO} CH_3\overset{O}{\overset{\|}{C}}\diagdown_{OOH} \qquad (2\text{-}85)$$

Oxidation catalysts often have a large proportion of ionic bonding, mostly with simple σ bonding of hard ligands (H_2O, ROH, RNH_2, OH^-, COO^-) to the metal ion. An example is the selective epoxidation of olefins with organic hydroperoxides (Eq. 2-86). The key step of this process is the nondissociative coordination of the hydroperoxide molecule by a hard–hard interaction of the type:

$$\begin{array}{c} \quad H \\ \quad | \\ >\!M\!<\quad O \\ |\quad \diagdown\!O\!_h \\ h\quad\; | \\ \phantom{>M<}R \end{array}$$

The metal center lowers the electron density on the peroxide oxygen atom, activating it towards nucleophilic attack of the olefin. Typical catalysts are Mo^{VI}, W^{VI}, and Ti^{IV} compounds.

2.2 Catalyst Concepts in Homogeneous Catalysis

$$M^{n+} + ROOH \rightleftharpoons [M^{n+}ROOH] \xrightarrow{>C=C<} \overset{O}{\underset{>C-C<}{\triangle}} + ROH + M^{n+} \qquad (2\text{-}86)$$

If the metal complex contains M=O groups (e.g., oxo complexes of molybdenum or vanadium), oxygen transfer from the metal hydroperoxide complex to the alkene proceeds via a cyclic transition state (Eq. 2-87).

$$\text{(cyclic transition state diagram)} \qquad (2\text{-}87)$$

As expected the catalytic effectivity increases with increasing Lewis acidity of the complex: $MoO_3 > WO_3$; electron-withdrawing ligands also increase the activity: $[MoO_2(acac)_2] > [MoO_2(diol)_2]$.

The oxirane process for the epoxidation of propene is of industrial importance. In this process, isobutane is oxidized with air to *tert*-butyl hydroperoxide, preferably with hard Mo^V and Mo^{VI} salts as catalysts. The hydroperoxide then oxidizes the propene.

Now let us turn our attention to hydrogenation reactions. Certain hydrogenation catalysts are highly substrate specific. While a combination of $CoCl_2$ and AlR_3 (hard) hydrogenates both α-olefins and dienes, in the presence of phosphine or phosphite, the diene is preferentially hydrogenated in the mixture (soft–soft interaction).

As a hard reagent, an aqueous hydrochloric acid solution of $RuCl_2$ catalyzes the hydrogenation of α,β-unsaturated carboxylic acids and amides, but not that of simple soft olefins.

Hard transition metal catalysts are also used in olefin polymerization. A prerequisite for polymerization is a rapid insertion reaction, which in turn requires high polarity of the metal–alkyl bond and positive polarization of the olefin. Therefore, in particular electropositive transition metals with low numbers of d electrons, such as Ti^{IV}, Ti^{III}, V^{III}, V^{II}, Cr^{II}, Zr^{IV}, are used as relatively hard catalysts here. In contrast, softer, electron-rich nickel^{II} complexes only lead to olefin dimerization or oligomerization. The reason is presumably the more facile β-hydride elimination reaction, which results in early chain termination.

2.2.3.2 Soft Catalysis with Transition Metal Compounds

Typical catalysts for the isomerization, hydrogenation, oligomerization, and carbonylation of olefins are characterized by a low oxidation state of the central atom, which is stabilized by σ–π interactions with soft ligands such as H^-, CO, $^tPR_3$, and X^- [7].

Numerous metal hydrides, such as $[HCo(CO)_4]$ and $[HRh(CO)(PPh_3)_3]$, or combinations of a metal complex and a hydride source (e.g., $[Co_2(CO)_8]/H_2$, $[Ni\{P(OEt_3)\}_4]/H_2SO_4$ catalyze the isomerization of 1-alkenes to 2-alkenes. In the industrial carbonylation of α-olefins, this double-bond isomerization is undesirable since the linear end products are of greater industrial importance.

Two mechanisms are discussed for the double-bond isomerization of olefins:

– The metal alkyl mechanism
– The metal allyl mechanism

The following examples illustrate the application of the HSAB concept to the above-mentioned possibilities. The addition of M–H to the double bond (Scheme 2-5) can proceed by a Markownikov (a) or anti-Markownikov route (b). Only after Markownikov addition is the 2-olefin formed by β-elimination.

Therefore, the isomerization depends crucially on the hydride character of the hydrogen atom. The hydride ligands of soft complexes such as $[HRh(CO)(PPh_3)_3]$ preferably undergo anti-Markownikow addition with the following polarization:

$$\begin{array}{c} \delta- \quad \delta+ \\ H\cdots CH-CH_2-R \\ | || \\ \delta+ M\cdots CH_2 \\ s \delta- \end{array}$$

With harder compounds such as $[HCo(CO)_4]$, in which the hydrogen atom has more protic than hydridic character, Markownikow addition is followed by isomerization (Eq. 2-88).

$$R-CH_2-\overset{\delta+}{CH}=\overset{\delta-}{CH_2} \longrightarrow R-CH_2-CH-CH_3 \longrightarrow R-CH=CH-CH_3 \atop + \atop HCo(CO)_x \quad (2\text{-}88)$$
$$(CO)_xCo\cdots H^{\delta+} \qquad (CO)_xCo$$

Thus the harder cobalt carbonyl compounds are more strongly isomerizing than the softer rhodium species. Furthermore, bulky, soft ligands like PPh_3 also favor anti-Markownikow addition for steric reasons.

An alternative reaction path for olefin isomerization involves metal alkyl intermediates (see also Section 2.1.2).

As the next example of soft catalysis, we shall discuss the dimerization of ethylene to 1-butene, which is catalyzed by rhodium complexes in a redox cycle (Scheme 2-6). The active Rh^I catalyst **A** undergoes oxidative addition of HCl and insertion of ethylene into the Rh–H bond to give the Rh^{III} alkyl complex **B**. The fol-

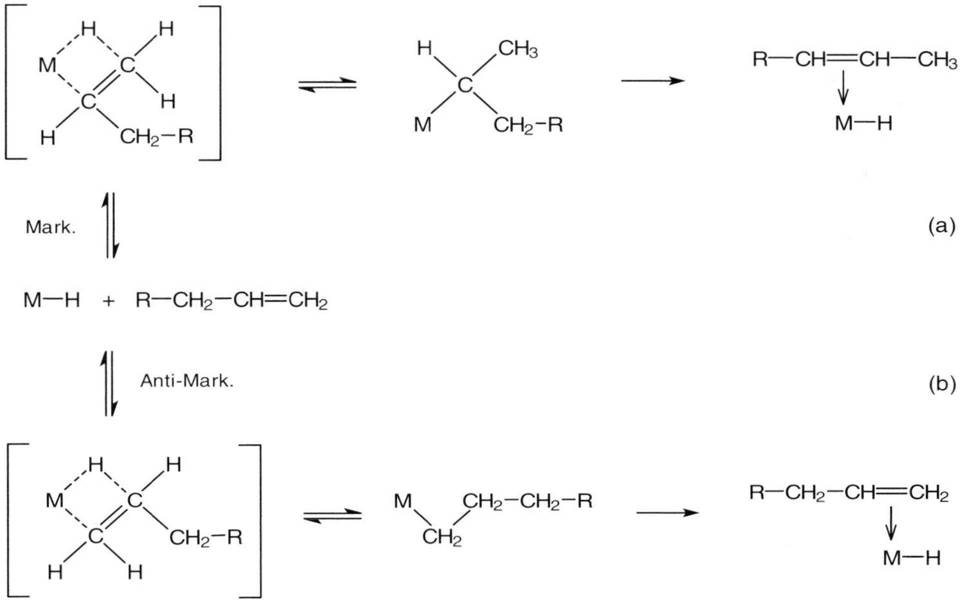

Scheme 2-5. Isomerization of α-olefins by the metal alkyl mechanism

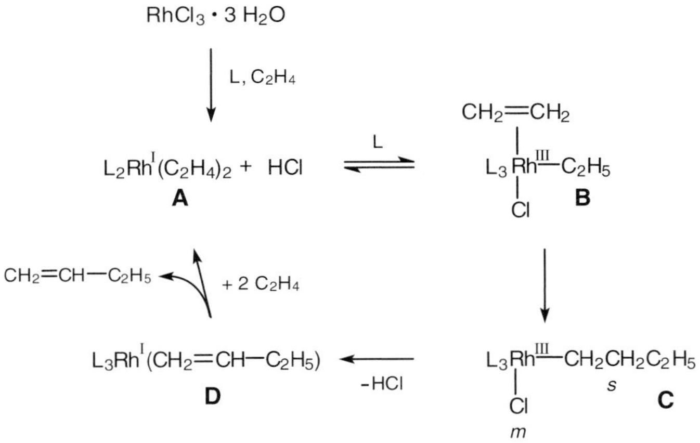

Scheme 2-6. Dimerization of ethylene to 1-butene with a rhodium catalyst (m = medium hard; s = soft)

lowing ethylene insertion reaction is the rate-determining step and is favored by the medium-hard Rh^{III} center. The resulting Rh^{III} butyl complex **C** has a hard–soft dissymmetry, and the system is stabilized by reductive elimination of HCl to give the soft Rh^{I} butene complex **D**, from which the desired product 1-butene is released in a displacement reaction with ethylene.

The homogeneously catalyzed hydrogenation of olefins and dienes has also been thoroughly investigated [12]. The advantage of the homogeneous reactions are the high selectivities that can be achieved in many cases. For example, with the weak catalysts [RhCl(PPh$_3$)$_3$], [RuCl$_2$(PPh$_3$)$_3$], and [RhH(CO)(PPh$_3$)$_3$], only alkene and alkyne groups are attacked, while other, harder unsaturated groups such as CHO, COOH, CN, and NO$_2$ remain unchanged [T11]. Wilkinson's catalyst [RhCl(PPh$_3$)$_3$] allows the hydrogenation of alkenes and alkynes to be carried out at 25 °C and 1 bar hydrogen pressure.

The rate-determining step in catalytic hydrogenation is believed to be the olefin–hydride migration (insertion reaction) to form a metal alkyl complex. This insertion reaction can regarded as the nucleophilic attack of a hydride ligand on an activated double bond. This explains why groups that increase the electron density on the hydrido group or lower the electron density in the olefinic double bond generally increase the reaction rate. In the hydrogenation of cyclohexene with [RhClL$_3$], the following ligand influence has been found:

L = I > Br > Cl
⟵
Rate of hydrogenation
Softness of σ donors

As an example of substrate effects, acrylonitrile and allyl acetate are more rapidly hydrogenated than unsubstituted 1-hexene.

As expected, soft catalyst systems such as [HCo(CN)$_5$]$^-$/CN$^-$ are particularly effective in hydrogenating soft substrates like conjugated dienes. In the case of butadiene, the CN$^-$ concentration can be used to control the selectivity for the end products 1-butene and 2-butene. Other soft homogeneous catalysts such as [Fe(CO)$_5$], [η^5-CpM(CO)$_3$H] (M = Cr, Mo, W), [Ru(H)Cl(PPh$_3$)$_3$], and *trans*-[Pt(SnCl$_3$)H(PPh$_3$)$_2$] also reduce conjugated dienes selectively to mono-enes.

The selectivity for the hydrogenation of dienes in the presence of mono-olefins depends on the stability of the π-allyl intermediates formed. For a hydrogenation mixture of diene, mono-olefin, and Pt/Sn catalyst, the competing reactions shown in Scheme 2-7 can be envisaged [T14].

Reaction route (b), in which the softer π-allyl complex is formed, is preferentially followed by soft catalyst systems. This is the case when excess ligand R$_3$P, CO, or [SnCl$_3$]$^-$ is present.

The reduction of harder substrates such as phenol requires harder catalysts (e.g., combinations with Lewis acids). The catalyst combination Co(2-ethyl hexanoate)$_2$/AlEt$_3$ allows the reduction of phenol to cyclohexanol to be carried out under mild conditions with over 90% selectivity [T11].

The most important homogeneously catalyzed industrial syntheses are the carbonylation reactions [T5]. Whereas hydroformylation of olefins with soft rhodium catalysts gives exclusively aldehydes as oxo products, with the harder cobalt catalysts alcohols can also be obtained. The initially formed aldehydes, which can be

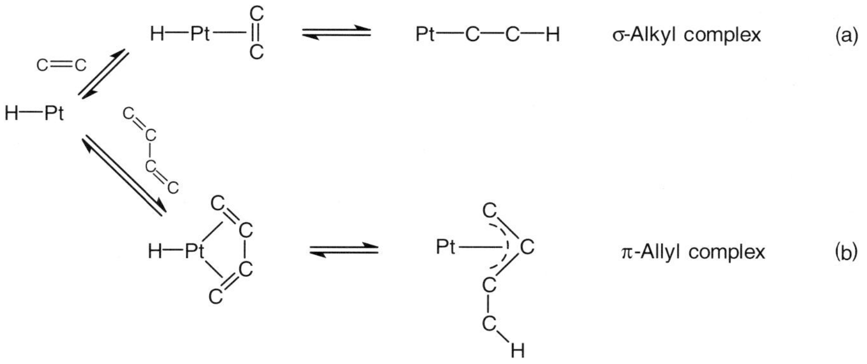

Scheme 2-7. Hydrogenation of dienes and monoolefins with Pt/Sn catalysts

regarded as relatively hard, are better able to form complexes with the hard cobalt center (Eq. 2-89).

$$R-CHO + HCo(CO)_nL_m \rightleftharpoons R-CH=O \rightarrow R-CH_2-O-Co(CO)_nL_m$$
$$\downarrow H-Co(CO)_nL_m \quad (2\text{-}89)$$
$$\xrightarrow{+H_2} R-CH_2OH + HCo(CO)_nL_m$$

The Reppe alcohol synthesis from olefins and CO/H_2O with hard iron/amine catalysts can be explained analogously: the end products are almost exclusively alcohols; the catalyst has a much higher hydrogenation activity than cobalt phosphine complexes.

The HSAB concept can also be applied to the related hydrocarboxylation reaction, in which carboxylic acids are produced from olefins, CO, water, and small amounts of hydrogen. With hard cobalt/*tert*-amine catalysts, the products are the hard carboxylic acids, whereas rhodium catalysts give mainly aldehydes. Rhodium makes the intermediate acyl complexes softer, and in the subsequent elimination step H_2, which is softer than H_2O, gives aldehyde as product.

Another carbonylation reaction of major industrial importance is the reaction of methanol with CO to give acetic acid, catalyzed by carbonyls of Fe, Co, and especially Rh in the presence of halides (Eq. 2-90).

$$CH_3OH + CO \rightarrow CH_3COOH \quad (2\text{-}90)$$

As in the case of hydroformylation, rhodium catalysts allow the process to be carried out at low temperatures and pressures (ca. 180 °C, 35 bar, Monsanto process). At the beginning of the reaction, iodide promotors convert the hard substrate methanol to the soft methyl iodide (Eq. 2-91).

$$CH_3OH + HI \rightleftharpoons CH_3I + H_2O \quad (2\text{-}91)$$

Rhodium(III) halide is used as catalyst precursor. Under the reaction conditions, it is reductively carbonylated to the active catalyst species, the anionic rhodium(I) complex $[Rh(CO)_2I_2]^-$. The reaction then proceeds as shown in Equation 2-92.

$$[Rh^I(CO)_2I_2]^- + CH_3I \longrightarrow [CH_3Rh^{III}(CO)_2I_3]^- \xrightarrow{+CO} [\underline{CH_3CO}-Rh^{III}(CO)_2I_3]^-$$
$$sss\phantom{I \longrightarrow [CH_3Rh^{III}(CO)_2I_3]^- \xrightarrow{+CO} [}h\phantom{\underline{CH_3CO}-Rh^{III}(CO)_2I_3]^-}ss$$
$$\mathbf{A}\mathbf{B}\mathbf{C}$$

$$\xrightarrow{-A} \underline{CH_3COI} \xrightarrow{+CH_3OH} \underline{CH_3COOH} + CH_3I$$
$$\phantom{\xrightarrow{-A}}hs\phantom{\xrightarrow{+CH_3OH}}hh$$

(2-92)

The soft Rh^I complex anion **A** readily undergoes oxidative addition of methyl iodide. Insertion of CO into the Rh–C bond of the resulting complex **B** then gives the acetyl rhodium complex **C**. Owing to a hard–soft dissymmetry, rapid elimination of acetyl iodide occurs. This initial product of the reaction is immediately solvolyzed by methanol to give acetic acid. The rate-determining step is believed to be the oxidative addition of methyl iodide to the Rh^I complex.

The experimental finding that bromide and chloride promoters are far less effective is explained by the fact that the rate of oxidative addition of RX to rhodium complexes decreases in the order $I > Br > Cl$, that is, with decreasing donor strength (softness) of the halide ligand.

The selective oxidation of ethylene to acetaldehyde with Pd^{II}/Cu^{II} chloride solutions has attained major industrial importance (Wacker process). This reaction can be regarded as an oxidative olefin substitution (oxypalladation). Once again the individual steps can be explained by applying the HSAB concept. The reaction mechanism in the presence of chloride has been studied in detail. The steps of interest here are shown in Scheme 2-8.

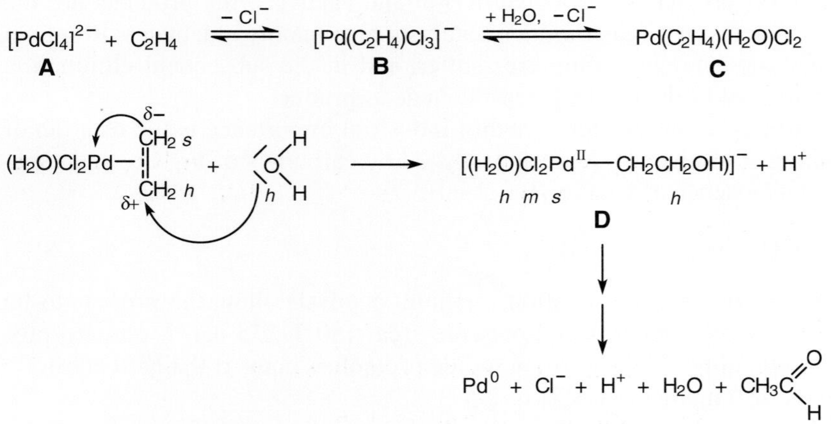

$$[PdCl_4]^{2-} + C_2H_4 \underset{}{\overset{-Cl^-}{\rightleftarrows}} [Pd(C_2H_4)Cl_3]^- \underset{}{\overset{+H_2O, -Cl^-}{\rightleftarrows}} Pd(C_2H_4)(H_2O)Cl_2$$
$$\phantom{[PdCl_4]^{2-}}\mathbf{A}\mathbf{B}\mathbf{C}$$

$$(H_2O)Cl_2Pd\!-\!\overset{\delta-\ CH_2\ s}{\underset{\delta+\ CH_2\ h}{\|}} + \overset{H}{\underset{H}{\diagdown O\diagup}} \longrightarrow [(H_2O)Cl_2Pd^{II}\!-\!CH_2CH_2OH)]^- + H^+$$
$$h\ m\ sh$$
$$\mathbf{D}$$

$$\downarrow$$

$$Pd^0 + Cl^- + H^+ + H_2O + CH_3C\!\!\overset{O}{\underset{H}{\diagup\!\!\diagdown}}$$

Scheme 2-8. Palladium-catalyzed oxidation of ethylene to acetaldehyde

After coordination of the ethylene to the tetrachloropalladate **A**, the strong *trans* effect of the ethylene ligand in complex **B** facilitates ligand substition to give the aquo complex **C**. The function of this neutral aquo complex is possibly that it exhibits less π backbonding from the metal to the olefin than the anionic complex, and the olefin therefore more readily undergoes nucleophilic attack in the former.

Newer investigations have shown that the complex undergoes nucleophilic attack by the hard reagent water [T18], whereas formerly insertion of ethylene into a palladium hydroxo species in an intramolecular step was assumed. The soft palladium(II) center in the hydroxyalkyl complex **D** is coordinated by several hard ligands, which explains the strong tendency towards elimination with release of the final product.

With the hard base water, oxidative olefin substitution leads to acetaldehyde; with acetic acid, vinyl acetate is formed. Finally, the metallic palladium is oxidized by atmospheric oxygen in the presence of Cu^{2+}, re-forming the starting complex.

The final example of a typical soft catalysis to be discussed here is the hydrocyanation of butadiene to adiponitrile (Eq. 2-93). Since both the substrate and the reagent HCN are very soft, soft Ni^0 complexes such as $[Ni\{P(OAr)_3\}_4]$ are preferred as catalysts.

$$\text{butadiene}_s + 2\ HCN_s \xrightarrow{Ni(0)} NC-(CH_2)_4-CN \qquad (2\text{-}93)$$

All the examples discussed here show that the selectivity of a homogeneously catalyzed reaction is decisively influenced by the central atom of the catalyst. Fine tuning can be made by modification of the ligands. The HSAB concept can be helpful in selecting catalysts, ligands, and solvents, as well as in planning test reactions.

Exercises for Section 2.2

Exercise 2.15

The acetylacetonate complex $[(acac)Rh(C_2H_4)_2]$ undergoes rapid ethylene exchange, as has been shown by NMR spectroscopy. In contrast, $[(\eta^5\text{-}C_5H_5)Rh(C_2H_4)_2]$ is inert. Explain these findings.

Exercise 2.16

The following rhodium complexes are important catalyst intermediates:

$[RhI_2(CO)_2]^-$ $RhCl(PPh_3)_3$ $H_2RhCl(PPh_3)_3$
A   **B**   **C**

a) What is the oxidation state of the metal in complexes **A**, **B**, and **C**?
b) Which of the complexes are coordinatively saturated?

Exercise 2.17

In the literature, the mechanism of the catalytic hydrogenation of ethylene with Wilkinson's catalyst [RhCl(PPh$_3$)$_3$] is given as follows:

P = PPh$_3$

Discuss the individual steps (**a–f**) of the reaction cycle.

Exercise 2.18

The thermodynamic stability of the complexes [PtX$_4$]$^{2-}$ increases in the series X = Cl < Br < I < CN. Explain these experimental findings.

Exercise 2.19

Certain carbonylation reactions can be carried out under mild conditions with the catalyst [PdCl$_2$(PPh$_3$)$_2$]. Addition of SnCl$_2$ in the presence of hydrogen gives even more stable and more active catalysts. Explain this in terms of the HSAB concept.

Exercise 2.20

Catalyst poisons for transition metal catalysts are often bases with P, As, Sb, Se, or Te in low oxidation states. Strong O and N bases such as amines and oxy anions are, however, not poisons. Give an explanation for this.

Exercise 2.21

a) Classify the following compounds according to the HSAB concept (acid, base; hard, soft, medium):

H^- Ir^+ N_2H_4 SO_2 Ti^{4+} CO_2 CO $CH_2=CH_2$

b) Apart from CO, which of the following fragments occur preferentially in carbonyl complexes?

OH^- $C_5H_5^-$ H_2O ROH NH_3 CN^- NR_3 PPh_3 C_6H_6

c) Which of the following pairs of compounds is harder (with reason)?

$Sn^{2+}-Sn^{4+}$

$P(C_2H_5)_3 - P(OC_2H_5)_3$

$[Co(CN)_5]^{2-} - [Co(NH_3)_5]^{3+}$

2.3 Characterization of Homogeneous Catalysts

In homogeneous catalysis, stoichiometric model reactions with well-defined transition metal complexes can be used to elucidate individual steps of the catalytic cycle. Other methods for testing the validity of an assumed reaction mechanism are the use of labelled compounds and the spectroscopic identification of intermediates [13]. An advantage of such investigations is that they can generally be carried out under mild conditions, for example, standard pressure and low temperatures.

Investigations of catalytically active systems is much more difficult. Complications here are the low catalyst concentration, the high reaction temperatures, and often also high pressures. Nevertheless, in some cases active catalysts can be isolated and analytically characterized. For example, catalytic processes can be terminated ("frozen"), or individual steps can be blocked by deliberate poisoning.

In the early years of homogeneous catalysis, it was thought that in situ spectroscopy (IR, NMR, ESR, Raman, etc.) would make a major contribution to the understanding of catalysis. However, experience has shown that this expectation has only partially been fulfilled. Infrared spectroscopy has proved useful in studying carbonyl complexes [5].

First of all, the postulated mechanism must be consistent with the kinetic measurements. Initially the rate law for the total process is of interest, but the rate laws for the individual steps of the reaction are also important. The influence of using different ligands in the catalyst and other substituents on the reactants, as well as solvent effects, provides further information.

Isotopic labelling allows element-transfer steps to be identified, and stereochemical studies provided support for certain reaction mechanisms.

The possible investigation methods are summarized in the following:

1) Deduction of reaction mechanisms
   – fundamental steps (key reactions, 16/18-electron rule)
   – electronic structure and stereochemistry of metal centers

2) Modelling of reaction steps
   – stoichiometric reactions
   – complex-formation equilibria of the metal complex
   – use of labelled compounds
   – spectroscopic methods
   – rate laws of the individual steps

3) Investigations performed on the catalytically active system
   – isolation of the catalyst
   – in-situ spectroscopy (IR, NMR, UV)
   – kinetics of the total reaction (e. g., gas consumption)
   – selectivity and stereospecificity

4) Special methods
   – influence of ligands
   – solvent effects
   – influence of substituents of the reactants

Here we will not deal with the individual analytical steps in detail, but instead give examples for the applicability of individual methods.

In the hydrogenation of olefins catalyzed by [RhCl(PPh$_3$)$_3$], the metal complexes have mostly been characterized by ^{1}H and ^{31}P NMR spectroscopy. Electronic and steric effects in ligand-exchange reactions involving phosphine ligands can also be studied by ^{31}P NMR spectroscopy. Infrared spectroscopy was used to identify metal carbonyl clusters in the rhodium-catalyzed production of ethylene glycol from synthesis gas [T11]. There are numerous examples for the use of IR spectroscopy in the literature, including high-pressure applications.

An example of the use of isotopically labelled compounds is the elucidation of the mechanism of the insertion of CO into σ-alkyl complexes to give acyl complexes. Such carbonylation reactions are often reversible. The carbonylation of methylmanganese pentacarbonyl with ^{14}CO was used as model reaction. None of the ^{14}C label was found in the acetyl group (Eq. 2-94).

$$CH_3Mn(CO)_5 + {}^*CO \longrightarrow CH_3COMn(CO)_4{}^*CO \qquad (2\text{-}94)$$

The reverse reaction (Eq. 2-95) shows that the labelled CO is incorporated as a ligand; no radioactivity was detectable in the gas phase.

$$CH_3{}^*COMn(CO)_5 \xrightarrow{\Delta} CH_3Mn(CO)_4{}^*CO + CO \qquad (2\text{-}95)$$

These experiments, together with kinetic and IR investigations, lead to the conclusion that carbonylation and decarbonylation are intramolecular processes. It was shown that instead of a carbonyl insertion into the metal–carbon bond, a methyl-group migration occurs (Eq. 2-96).

$$\underset{\substack{\diagup\;\;|\\OC\;\;OC}}{\overset{\overset{\displaystyle CH_3\;\;CO}{\diagdown\;\;\diagup}}{OC-Mn-CO}} \xrightarrow{+CO} \underset{\substack{\|\;\;\diagup\;\;\diagdown\\O\;\;CO\;\;CO}}{\overset{\overset{\displaystyle OC\;\;CO}{\diagdown\;\;\diagup}}{CH_3-C-\!-\!-Mn-CO}} \qquad (2\text{-}96)$$

Extensive investigations have been carried out on the Rh/iodide-catalyzed carbonylation of methanol to acetic acid. The most important results are summarized in the following:

Reaction: $CH_3OH + CO \longrightarrow CH_3COOH$

Catalyst: RhX_3/CH_3I

Kinetics: $r = k[Rh][I]$

$E_a = 61.5$ kJ/mol

IR: bands at 1996 and 2067 cm^{-1} at 100 °C and 6 bar, typical of $[RhI_2(CO)_2]^-$

Isolated: $[Rh_2(COCH_3)_2(CO)_2I_6]^{2-}$ (X-ray structure)

Model reactions on Rh and Ir complexes.

A further example of practical catalyst development is the hydroformylation of long-chain α-olefins with various copper(I) complex catalysts [6]. Modification of the catalysts with tertiary phosphines and amines led to aldehydes as products in varying yields, with alcohols and alkanes as byproducts. It was found that defined copper complexes have only a low catalytic activity. Only after the introduction of tertiary amines as solvents and catalyst components were better results obtained.

Since apart from the catalyst components, their stoichiometry and the reaction conditions can be varied, there is a wide range of possibilities for optimization experiments. The course of the reaction was followed by a simple high-pressure IR system. Sample spectra are shown in Figure 2-1.

Immediately after application of synthesis gas pressure, a band (1) is observed for dissolved CO at 2130 cm^{-1}. The peak at 2060 cm^{-1} indicates a mononuclear copper carbonyl complex. The complex of type [Cu(CO)L] (L = ligand), formed in situ, is the active catalyst. After a reaction time of 140 min, an aldehyde band appears at 1720 cm^{-1} (3), while the sharp olefin peak at 1640 cm^{-1} (4) continually decreases in intensity. The catalyst is only effective in the temperature range 160–180 °C and rapidly decomposes above 180 °C.

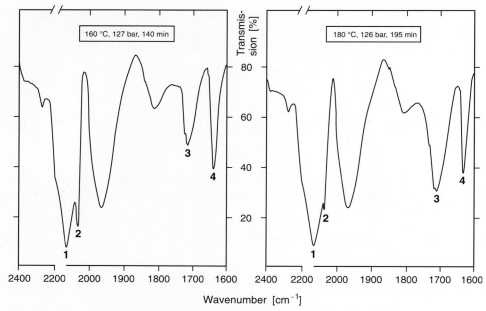

Fig. 2-1. Carbonylation of 1-decene in a high-pressure IR apparatus; catalyst [(PPh$_3$)$_3$CuCl]/ tetramethylethylenediamine, solvent THF.
Bands: 1) 2130 cm^{-1}, dissolved CO
2) 2060 cm^{-1}, Cu(CO) complex
3) 1710–1720 cm^{-1}, aldehyde
4) 1640 cm^{-1}, 1-decene

The high-pressure IR system is shown schematically in Figure 2-2 [8]. The autoclave is equipped with a magnetic piston stirrer that also acts as a displacement pump with a teflon ball valve. The reaction solution is pumped through the steel capillary, a microfilter, a nonreturn valve, and finally the high-pressure cell with 15 mm thick salt windows (NaCl or CaF$_2$). The solution is then returned to the autoclave. Figure 2-3 shows the complete mobile IR unit.

An IR unit of this type offers the following possibilities:

– Detection of intermediates and reaction products under test conditions [5]
– Performing kinetic measurements as a prerequisite for reactor design
– Investigation and characterization of catalyst species for optimization of the test conditions

In the final example, we shall consider kinetics and ligand effects in the cobalt-catalyzed hydroformylation of olefins. The unmodified cobalt catalyst in this case is [HCo(CO)$_4$], which dissociates with loss of CO in an equilibrium reaction (Eq. 2-97).

$$HCo(CO)_4 \rightleftharpoons HCo(CO)_3 + CO \tag{2-97}$$

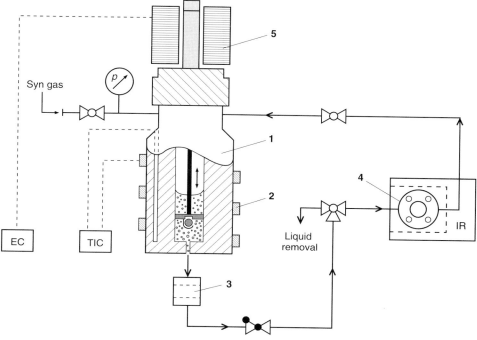

Fig. 2-2. Schematic of the high-pressure IR apparatus (FH Mannheim)
1) Magnetic-piston autoclave with recirculating pump;
2) Heating strip
3) Microfilter
4) High-pressure IR cuvette
5) Magnetic coil

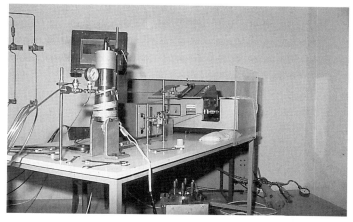

Fig. 2-3. IR high-pressure plant for homogeneous catalysis (high-pressure laboratory, FH Mannheim)

The product is formed in the rate-determining step by hydrogenolysis of the metal acyl complex, which plays a key role in this reaction.

Kinetics: $r = k[\text{olefin}][\text{Co}]\, p_{H_2} (p_{CO})^{-1}$

Increasing partial pressure of CO: higher selectivity for linear aldehydes

IR: $[(CO)_4Co-CO-CH_2CH_2R]$ ia a cobalt/1-octene system at 150 °C and 250 bar

The higher content of linear aldehydes in the reaction mixture is explained by the lower steric hindrance of the CO insertion reaction for a linear acyl complex compared to the branched isomer. However, kinetic measurements showed that the reaction rate is inversely proportional to the CO partial pressure, which can be explained by the equilibrium reactions (2-97) and (2-98), the latter preceding oxidative addition of hydrogen.

$$(CO)_4Co-CO-R \rightleftharpoons (CO)_3Co-CO-R + CO \qquad (2\text{-}98)$$

In both cases the tetracarbonyl species are inactive, and their formation is favored by high CO pressure.

Under normal oxo synthesis conditions, a small fraction of the aldehyde product is hydrogenated to alcohol (Eq. 2-99).

$$R-CHO + H_2 \longrightarrow R-CH_2OH \qquad (2\text{-}99)$$

In this reaction, too, the active catalyst is the hydrido tricarbonyl complex $[HCo(CO)_3]$. This cobalt-catalyzed hydrogenation of aldehydes is even more strongly inhibited by CO [T11].

Kinetics: $r = k[RCHO][\text{Co}]\, p_{H_2} (p_{CO})^{-2}$

This explains the low hydrogenation activity under hydroformylation conditions where CO partial pressures can exceed 100 bar.

In the case of phosphine-modified cobalt catalysts, ligand effects have been thoroughly investigated. The influence of ligand basicity can be represented by the following equilibrium reaction (Eq. 2-100).

$$HCo(CO)_4 + L \rightleftharpoons HCo(CO)_3L + CO \qquad (2\text{-}100)$$
$$L = \text{phosphine ligand}$$

With donors such as triphenylphosphine the equilibrium lies well to the left; with increasing ligand basicity it is displaced to the right. In the more stable catalysts $[HCo(CO)_3L]$, strongly basic trialkylphosphine ligands increase the electron density at the metal center and thus on the hydride ligand. This facilitates the migration of the hydride ligand to the acyl carbon atom and promotes the oxidative addition of hydrogen. Complexes containing tertiary phosphines also have higher hydrogenation activity. The following catalyst properties are influenced by the σ-donor strength:

Ligand influences in hydroformylation with $HCo(CO)_3L$

L = PEt_3 > $P(^nBu)_3$ > PEt_2Ph > $PEtPh_2$ > PPh_3

←──────────────────────────
σ-Donor strength, selectivity for
linear products, hydrogenation activity
(ratio $RCH_2OH/RCHO$)

──────────────────────────→
Catalyst activity

Such *tert*-phosphine-modified catalysts are used industrially in the Shell hydroformylation process. This is one of many examples of the influence of auxiliary ligands (cocatalysts) on homogeneous catalysis.

Exercises for Section 2.3

Exercise 2.22

a) Discuss the CO stretching frequencies of the following transition metal complexes:

$[Ni(CO)_4]$ 2060 cm^{-1}
$[Mn(CO)_6]^+$ 2090 cm^{-1}
$[V(CO)_6]^-$ 1860 cm^{-1}

b) For molybdenum carbonyl complexes, the following CO bands are found in the IR:

$[(PPh_3)_3Mo(CO)_3]$ 1910, 1820 cm^{-1}
$[(PCl_3)_3Mo(CO)_3]$ 2040, 1960 cm^{-1}

Explain the position of the CO bands.

Exercise 2.23

In many carbonylation reactions, cobalt carbonyl hydride is regarded as the active catalyst. The following CO stretching frequencies were measured:

$Co(CO)_4^- + H^+$ ⟶ $HCo(CO)_4$
1892 cm^{-1} 2067 cm^{-1}

Explain this finding.

Exercise 2.24

For the catalyst octacarbonyldicobalt different CO stretching frequencies were measured in the regions A and B:

A 2150–1900 cm^{-1}
B 1850–1700 cm^{-1}

Which structure can be deduced for the complex?

3 Homogeneously Catalyzed Industrial Processes

3.1 Overview

In the last three decades homogeneous catalysis has undergone major growth. Many new processes with transition metal catalysts have been developed, and many new products have become available. Although heterogeneous catalysis is still of much greater economic importance in industrial processes, homogeneous catalysis is continually increasing in importance. The share of homogeneous transition metal catalysis in catalytic processes is currently estimated at 10–15 % [8]. Economic data on homogeneous catalysis are difficult to obtain. Homogeneous catalysts are often used internally in a company without this fact being made public. In many cases the catalysts are prepared in situ from metal compounds.

Homogeneous transition metal catalyzed reactions are now used in nearly all areas of the chemical industry, as shown in Scheme 3-1 [10].

Homogeneous hydrogenation is used in polymer synthesis, the hydrogenation of aldehydes to alcohols (oxo process), in asymmetric hydrogenation (L-dopa, Monsanto), and for the hydrogenation of benzene to cyclohexane (Procatalyse).

The most important industrial application of homogeneous catalysts is the oxidation of hydrocarbons with oxygen or peroxides. Mechanistically, a distinction is made between:

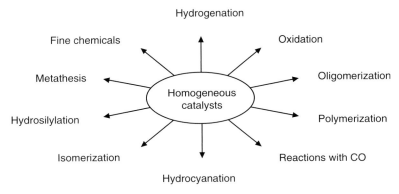

Scheme 3-1. Homogeneous transition metal catalyzed reactions carried out industrially [10]

- Homolytic processes: the transition metals react with formation of radicals, and the oxidation or reduction steps are one-electron processes.
- Heterolytic processes: normal two-electron steps of coordination chemistry.

Oligomerization reactions involve mono-olefins and dienes; polymerization reactions are mechanistically similar. Polymerization or copolymerization with soluble or insoluble transition metal catalysts is used to produce:

- Polyethylene and polypropylene (Ti- and Zr-based metallocene catalysts)
- Ethylene–butadiene rubber
- Poly(*cis*-1,4-butadiene)
- Poly(*cis*-1,4-isoprene)

Polymers prepared with transition metal complexes have different physical properties to those prepared by radical polymerization.

Reactions with CO are one of the most important areas of application of homogeneous catalysis [T5]. They belong to the earliest industrial processes and are associated with the names Walter Reppe (BASF, Ludwigshafen) and Otto Roelen (Ruhrchemie, Oberhausen).

The hydrocyanation of butadiene with two moles of HCN in the presence of nickel complexes to give adiponitrile with high regioselectivity has been developed to industrial scale by DuPont.

Isomerization processes involving homogeneous catalysts are mostly intermediate steps in industrial processes. For example, in the Shell oxo process, inner olefins are converted to primary alcohols. The isomerization occurs prior to CO insertion. The key step in the above mentioned DuPont process is the isomerization of 2-methyl-3-butenenitrile to a linear nitrile. A further example is the Cu_2Cl_2 catalyzed isomerization of dichlorobutenes [10].

Metathesis of mono- and diolefins can be performed with both homogeneous and heterogeneous catalysis. The most important processes involving metathesis steps, the SHOP process and the Phillips triolefin process, are based on heterogeneous catalysts. Homogeneous catalysts are used in the ring opening metathesis of norbornene (Norsorex, CDF-Chemie) and cyclooctene (Vestenamer, Hüls) [7].

Homogeneous catalysis is also used in the manufacture of low-scale but high-value products such as pharmaceuticals and agrochemicals. A rapidly growing area is the synthesis of fine chemicals [16]. Table 3-1 summarizes the most important industrial processes involving homogeneous catalysts [8]. Production data are listed in Table 3-2.

Table 3-1. Industrial processes with homogeneous transition metal catalysis [8]

Unit operation	Process/products
Dimerization of olefins	dimerization of monoolefins (Dimersol process); synthesis of 1,4-hexadiene from butadiene and ethylene (DuPont)
Oligomerization of olefins	trimerization of butadiene to cyclododecatriene (Hüls); oligomerization of ethylene to α-olefins (SHOP, Shell)
Polymerization	polymers from olefins and dienes (Ziegler-Natta-catalysis)
CO reactions	carbonylations (hydroformylation, hydrocarboxylation, Reppe reactions); carbonylation of methanol to acetic acid (Monsanto)
Hydrocyanation	adiponitrile from butadiene and HCN (DuPont)
Oxidation	cyclohexane oxidation; production of carboxylic acids (adipic and terephthalic acid); epoxides (propylene oxide, Halcon process); acetaldehyde (Wacker-Hoechst)
Isomerization	isomerization of double bonds; conversion of 1,4-dichloro-2-butene to 3,4-dichloro-1-butene (DuPont)
Metathesis	octenenamer from cyclooctene (Hüls)
Hydrogenation	asymmetric hydrogenation (L-dopa, Monsanto); benzene to cyclohexane (Procatalyse)

Table 3-2. Production of chemicals by homogeneous catalysis

Process	10^6 t/a
Oxidation	14.0
Reactions with CO	8.0
Hydrogenation	1.4
Oligomerization	0.8
Hydrocyanation	0.4

3.2 Examples of Industrial Processes

In this chapter, we will take a closer look at some large-scale industrial processes that involve homogeneous transition metal catalysts [3, 6].

3.2.1 Oxo Synthesis

Oxo synthesis, or more formally hydroformylation, is an olefin/CO coupling reaction which in the presence of hydrogen leads to the next higher aldehyde. The process was discovered in 1938 by Otto Roelen at Ruhrchemie, where it was first commercialized [4]. This reaction is the most important industrial homogeneous catalysis in terms of both scale and value. The most important olefin starting material is propene, which is mainly converted to 1-butanol and 2-ethylhexanol via the initial product butyraldehyde (Eq. 3-1).

$$CH_3CH=CH_2 + CO + H_2 \longrightarrow CH_3CH_2CH_2CHO \begin{array}{c} \xrightarrow{H_2} CH_3CH_2CH_2CH_2OH \\ \xrightarrow[2.\ H_2]{1.\ Base} CH_3(CH_2)_3-\underset{C_2H_5}{CH}-CH_2OH \end{array}$$
$$\left(+ CH_3-\underset{CHO}{CH}-CH_3\right)$$

(3-1)

The most important location for this reaction is Germany, with the plants of Hoechst (Ruhrchemie works in Oberhausen) and BASF in Ludwigshafen. Approximately 50% of world capacity is located in Europe and about 30% in the USA. Numerous industrial variants of oxo synthesis are known. Cobalt and rhodium catalysts are used, the latter now being preferred [3].

The original catalyst was $[Co_2(CO)_8]$, which was modified with phosphines to increase the yield of the industrially more important linear aldehydes. A breakthrough was achieved in 1976 at Union Carbide with the introduction of rhodium catalysts such as $[HRh(CO)(PPh_3)_3]$. The rhodium-catalyzed process operates at ca. 100 °C and 10–25 bar and gives a high ratio of linear to branched products.

The low pressure allows the synthesis gas to be used directly under its normal production conditions, so that investments for compressors and high-pressure reactors can be saved. However, the economic advantages are strongly dependent on the lifetime of the expensive catalysts, and loss-free catalyst recovery is of crucial importance [14].

The mechanisms of hydroformylation with rhodium and cobalt catalysts have been studied in detail and are very similar. We have already learnt that for cobalt

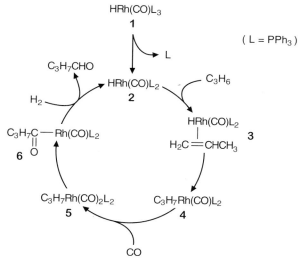

Scheme 3-2. Mechanism of the hydroformylation of propene with [HRh(CO)(PPh$_3$)$_3$]

the active catalyst precursor is [HCo(CO)$_4$]; in the case of the modified rhodium catalyst it is the complex [HRh(CO)(PPh$_3$)$_3$].

The catalytic cycle for the rhodium system in the presence of excess triphenylphosphine as cocatalyst is shown in Scheme 3-2 [T14].

Dissociation of a phosphine ligand leads to a coordinatively unsaturated complex **2**, to which the olefin coordinates. This is followed by the familiar steps of olefin insertion, CO insertion to give the Rh acyl complex **6**, and hydrogenolysis of the acyl complex with liberation of the aldehyde, which completes the cycle.

The Shell process is a variant of the cobalt-catalyzed process in which phosphine-modified catalysts of the type [HCo(CO)$_3$(PR$_3$)] are used. Such catalysts, which are stable at low pressures, favor the hydrogenation of the initially formed aldehydes, so that the main products are oxo alcohols. However, a disadvantage is the lower catalyst activity and increased extent of side reactions, especially the hydrogenation of the olefin starting material. The superiority of the low-pressure rhodium process can be seen from the process data listed in Table 3-3.

The advantages of the rhodium catalysis can be summarized as follows:

1) Rhodium is about 1000 times more active than cobalt as a hydroformylation catalyst.
2) The large excess of PPh$_3$ allows high aldehyde selectivity and a high fraction of linear product to be achieved and at the same time inhibits hydrogenation reactions.

3) The presence of PPh$_3$ dramatically increases the stability of the catalyst and prolongs its life. The low volatility of the catalyst allows the product to be distilled from the reactor with minimal rhodium losses (<1 ppm).
4) Efficient purification of the reactants avoids catalyst poisons and prolongs catalyst life [14].

Table 3-3. Industrial propene hydroformylation processes [14]

	Catalysts		
	Co	Co/phosphine	Rh/phosphine
Reaction pressure (bar)	200–300	50–100	7–25
Reaction temperature (°C)	140–180	180–200	90–125
Selectivity C$_4$ (%)	82–85	>85	>90
n/iso-Aldehyde	80/20	up to 90/10	up to 95/5
Catalyst	[HCo(CO)$_4$]	[HCo(CO)$_3$(PBu$_3$)]	[HRh(CO)(PPh$_3$)$_3$]/ PPh$_3$ up to 1 : 500
Main products	aldehydes	alcohols	aldehydes
Hydrogenation to alkane (%)	1	15	0.9

The costs of the rhodium process are, however, higher owing to the required work up, catalyst recycling, and corrosion problems. Therefore, intensive research is being carried out to develop heterogeneous rhodium catalysts. However, this has so far been thwarted by the low stability of the catalysts.

A recent breakthrough has been the use of two-phase technology, commercialized in the Ruhrchemie/Rhône Poulenc process, which uses a new water-soluble rhodium complex with polar SO$_3$Na groups on the phenyl rings of the phosphine (TPPTS) [1].

The Ruhrchemie works of Hoechst AG in Oberhausen produces over 300 000 t/a of butyraldehyde using a two-phase water/organic phase system. The process gives improved product selectivity (n/i ratio > 95/5), and the separation of the catalyst and its recycling are straightforward. Figure 3-1 shows a flowsheet of the process. Such two-phase processes in which the reaction occurs at the phase boundary are expected to be of major future importance in industrial chemistry [16].

3.2 Examples of Industrial Processes

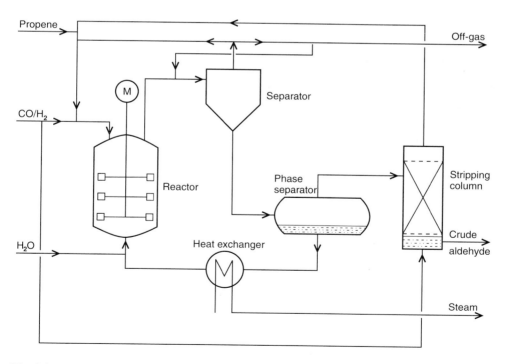

Fig. 3-1. Ruhrchemie/Rhône-Poulenc process for the hydroformylation of propene

3.2.2 Production of Acetic Acid by Carbonylation of Methanol

Another industrially important process with soluble rhodium catalysts is the direct carbonylation of methanol to acetic acid (Eq. 3-2) [T5].

$$CH_3OH + CO \xrightarrow{[RhI_2(CO)_2]^-} CH_3COOH \qquad (3\text{-}2)$$

The process was commercialized by Monsanto and has replaced the original high-pressure cobalt-catalyzed BASF process. The catalytic cycle of the Monsanto process is shown in Scheme 3-3 [T18].

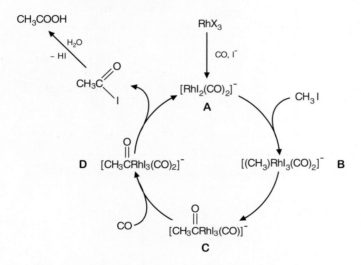

Scheme 3-3. Carbonylation of methanol to produce acetic acid (Monsanto process)

The rate-determining step is oxidative addition of methyl iodide to the four-coordinate 16-electron complex $[RhI_2(CO)_2]^-$ **A** to give the six-coordinate 18-electron complex **B**. This is followed by CO insertion to give the 16-electron acyl complex **C**. Further coordination of CO to the metal center leads to the 18-electron complex **D**, which undergoes reductive elimination of acetyl iodide, re-forming the active catalyst **A**. The acetyl iodide is then hydrolyzed by water to acetic acid and HI (Eq. 3-3).

$$CH_3COI + H_2O \longrightarrow CH_3COOH + HI \qquad (3\text{-}3)$$

The strong acid HI converts the methanol starting material to methyl iodide (Eq. 3-4).

$$CH_3OH + HI \rightleftharpoons CH_3I + H_2O \quad (3\text{-}4)$$

This reaction mechanism is supported by model studies. Paricularly advantageous are the mild reaction conditions (30–40 bar, 150–200 °C) and the high selectivity with respect to methanol (99%) and CO (>90%) compared to the older cobalt process. Methanol carbonylation is one of the few industrially important catalytic reactions whose kinetics are known in full [7].

Since the reaction is zero order with respect to the reactants, stirred tank reactors have no disadvantages relative to tubular reactors. In fact, stirred vessels allow better heat and material transfer in the gas–liquid reaction. Since the intermediates are anionic, the reaction is carried out in polar solvents.

Industrially, processes in which the products are separated by distillation predominate. Numerous columns are necessary because the boiling point of acetic acid lies between those of the low-boiling components (unchanged CO, CH_3I, and the byproduct dimethyl ether) and that of the higher boiling rhodium complex.

The economics of the process depend on loss-free rhodium recycling, which is now readily achievable. A disadvantage is the corrosivity of the iodide, which requires the use of expensive stainless steels for all plant components. Up to now, alternatives such as replacement of the halogen or immobilization of the catalyst have not proved feasible.

Today, methanol carbonylation is carried out exclusively in plants using the Monsanto process, which has been licensed worldwide.

3.2.3 Selective Ethylene Oxidation by the Wacker Process

The Wacker process was the first organometallic catalytic oxidation [15, 16]. It was developed 1959 by Smidt and co-workers at the Wacker Consortium for Industrial Electrochemistry in Munich and is mainly used for the production of acetaldehyde from ethylene and oxygen (Eq. 3-5)

$$CH_2=CH_2 + 1/2\, O_2 \xrightarrow{\text{Pd cat.}} CH_3CHO \quad (3\text{-}5)$$

The process proceeds by homogeneous catalysis on $PdCl_2$. It had been known much ealier that solutions of Pd^{II} complexes stoichiometrically oxidize ethylene to acetaldehyde, but the crucial discovery was the exploitation of this reaction in a catalytic cycle. A closed-cycle process was developed in which an excess of the oxidizing agent Cu^{2+} re-oxidizes the palladium formed in the process without its depositing on the reactor walls. The Cu^+ formed in the redox process is re-oxidized to Cu^{2+} by oxygen. The reaction steps are described by Equations 3-6 to 3-8.

$$CH_2=CH_2 + H_2O + PdCl_2 \rightarrow CH_3CHO + Pd + 2\, HCl \quad (3\text{-}6)$$

$$\text{Pd} + 2\ \text{CuCl}_2 \longrightarrow \text{PdCl}_2 + 2\ \text{CuCl} \tag{3-7}$$

$$2\ \text{CuCl} + \tfrac{1}{2}\text{O}_2 + 2\ \text{HCl} \longrightarrow 2\ \text{CuCl}_2 + \text{H}_2\text{O} \tag{3-8}$$

The complete catalytic process is depicted in Scheme 3-4.

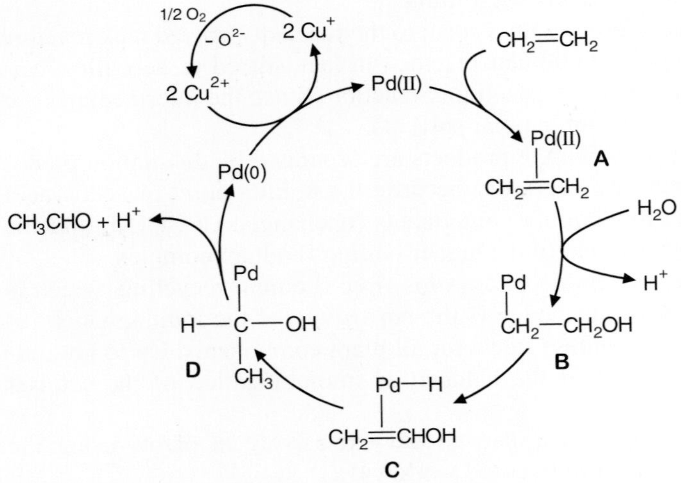

Scheme 3-4. Mechanism for the oxidation of ethylene to acetaldehyde in the Wacker process (chloride ligands omitted)

A mechanistic study of the Wacker process involving detailed stereochemical investigations showed that CO bond formation occurs with *trans* stereochemistry; that is, the ethylene molecule is not attacked intramolecularly by a coordinated water molecule. Instead, an additional, uncomplexed water molecule attacks the double bond.

The formation of **B** by addition of water is followed by two further steps in which the coordinated alcohol is isomerized. First, a β-hydride elimination gives **C**, and then an insertion reaction forms **D**. The elimination of the product acetaldehyde and H^+ gives Pd^0, which is oxidized back to Pd^{2+} by $\text{Cu}^{2+}/\text{O}_2$. With the exception of this last step, the oxidation state of palladium in all steps of the cycle is +2 [7].

In industry, bubble column reactors are used to react the gaseous starting materials ethylene and air (or oxygen) with the aqueous hydrochloric acid solution of the catalyst. Two process variants compete with one another.

In the one-step process, reaction and regeneration with oxygen are carried out simultaneously, while in the two-step process they are carried out separately. In the latter case, air can be used for regeneration, and complete ethylene conversion is achieved. A disadvantage is the higher energy requirement for catalyst circulation compared to the gas circulation used in the one-stage process. In addition, the double reactor design for higher pressures and the use of corrosion-resistant materials lead to higher investment costs.

The two-step process operates at 100–110 °C and 10 bar; catalyst regeneration is carried out at 100 °C/10 bar. Selectivities of 94% are attained. Side products, such as acetic acid and crotonaldehyde, and chlorinated compounds are removed by two-stage distillation, and the crude aldehyde is concentrated (Fig. 3-2). This process accounts for about 85% of total acetaldehyde production.

In analogous processes, the oxidation of ethylene in the presence of acetic acid produces vinyl acetate, and in the presence of alcohols, vinyl ethers. In this case heterogeneous catalysts are mainly used.

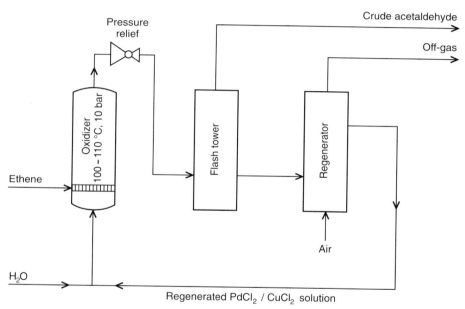

Fig. 3-2. Acetaldehyde production in the two-stage Wacker–Hoechst process

3.2.4 Oxidation of Cyclohexane

The chemistry of the Wacker process is atypical for an oxidation reaction. Normally, catalytic oxidations proceed by chain reactions initiated by radical intermediates. Well-known products of such reactions are hydroperoxides, which themselves often undergo further reaction to give other products.

Radical reactions are characterized by complex product distributions since oxygen exhibits high reactivity towards organic reactants, metal centers, and many ligands. Metals play an important role as initiators for radical chain reactions. Radicals are often generated by metal-catalyzed decomposition of organic hydroperoxides [15].

An industrially important example of such a process is the oxidation of cyclohexane to cyclohexanone and cyclohexanol. Cobalt salts are used as multifunctional catalysts. Cyclohexane is generally oxidized in the presence of about 20 ppm of a soluble cobalt salt such as cobalt naphthenate in the liquid phase at 125–165 °C and 8–15 bar up to a conversion of 10–12%.

Higher conversions are undesirable as the selectivity decreases because the products are more reactive than cyclohexane. Sometimes boric acid is added to stabilize the oxidation mixture. The selectivities with respect to cyclohexanone and cyclohexanol are 80–85%. Unreacted cyclohexane is removed by distillation and recycled. The high-boiling components, mainly cyclohexanone and cyclohexanol, are purified by distillation [12]. The most important intermediate in cyclohexane oxidation is cyclohexyl hydroperoxide; a proposed mechanism is shown in Scheme 3-5.

The radical process begins with the radical-transfer agents $R^\bullet$ and $ROO^\bullet$ ($R = C_6H_{11}$). Cobalt acts as an electron-transfer catalyst and redox initiator in the process. In a one-electron step, the oxidation state of the metal varies between +2 and +3, and radicals are released from the cyclohexane hydroperoxide. Since the cobalt is also involved in a cyclic process, its function is purely catalytic, and thus only small amounts of catalyst are required. Other metals such as V, Cr, Mo, Mn can also be used. Industrial variants of the process have been developed by companies such as BASF, Bayer, DuPont, ICI, Inventa, Scientific Design, and Vickers-Zimmer [T9].

The mixture of cyclohexanone and cyclohexanol can be converted to adipic acid in a second step by oxidation with nitric acid in the presence of metal compounds such as Cu^{II} or V^V salts as homogeneous catalysts.

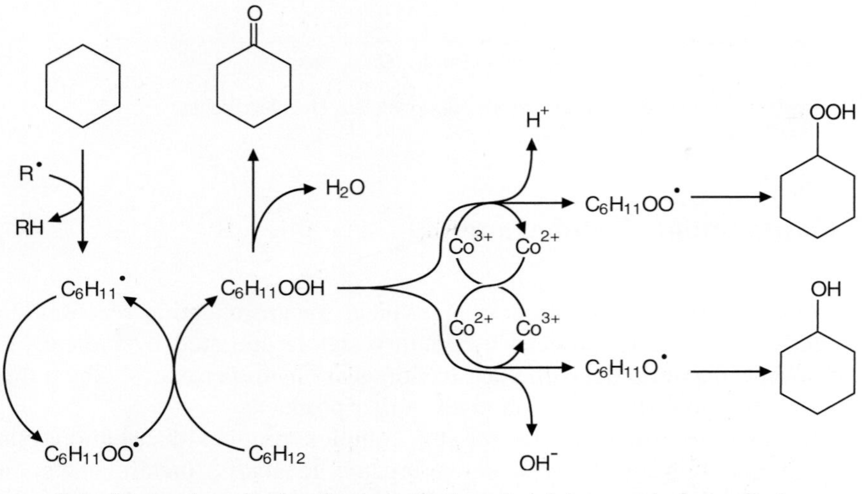

Scheme 3-5. Proposed mechanism for the oxidation of cyclohexane via free radicals.

3.2.5 Asymmetric Hydrogenation: Monsanto L-Dopa Process

As we have already seen, rhodium(I) phosphine complexes are particularly active hydrogenation catalysts. The most intensively investigated catalysts are [RhCl(PPh$_3$)$_3$] (Wilkinson's catalyst) and [HRh(CO)(PPh$_3$)$_3$], both of which have long been commercially available.

Wilkinson's catalyst is very sensitive to the nature of the phosphine ligand and the alkene substrate. It is used for laboratory-scale organic syntheses and for the production of fine chemicals.

One of the most elegant applications of homogeneous catalytic hydrogenation is the Monsanto process for the synthesis of L-dopa, a chiral amino acid used in the treatment of Parkinson's disease.

For the synthesis of such optically active products in enantioselective reactions, rhodium(I) catalysts similar to Wilkinson's catalyst but with optically active phosphine ligands were developed. A requirement is that the alkenes to be hydrogenated must be prochiral, that is, they must have a structure that on complexation to the metal center leads to (R) or (S) chirality [2].

In the Monsanto process, the acetamidocinnamic acid derivative **A** is asymmetrically hydrogenated to give a levorotatory precursor of L-dopa (3,4-dihydroxyphenylalanine). L-Dopa is formed by removing the acetyl protecting group from the nitrogen atom (Eq. 3-9). The asymmetry is introduced by a cationic rhodium complex containing optically active phosphine ligands. Asymmetric chelate ligands are particularly effective in forming an asymmetric coordination center for the complexation of an olefin. The resulting complex can exist in two diastereomeric forms that differ in the way the alkene is coordinated [T18].

Diastereomeric complexes generally have different thermodynamic and kinetic stabilities, and in favorable cases one of these effects can lead to enantioselective product formation.

Variation of the ligands in the rhodium complex eventually led to the chiral phosphine DIPAMP.

DIPAMP

Rhodium catalysts with this ligand can hydrogenate amino acid precursors to give optically active amino acids with enantiomeric excesses up to 96 % (enantiomer ratio of 98:2). In the meantime, other chiral disphosphine ligands have become available for use as catalyst ligands in enantio- and diastereoselective catalysis [7].

Particularly interesting in this process is that the diastereomer that is present in lower concentration leads to the desired product. This is explained by its lower activation energy, which makes a higher turnover rate possible.

Asymmetric hydrogenation is a good example of the tailoring of catalysts by modification of the ligands [16].

3.2.6 Oligomerization of Ethylene (SHOP Process)

Long-chain α-olefins are of major industrial importance in the production of detergents, plasticizers, and lubricants. Today such α-olefins are mainly produced by oligomerization of ethylene (Eq. 3-10). Numerous homogeneous transition metal catalyst on the basis of Co, Ti, and Ni have been described for this reaction.

$$n\,CH_2=CH_2 \longrightarrow CH_3CH_2-(CH_2CH_2)_{(n-2)}-CH=CH_2 \qquad (3\text{-}10)$$

The nickel-catalyzed Shell higher olefin process (SHOP) is of major industrial importance [9,11]. Ethylene is converted to α-olefins with a statistical distribution in which the lower oligomers are favored (so-called Schulz–Flory distribution). This is carried out at 80–120 °C and 70–140 bar in the presence of a nickel catalyst with phosphine ligands such as Ph_2PCH_2COOK. The product mixture is separated into C_{4-10}, C_{12-18}, and C_{20+} fractions by distillation.

The C_{12-18} fraction contains α-olefins with the desired chain length for the detergent industry. The top and bottom olefins are subjected to a combination of double-bond isomerization and metathesis. Isomerization gives a mixture of inner olefins with a statistical distribution of the double bond, metathesis of which gives a new mixture of inner olefins from which the C_{10-14} olfins can be separated by distillation. The process is depicted schematically in Equation 3-11.

$$\begin{array}{c}
C_{18}-C=C \xrightarrow{\text{Isom.}} C_9-C\dot{=}C-C_9 \\
C-C-C=C \xrightarrow{\text{Isom.}} C-C\dot{=}C-C \\
\text{Cat.} \downarrow \text{metathesis} \\
2\ C_9-C=C-C
\end{array}$$
(3-11)

If, however, the inner olefins are cleaved with ethylene over heterogeneous catalysts (e.g., $Re_2O_7/Sn(CH_3)_4/Al_2O_3$), a mixture of unbranched terminal olefins is obtained. Undesired higher and lower olefins are recycled. The products consist of 94–97% n,α-olefins and >99.5% monoolefins. A schematic of the SHOP process is shown in Scheme 3-6.

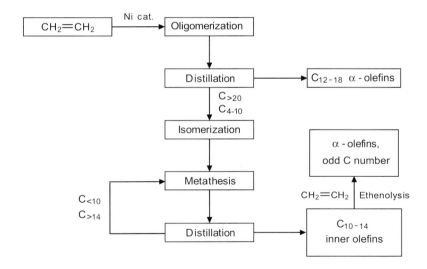

Scheme 3-6. Block schematic of the SHOP process

The combination of isomerization and metathesis with distillation and recycling offers a unique technology for obtaining a desired carbon-number distribution [7].

Mechanistic investigations with special nickel complex catalysts have shown that nickel hydrides with chelating P–O groups are the catalytically active species. The metal hydride reacts with ethylene to give alkylnickel intermediates, which can

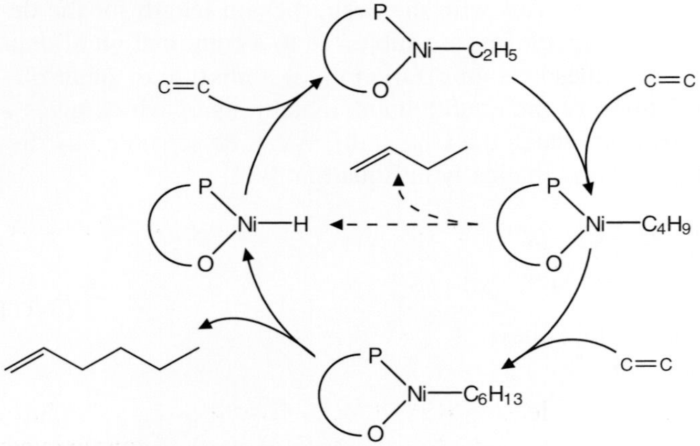

Scheme 3-7. Schematic of ethylene oligomerization with nickel complex catalysts [9]

grow further by ethylene insertion or eliminate the corresponding α-olefins. A simplified mechanism is shown in Scheme 3-7 [9].

The SHOP process first came on-stream in Geismar (USA) in 1979 and has since reached a capacity of 600 000 t/a. Further plants were built in the UK, the Netherlands, and France. The major advantage of the process is the ability to adjust the α-olefin products in response to market demands.

The products 1-hexene and 1-octene are copolymerized with ethylene to give high tensile strength polyethylenes for use in packaging materials. 1-Decene is used for producing high-temperature motor oils, and the higher olefins are converted to tensides.

Prior to introduction of the SHOP process α-olefins were produced by pyrolysis of waxes above 500 °C (e.g., Chevron process) or by olefin oligomerization with triethylaluminium (Gulf process). However, both produce olefins that are less suited to market requirements [17].

Exercises for Chapter 3

Exercise 3.1

The homogeneous catalytic hydrogenation of 1,3-butadiene with dihydrido platinum complexes gives a mixture of 1-butene, *cis*-2-butene, and *trans*-2-butene. Which intermediates are involved in the process?

Exercise 3.2

How is heptanal produced industrially?

Exercise 3.3

Acetic acid can be produced by two homogeneous catalytic processes. Name the two routes and the catalysts involved.

Exercise 3.4

The stereoselective coupling of butadiene with ethylene to give *trans*-1,4-hexadiene is described by the cyclic process shown in Scheme 3-8. Discuss the catalytic cycle and the individual intermediates **A** to **D**.

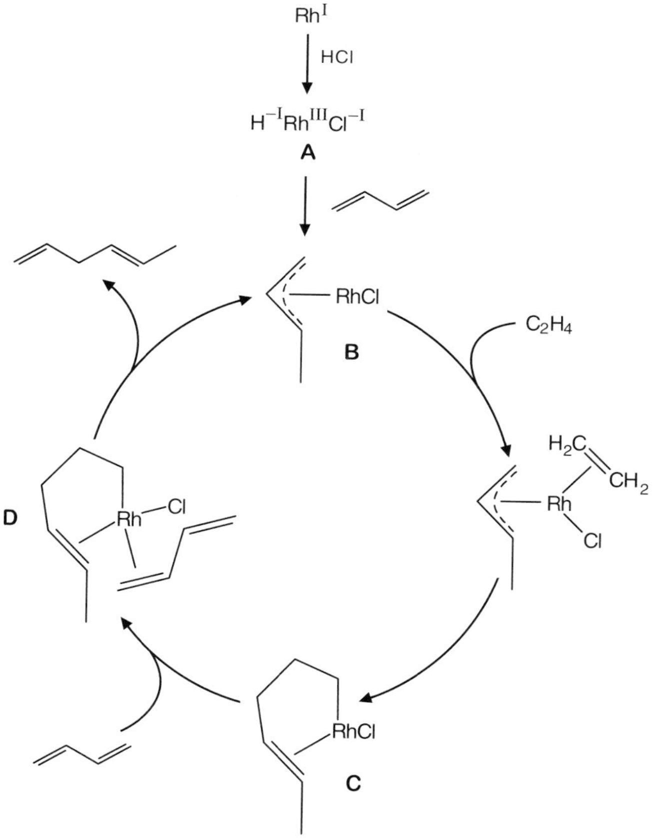

Scheme 3-8. Rhodium-catalyzed coupling of ethylene and butadiene

Exercise 3.5

The low-pressure polymerization of ethylene with Ziegler catalysts (Ti^{III} compounds/Al alkyls) is depicted in Scheme 3-9. Explain the mechanism of the polymerization.

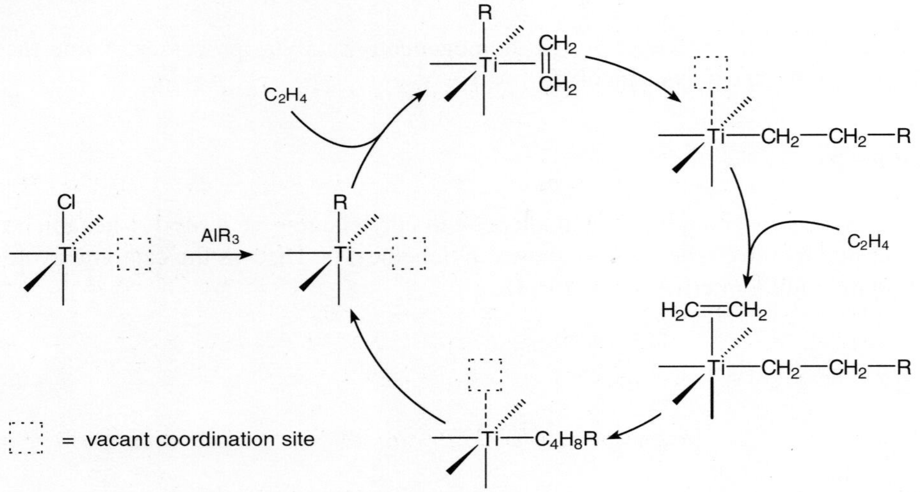

Scheme 3-9. Ziegler polymerization of ethylene with Ti/Al catalysts

Exercise 3.6

Why is it not to be expected that modification of the surface of heterogeneous catalysts with optically active substances will lead to asymmetric hydrogenations with high optical yields?

4 Heterogeneous Catalysis: Fundamentals

4.1 Individual Steps in Heterogeneous Catalysis

Heterogeneously catalyzed reactions are composed of purely chemical and purely physical reaction steps. For the catalytic process to take place, the starting materials must be transported to the catalyst. Thus, apart from the actual chemical reaction, diffusion, adsorption, and desorption processes are of importance for the progress of the overall reaction.

We will now consider the simplest case of a catalytic gas reaction on a porous catalyst. The following reaction steps can be expected (Fig. 4-1) [T20, T26]:

1) Diffusion of the starting materials through the boundary layer to the catalyst surface.
2) Diffusion of the starting materials into the pores (pore diffusion).
3) Adsorption of the reactants on the inner surface of the pores.
4) Chemical reaction on the catalyst surface.
5) Desorption of the products from the catalyst surface.
6) Diffusion of the products out of the pores.
7) Diffusion of the products away from the catalyst through the boundary layer and into the gas phase.

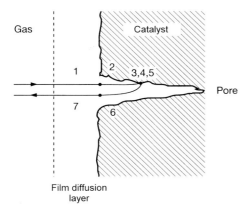

Fig. 4-1. Individual steps of a heterogeneously catalyzed gas-phase reaction

In heterogeneous catalysis chemisorption of the reactants and products on the catalyst surface is of central importance, so that the actual chemical reaction (step 4) can not be considered independently from steps 3 and 5. Therefore, these steps must be included in the microkinetics of the reaction. In cases where the other transport processes discussed above play a role, the term macrokinetics is used.

The measured reaction rate, known as the effective reaction rate, is determined by the most strongly inhibited and therefore slowest step of the reaction sequence. This rate-determining step also determines the reaction order.

The effective reaction rate r_{eff} is influenced by many parameters, including the nature of the phase boundary, the bulk density of the catalyst, the pore structure, and the transport rate in the diffusion boundary layer. If the physical reaction steps are rate determining, then the catalyst capacity is not fully exploited.

If one wishes to determine the mechanism and to describe it exactly in terms of rate equations, then one must ensure that only steps 3–5 are rate determining.

For example, the film diffusion resistance can be suppressed by increasing the gas velocity in the reactor. If pore diffusion is of desicive influence, then the ratio of the outer to the inner surface area is too small. In this case, lowering the particle size of the catalyst shortens the diffusion path, and the reaction rate increases until it is no longer dependent on pore diffusion.

Plotting concentration against position in the pore provides information about the ratio of the reaction rate to the transport rate. First we shall discuss this qualitatively. As shown in Figure 4-2, the following regions can be distinguished:

a) Film diffusion region: the reaction is fast compared to diffusion in the film layer and to diffusion in the pores.
b) Pore diffusion region: the reaction is fast compared to diffusion in the pores, but slow compared to film diffusion.

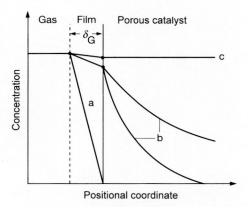

Fig. 4-2. Concentration–position curves in the film diffusion region (a), the pore diffusion region (b), and the kinetic region (c)

c) Kinetic region: the reaction is slow compared to diffusion in the pores and through the gas film.

Changing the temperature changes the ratio of reaction to transport rate (Fig. 4-3). In the kinetic region, the reaction rate increases rapidly with increasing temperature, as in a homogeneous reaction obeying the Arrhenius law. In the pore diffusion region, the reaction rate also increases according to the Arrhenius law, but at the same time the concentration profile becomes steeper, so that an ever decreasing fraction of the catalyst is active. This results in a less rapid increase of the reaction rate than in the kinetic region.

In the film diffusion region, r_{eff} increases slowly with increasing temperature because the diffusion has only a slight temperature dependence. There is practically no reaction resistance, and the gas already undergoes complete conversion on the outer surface of the catalyst.

Mathematical treatment of the total catalytic process is complicated by strong coupling of the physical and chemical reaction steps, and by the heat of reaction of the chemical reactions. This leads to temperature and pressure gradients that are difficult to solve mathematically.

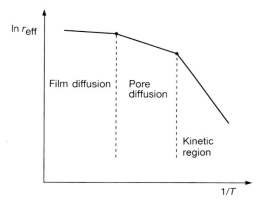

Fig. 4-3. Dependence of effective reaction rate on temperature

4.2 Kinetics and Mechanisms of Heterogeneously Catalyzed Reactions [T20, T32]

Knowledge concerning the kinetic parameters of a catalytic reaction is of major practical importance:

1) Knowledge of the reaction order with respect to the reactants and products is a prerequisite for studying the mechanism of the reaction. A precise reaction mechanism allows the catalyst to be optimized on a scientific basis.
2) The design of the reactor, including the size and shape of the catalyst particles, depends directly on the reaction order of the reactants and the thermodynamic conditions (see also Chap. 9).
3) The influence of temperature on the reaction rate can give helpful indications as to which is the slowest step of the total catalytic process.

As we have already seen in the preceding chapter, the adsorption steps that precede and follow the chemical reaction are part of the microkinetics. For this reason we shall now deal with the phenomenon of adsorption in more detail.

4.2.1 The Importance of Adsorption in Heterogeneous Catalysis [T42, T43]

First we must distinguish between physical adsorption (physisorption) and chemical adsorption (chemisorption).

Physisorption is the result of van der Waals forces, and the accompanying heat of adsorption is comparable in magnitude to the heat of evaporation of the adsorbate. In chemisorption, chemical bonds are formed between the the catalyst and the starting material. The resulting surface molecules are much more reactive than free adsorbate molecules, and the heats of adsorption are comparable in magnitude to heats of chemical reaction. This is demonstrated by the following example: the heat of adsorption of oxygen on carbon is ca. 330 kJ/mol, which is almost as high as the heat of combustion of carbon (394 kJ/mol).

One might be tempted to believe that highly effective adsorbents are also good catalysts, but in reality the situation is not so simple, because catalytic reactions proceed highly specifically. Today it is known that adsorption is a necessary but not sufficient prerequisite for molecules to react with one another under the influence of a solid surface. Furthermore, it is important that a distinction be made between the amount of adsorbed substance and the rate of adsorption.

Since both types of adsorption are exothermic, raising the temperature generally decreases the equilibrium quantity of adsorbate. Physisorption is fast, and equilibrium is rapidly reached, even at low temperature. Chemisorption generally requires high activation energies. The rate of adsorption is low at low temperatures, but the process can be rapid at higher temperatures.

The rate of both types of adsorption is strongly dependent on pressure. Chemisorption leads only to a monolayer, whereas in physisorption multilayers can form. Table 4-1 compares the two types of adsorption.

Table 4-1. Comparison of physisorption and chemisorption

	Physisorption	Chemisorption
Cause	van der Waals forces, no electron transfer	covalent/electrostatic forces, electron transfer
Adsorbents	all solids	some solids
Adsorbates	all gases below the critical point, intact molecules	some chemically reactive gases, dissociation into atoms, ions, radicals
Temperature range	low temperatures	generally high temperatures
Heat of adsorption	low, ≈ heat of fusion (ca. 10 kJ/mol), always exothermic	high, ≈ heat of reaction (80–200 (600) kJ/mol), usually exothermic
Rate	very fast	strongly temperature dependent
Activation energy	low	generally high (unactivated: low)
Surface coverage	multilayers	monolayer
Reversibility	highly reversible	often reversible
Applications	determination of surface area and pore size	determination of surface concentrations and kinetics, rates of adsorption and desorption, determination of active centers

The surface also has a major influence on adsorption. Whereas in physisorption only the magnitude of the surface area is important, chemisorption is highly specific. For example, hydrogen is chemisorbed by nickel but not by alumina, and oxygen by carbon but not by MgO. Some examples of chemisorption processes are given in Table 4-2.

The type of surface also has considerable influence on chemisorption, with surface irregularities such as corners, edges, and lattice defects playing a major role. In particular, raised areas, generally atoms with free valences, are referred to as active centers. The number of active centers is shown by the example of cumene cracking, in which the active-center concentration is 3.6×10^{19}/g catalyst or 1.2×10^{17}/m² catalyst surface.

Table 4-2. Examples of chemisorption processes

System	Heat of chemisorption [kJ/mol]	Activation energy [kJ/mol]
H_2 on graphite	189	25
CO on Cr_2O_3	38–63	0.8–3
N_2 on Fe (with promoters Al_2O_3, K_2O)	147	67
CO on Pd	72–76	9.6–38.0
H_2 on W powder	84–315	42–105

Finally, let us summarize the most important factors influencing the reaction kinetics:

1) Adsorption is a necessary step preceding the actual chemical reaction on solid catalyst surfaces.
2) Heterogeneous catalysis involves chemisorption, which has the characteristics of a chemical reaction in that the molecules of the starting material react with the surface atoms of the catalyst.
3) Catalyst surfaces have heterogeneous structures, and chemisorption takes place preferentially at active sites on the surface.

In the following we shall consider the fundamental laws of adsorption, which provide the basis for the rate expressions of heterogeneously catalyzed reactions [15, T32].

Adsorption equilibria are normally described empirically. The Freundlich equation (Eq. 4-1) describes general practical cases of adsorption.

$$c_A = a\, p_A^n \qquad (4\text{-}1)$$

c_A = concentration of the adsorbed gas
p_A = partial pressure of the adsorbed gas under equilibrium conditions
a = empirical constant
n = fraction between 0 and 1

Experimentally it is found that the amount of gas adsorbed by a solid increases with increasing total pressure P. Langmuir expressed the concentration of the adsorbed gas as a function of the partial pressure p_A and two constants (Eq. 4-2).

$$c_A = \frac{a\,b\,p_A}{(1 + b\,p_A)} \qquad a, b = \text{empirical constants} \qquad (4\text{-}2)$$

The adsorption isotherm rises up to a quantity of adsorbed substance corresponding to mononuclear coverage of the boundary layer (Fig. 4-4).

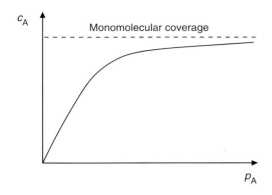

Fig. 4-4. Langmuir isotherm

Consideration of the boundary conditions shows that for large values of b or p_A, $c_A = a$, an expression identical to the Freundlich equation with $n = 0$. At very low partial pressure p_A or very small values of b, Equation 4-2 becomes the Freundlich equation with $n = 1$ (Eq. 4-3).

$$c_A = abp_A \tag{4-3}$$

The intermediate pressure range can therefore be described by the Freundlich equation with $n = 0-1$.

For a better understanding of catalysis we shall now derive the Langmuir equation on the basis of chemisorption on the active centers of the catalyst. Langmuir assumed the simple case of an energetically homogeneous catalyst surface, so that the adsorption enthalpy is independent of the degree of coverage of the surface θ_A.

For the reaction of a molecule A from the gas phase with a free site of the catalyst surface F:

$$A + F \rightleftharpoons AF \tag{4-4}$$

the law of mass action is

$$K_A = \frac{c_{AF}}{c_F p_A} \tag{4-5}$$

c_{AF} = effective concentration of chemisorbed A per unit mass of catalyst
c_F = effective concentration of active centers on the surface of the catalyst per unit mass of adsorbent

Introducing the degree of coverage of the surface we obtain

$c_{AF} = \theta_A$ θ_A = Degree of coverage of starting material A
$c_F = (1 - \theta_A)$

and Equation 4-5 becomes Equation 4-6

$$K_A = \frac{\theta_A}{(1-\theta_A)p_A} \tag{4-6}$$

which can be rearranged in terms of the degree of coverage θ_A (Eq. 4-7)

$$\theta_A = \frac{K_A p_A}{1 + K_A p_A} \tag{4-7}$$

The Langmuir isotherms derived therefrom provide the basis for the formulation of rate equations.

Consider a mononuclear gas-phase reaction A→C in which A is adsorbed without dissociation and the product C is not adsorbed. The reaction rate with respect to A thus depends only on the concentration of adsorbed A, that is, on its degree of coverage (Eq. 4-8).

$$-\frac{dp_A}{dt} = k\,\theta_A = \frac{k\,K_A p_A}{1 + K_A p_A} \tag{4-8}$$

Considering the boundary conditions shows that:

1) If K_A or p_A becomes so small that the product $K_A p_A \ll 1$, then $\theta_A \approx K_A p_A$ and the reaction is first order in A. Under these conditions the degree of coverage is low.
2) If K_A or p_A becomes so large that the product $K_A p_A \gg 1$, then θ_A becomes independent of p_A and the reaction is zero order in A. This is the case when the degree of coverage is near unity.

If neither of these approximations applies, the reaction order in A must lie between 0 and 1. If it is possible to follow the reaction order of such a reaction over a wide pressure range, then at low pressure a reaction order of unity would be observed, which at higher pressures eventually drops to zero.

Let us now consider another widely occurring situation: mixed adsorption. In this case two gases A and B compete for free sites on the catalyst surface. The number of free sites is now $1 - \theta_A - \theta_B$, and Equations (4-9) and (4-10) are obtained for the degrees of coverage of the two starting materials.

$$\theta_A = \frac{K_A p_A}{1 + K_A p_A + K_B p_B} \tag{4-9}$$

$$\theta_B = \frac{K_B p_B}{1 + K_A p_A + K_B p_B} \tag{4-10}$$

Some molecules undergo dissociative chemisorption on the surface, as we shall see below. For the reaction:

$$\tfrac{1}{2}A-A + F \rightleftharpoons AF \tag{4-11}$$

we obtain the expression

$$\theta_A = \frac{K_A \sqrt{p_A}}{1 + K_A \sqrt{p_A}} \quad (4\text{-}12)$$

Other possibilities for the adsorption of gas molecules can be discussed in an analogous manner, and the derived relationships serve as the basis for rate equations and for understanding the mechanisms of heterogeneously catalyzed reactions [2].

4.2.2 Kinetic Treatment [8, T26]

A prerequisite for the design and operation of chemical reactors is knowledge of the dependence of the reaction rate r on the process parameters. It has proved useful to make a distinction between micro- and macrokinetics. Whereas the true reaction rate (microkinetics) depends only on the concentration of the reactants, the temperature, and the catalyst, the macrokinetics in industrial systems are additionally influenced by mass- and heat-transfer processes in the reactor.

According to Equation 1-2, the reaction rate can depend on the concentration of all the reactants, but also on the concentration of the catalyst. It should be noted that a rate equation as a time law, the so-called formal reaction kinetics, doe not describe the reaction mechanism of a chemical conversion. A strict distinction must be made between molecularity (i.e., the number of molecules involved in an elementary step) and reaction order.

As we have already seen, there are reactions for which a constant reaction order can not be given, that is, the reaction rate can not be expressed in terms of a power of the concentration. This is often the case for heterogeneous reactions.

In the case of heterogeneous reactions the reaction rate can be expressed relative to the specific surface area S of the catalyst (m^2/kg) instead of the reaction volume in Equation 1-2 (Eq. 4-13).

$$r_{A,S} = -\frac{1}{S}\frac{dn_A}{dt} = k\,f(c_A) \text{ kmol kg m}^{-2}\text{ s}^{-1} \quad (4\text{-}13)$$

The most practical approach, however, is to express the reaction rate relative to the mass of catalyst m_{cat} to give an expression for the effective reaction rate $r_{A,eff}$ (Eq. 4-14).

$$r_{A,eff} = -\frac{1}{m_{cat}}\frac{dn_A}{dt} = k\,f(c_A) \text{ kmol kg}^{-1}\text{ s}^{-1} \quad (4\text{-}14)$$

It should again be emphasized that the effective reaction rate in heterogeneous reactions depends not only on the temperature and the concentration of the reactants, but also on macrokinetic parameters such as phase boundary, bulk density, and particle size of the catalyst; pore structure; and rate of diffusion.

In the following we will deal with setting up rate equations for simple heterogeneously catalyzed gas-phase reactions [T20,T26].

Consider the gas-phase reaction

A + B ⟶ C

The dependence of the reaction rate on the partial pressure of the components can in general form be expressed as a power law of the type

$$r = k p_A^a p_B^b p_C^c \tag{4-15}$$

where r is the effective reaction rate per unit mass of catalyst and k is the rate coefficient, the dimensions of which depend on the values of the exponents a, b, and c. The exponents are generally not equal to unity. It is noteworthy that in homogeneous reactions the product does not normally appear in the rate equation (i.e., c is generally zero). In heterogeneous reactions the product can remain adsorbed on the surface and thus influence the reaction rate.

The applicability of such formal approaches is limited by the fact that the exponents are not always constants and may be dependent on temperature and pressure. Such treatments are generally restricted to narrow pressure ranges and are therefore not particularly meaningful.

A better basis for developing rate equations can often be obtained by modelling the adsorption and desorption of the reaction partners on active centers. The rate equations then contain the partial pressures of the components of the reaction mixture. The effective reaction rate can be expressed as the ratio of the product of the kinetic term and the driving force (or distance from equilibrium) to the resistance term (Eq. 4-16)

$$r_{\text{eff}} = \frac{(\text{kinetic term}) \cdot (\text{driving force})}{(\text{resistance term})^n} \tag{4-16}$$

The exponent n usually has the value 1 or 2 and depends on the number of catalytically active centers of the catalyst surface that are involved in the rate-determining step. The resistance term can also be referred to as the chemisorption term.

The terms in Equation 4-16 contain the relative adsorptivity of the catalyst for the individual components of the reaction mixture. For a complete derivation of the kinetics of a catalytic reaction, that is, the functional relationship between r and the variables concentration, temperature, and pressure, the reaction mechanism must be known. It often sufficient to formulate the kinetic equation in terms of the slowest, rate-determining elementary step [2]. In this way, multiparameter equations can often be replaced by equivalent rate expressions that describe the influence of the most important experimental variables with sufficient accuracy. For irreversible reactions in which the rate of mass transport is decisive, simple expressions of the type shown in Equation 4-17 are often sufficient.

$$r_{\text{eff}} = k\, p_A^n \tag{4-17}$$

or, for the reaction A + B ⟶ R + S

$$r_{\text{eff}} = \frac{k\, p_A\, p_B}{(1 + K_A\, p_A + K_R\, p_R)^n} \tag{4-18}$$

Rate expressions such as those of Equations 4-17 and 4-18 are based on the theory of active centers.

The methods for determining reliable rate equations that describe the mechanisms of heterogeneously catalyzed reactions, some of which are quite laborious, will not be described in further detail here. Chemical engineers are interested in the kinetics of a reaction in so far as they can be used in reactor design.

4.2.3 Mechanisms of Heterogeneously Catalyzed Gas-Phase Reactions [15, 34, T35]

In this chapter we shall deal with bimolecular gas-phase reactions, which occur widely in heterogeneous catalysis. Two mechanisms are often discussed for reactions of the type:

$$A_G + B_G \longrightarrow C_G \tag{4-19}$$

Langmuir–Hinshelwood Mechanism (1921)

This mechanism is based on the following assumption: both reaction partners are adsorbed without dissociation at different free sites on the catalyst surface. This is then followed by the actual surface reaction between neighboring chemisorbed molecules to give the product C, adsorbed on the surface. In the final step the product is desorbed. The reaction sequence is thus:

$A_G \rightleftharpoons A^*$ and $B_G \rightleftharpoons B^*$

$A^* + B^* \rightleftharpoons C^*$

$C^* \rightleftharpoons C_G$ * adsorbed molecules

The Langmuir–Hinshelwood mechanism can be depicted as shown in Figure 4-5.

Each of the above-mentioned steps can be rate determining, but here we shall only discuss the case in which the surface reaction between the two adsorbed mole-

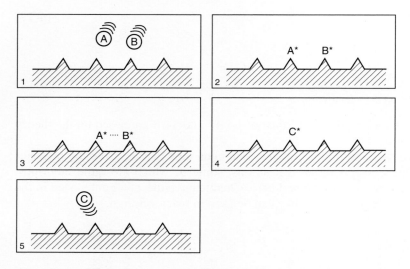

Fig. 4-5. Langmuir–Hinshelwood mechanism (schematic)

cules is the rate-determining step. On the basis of the relationship for mixed adsorption, the following rate equation can be formulated (Eq. 4-20).

$$r_{eff} = \frac{dp_C}{dt} = k\,\theta_A\,\theta_B = \frac{k\,K_A\,p_A\,K_B\,p_B}{(1 + K_A\,p_A + K_B\,p_B)^2} \qquad (4\text{-}20)$$

Of the numerous boundary conditions that are possible, we will consider only two in more detail here:

1) When both starting materials are only weakly adsorbed, then both K_A and $K_B \ll 1$ and the rate equation becomes $r_{eff} = k'\,p_A\,p_B$ and $k' = k K_A K_B$. The reaction is first order in both reactants and second order overall.
2) When A is weakly and B strongly adsorbed, $K_A \ll 1 \ll K_B$ and the rate equation reduces to

$$r_{eff} = \frac{k''\,p_A}{p_B} \quad \text{where} \quad k'' = k\frac{K_A}{K_B}$$

The reaction order is one with respect to A and minus one with respect to B.

Let us consider the reaction rate as a function of the partial pressure of component A, that is, at constant partial pressure p_B:

1) At low partial pressure p_A, the product $K_A\,p_A$ in the denominator of Equation 4-20 is negligible compared to $(1 + K_B\,p_B)$ and it follows that

$$r_{eff} \approx k\, K_A p_A \frac{K_B p_B}{1 + K_B p_B} \approx k' p_A$$

Thus the reaction rate in this case is proportional to p_A.
2) The reaction rate reaches a maximum when $\theta_A = \theta_B$ or $K_A p_A = K_B p_B$.
3) At high partial pressure p_A, the term $(1 + K_B p_B)$ in the denominator of Equation 4-20 is negligible compared to $K_A p_A$ and it follows that

$$r_{eff} \approx \frac{k''}{K_A p_A} \approx \frac{1}{p_A}$$

Hence the reaction order with respect to component A is -1.

Figure 4-6 depicts the three cases qualitatively [15].

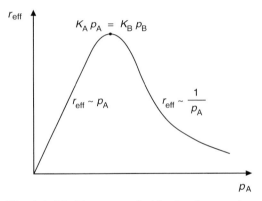

Fig. 4-6. Limiting cases of a bimolecular gas-phase reaction according to the Langmuir–Hinshelwood mechanism

At low partial pressure of component A, the degree of coverage θ_A is low, and all the chemisorbed molecules can react with component B. The reaction rate increases to a maximum where the surface is covered to an equal extent with A and B (i.e., $\theta_A = \theta_B$). With increasing partial pressure of component A, the surface becomes increasingly occupied by A, and the probability of reaction with chemisorbed B decreases. Thus it could be said that the surface is blocked by A.

The Langmuir–Hinshelwood mechanism has been proven for many reactions, including some carried out on an industrial scale, for example:

1) Oxidation of CO on Pt catalysts

$$2\,CO + O_2 \longrightarrow 2\,CO_2$$

2) Methanol synthesis on ZnO catalysts

$$CO + 2\,H_2 \longrightarrow CH_3OH$$

3) Hydrogenation of ethylene on Cu catalysts

   $C_2H_4 + H_2 \longrightarrow C_2H_6$

4) Reduction of N_2O with H_2 on Pt or Au catalysts

   $N_2O + H_2 \longrightarrow N_2 + H_2O$

5) Oxidation of ethylene to acetaldehyde on Pd catalysts

   $CH_2{=}CH_2 + O_2 \longrightarrow CH_3CHO$

Eley–Rideal Mechanism (1943)

In this mechanism only one of the gaseous reaction partners (e.g., A) is chemisorbed. Component A then reacts in this activated state with starting material B from the gas phase to give the chemisorbed product C. In the final step the product is desorbed from the catalyst surface. The reaction sequence is thus:

$A_G \rightleftharpoons A^*$

$A^* + B_G \rightleftharpoons C^*$

$C^* \rightleftharpoons C_G$

In this case only the degree of coverage of the gas A is decisive for the reaction kinetics, and on the basis of the Langmuir isotherm (Eq. 4-7), the following rate equation can be formulated:

$$r_{\text{eff}} = k\,\theta_A\, p_B = k \frac{K_A p_A}{(1 + K_A p_A)} p_B \tag{4-21}$$

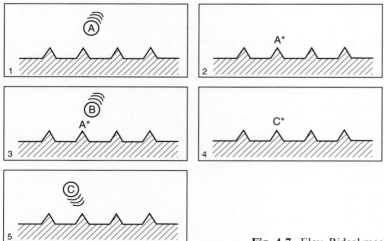

Fig. 4-7. Eley–Rideal mechanism (schematic)

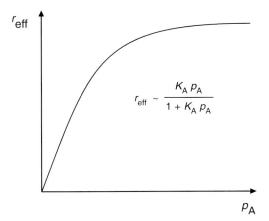

Fig. 4-8. Bimolecular gas-phase reaction with the Eley–Rideal mechanism

The Eley–Rideal mechanism is depicted schematically in Figure 4-7.

If we observe the reaction rate as a function of the partial pressure of component A at constant p_B, we see that it follows the isotherm for p_A and eventually reaches a constant final value (Fig. 4-8).

Several examples of reactions that follow the Eley–Rideal mechanism can be given:

1) Oxidation of ethylene to ethylene oxide:

$$C_2H_4 + O_2{}^* \longrightarrow \underset{O}{CH_2\text{—}CH_2}$$

In this industrially important oxidation reaction, it has been shown that in the initial stages molecularly adsorbed oxygen reacts with ethylene from the gas phase to give ethylene oxide. However, at the same time O_2 is dissociatively adsorbed as highly reactive atomic oxygen, which in an undesired side reaction gives rise to the combustion products CO_2 and H_2O.

2) Reduction of CO_2 with H_2:

$$CO_{2,G} + H_2{}^* \longrightarrow H_2O + CO$$

3) Oxidation of ammonia on Pt catalysts:

$$2\,NH_3 + 3/2\,O_2{}^* \longrightarrow N_2 + 3\,H_2O$$

4) Hydrogenation of cyclohexene:

$$\text{C}_6\text{H}_{10} + \text{H}_2^* \longrightarrow \text{C}_6\text{H}_{12}$$

5) Selective hydrogenation of acetylene on Ni or Fe catalysts:

$$HC \equiv CH + H_2^* \longrightarrow H_2C=CH_2$$

The two above-mentioned mechanisms are relatively straightforward. In the literature, however, up to a hundred different mechanisms and their rate equations are described. Knowledge of the mechanism of a heterogeneously catalyzed reaction is a prerequisite for obtaining functional relationships between the reaction rate and the variables on which it depends.

For practical reactor calculations, however, it is generally sufficient to use a kinetic approach based on the rate-determining elementary step. In many cases, an empirical rate equation that describes the influence of the most important variables wth sufficient accuracy in the chosen operating range is adequate.

Mathematical modelling of the reaction kinetics on the basis of statistical methods allows one to choose between different models and to obtain the best possible rate expression, but the effort required is considerable.

Exercises for Section 4.2

Exercise 4.1

The adsorption of CO on activated carbon was followed experimentally at 0 °C. At the given pressures, the following quantities of adsorbed gas were measured (corrected to a standard pressure of 1 bar):

p [mbar]	133	267	400	533	667	800	933
V [cm^3]	10.3	19.3	27.3	34.1	40.0	45.5	48.0

Determine whether the measurements conform to the Langmuir isotherm and calculate

a) the constant K_A and
b) the volume corresponding to complete coverage

Exercise 4.2

The decomposition of phosphine PH_3 on tungsten catalysts is first order at low pressures but zero order at high pressures. Interpret these findings.

Exercise 4.3

The reduction of CO_2 with H_2 on Pt catalysts is described by the equation:

$$CO_{2(G)} + H_{2(ads)} \longrightarrow H_2O + CO$$

a) According to which well-known mechanism does this hydrogenation proceed (with explanation)?
b) What is the name of a general kinetic treatment for such reactions, based on adsorption theory?

Exercise 4.4

The kinetics of isobutene oligomerization on macroporous polystyrene sulfonic acid is described as follows:
At low concentrations of isobutene (IB)

$$r = k_1 c_{IB}^2,$$

and at high concentrations

$$r = k_2 c_{IB}$$

Which simple model can be used to explain this?

Exercise 4.5 [T24]

When organosulfur compounds react with H_2 in the presence of a sulfur-containing Ni-Mo/γ-Al_2O_3 supported catalyst, the reaction is much faster than that of organo-nitrogen compounds under the same conditions. However, when a mixture of the same sulfur and nitrogen compounds is hydrogenated, then the nitrogen compounds react faster, regardless of the concentration ratio. Explain these observations with the aid of a kinetic model.

Exercise 4.6

The oxidation of SO_2 on Pt catalysts proceeds in two steps:

1. $O_2 + 2* \rightleftharpoons 2\,O*$
2. $SO_2 + O* \longrightarrow SO_3 + *$

   $\overline{2\,SO_2 + O_2 \longrightarrow 2\,SO_3}$

Explain these reaction steps. What mechanism is involved?

Exercise 4.7

In carrying out heterogeneously catalyzed reactions, a distinction is made between microkinetics and macrokinetics. Which steps have to be taken into account in the case of microkinetics?

Exercise 4.8

A methanation reaction was investigated on a commercial supported catalyst 0.5 % Rh/γ-Al$_2$O$_3$:

$$CO + 3 H_2 \longrightarrow CH_4 + H_2O$$

The degree of dispersion D of the the catalyst was found to be 42 % by means of chemisorption measurements with H_2. At 10 bar and 300 °C a catalyst turnover number of 0.16 s^{-1} was determined for methane. Calculate the rate of formation of methane r'_{CH_4} in mol s^{-1} g(cat.)$^{-1}$ (metal + support).

Exercise 4.9

Distinguish between chemisorption and physisorption according to the following criteria:

	Chemisorption	Physisorption
Cause		
Adsorption heat (magnitude)		
Temperature range		
Number of adsorbed layers		

4.3 Catalyst Concepts in Heterogeneous Catalysis

4.3.1 Energetic Aspects of Catalytic Activity [8, T38]

If a molecule is to enter a reactive state, it must undergo activated adsorption on the catalyst surface. Hence the catalyst must chemisorb at least one of the reaction partners, as we have already seen.

The strength of adsorption of the molecules is decisive for effective catalysis: neither too strong nor too weak binding of the reactants can induce the required reactivity; a certain medium binding strength is optimum.

Thus chemisorption and the associated energetic aspects play a crucial role in understanding heterogeneous catalysis [10]. The active centers on the catalyst surface are probably the result of free valences or electron defects, which weaken the bonds in the adsorbed molecules to such an extent that a reaction can readily occur. The course of a heterogeneously catalyzed reaction is compared to that of an uncatalyzed reaction in Figure 4-9.

In Figure 4-9 the three elementary steps on the catalyst surface are depicted qualitatively together with the corresponding energies. For the catalyzed reaction a distinction should be made between the apparent activation energy, starting from the ground state of the gaseous molecule, and the true activation energy, relative to the chemisorbed state. The latter, also known as catalytic activation energy, is more important.

Sometimes the product or transition state being formed may be so strongly bound on the surface that its desorption or further reaction is hindered. In this case the catalyst is poisoned by the product and becomes inactive.

For a deeper understanding of the catalytic reaction mechanism, knowledge regarding the structure and stability of the adsorbed intermediates is particularly important. In many cases a simple qualitative view of the chemisorption is sufficient.

The chemisorption of gases on metals has been the subject of particularly intensive investigations, and the available data allow the catalytic properties of metals to be explained well. Experimentally determined, qualitative orders of catalytic effec-

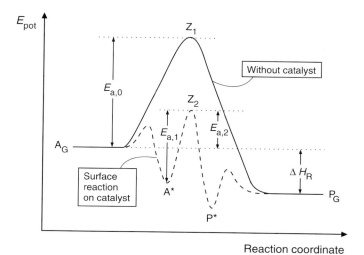

Fig. 4-9. Course of a heterogeneously catalyzed gas-phase reaction $A_G \rightarrow P_G$
$E_{a,0}$ = activation energy of the homogeneous uncatalyzed gas-phase reaction
$E_{a,1}$ = true activation energy
$E_{a,2}$ = apparent activation energy of the catalyzed reaction
Z_1 = transition state of the gas-phase reaction
Z_2 = transition state of the surface reaction
ΔH_R = reaction enthalpy

tiveness are often found in the literature. For example, for the adsorption of hydrocarbons:

acetylenes > dienes > alkenes > alkanes
polar substances > nonpolar substances

For the strength of chemisorption on many metals, the following sequence is given:

$O_2 > C_2H_2 > C_2H_4 > CO > H_2 > CO_2 > N_2$

The following explanation can be given: the reactivity of metal surfaces towards the above gases differs widely, depending on the chemical structure of the metal. As early as the 1950s, the metals were classified according to their chemisorption capabilities [T20]. Table 4-3 lists the metals in groups A to E in order of decreasing activity. The highest activity is found for the transition metals, although there are a few exceptions. In general the activity first increases along a transition metal period and then declines again at the end. The metals of group A chemisorb all seven gases, including nitrogen, which is generally the most difficult to activate.

Table 4-3. Classification of the metals according to their chemisorption properties [T20]

Metal groups		Gases						
		O_2	C_2H_2	C_2H_4	CO	H_2	CO_2	N_2
(A)	Ti, Zr, Hf, V, Nb, Ta, Cr, Mo, W, Fe, Ru, Os	+	+	+	+	+	+	+
(B_1)	Ni, Co	+	+	+	+	+	+	−
(B_2)	Rh, Pd, Pt, Ir	+	+	+	+	+	−	−
(B_3)	Mn, Cu	+	+	+	+	±	−	−
(C)	Al, Au	+	+	+	+	−	−	−
(D)	Li, Na, K	+	+	−	−	−	−	−
(E)	Mg, Ag, Zn, Cd, In, Si, Ge, Sn, Pb, As, Sb, Bi	+	−	−	−	−	−	−

+ strong chemisorption; ± weak chemisorption; − no chemisorption

The elements of lowest activity chemisorb only oxygen, which is the most easily activated. In between are the metals of medium activity, which activate only molecules from O_2 to CO or H_2. Metals that adsorb several gases can be classified according to various criteria:

− The adsorption coefficient, which reflects the strength of adsorption
− Exchange of one bound gas with another
− The heat of adsorption

A criterium for adsorption is whether it is volumetrically measurable at 10^{-3} bar at room temperature. In some cases the precise classification of a metal depends on its purity or its physical state. For example, technical-grade copper weakly adsorbs hydrogen, but pure copper not at all.

Let us now attempt to explain the above classification of the metals in terms of their atomic structure. The metals of class A belong to groups 4–8 of the periodic table, class B_1 contains the nonnoble metals of groups 9 and 10, and class B_2 the noble metals of these groups. Class B_3 contains manganese and copper, two metals of the first transition metal period with anormal behavior. All other metals of classes C, D, and E precede or follow the transition metals in the periodic table.

Thus the electronic structure of the metals is decisive for their catalytic activity. The transition metals, with their partially filled d orbitals, are particularly good catalysts. These orbitals are responsible for the covalent binding of gases on metal surfaces in chemisorption and catalysis. Whereas transition metals have one or more unpaired d electrons in the outer electron shell, the weakly chemisorbing main group elements have only s or p electrons. It is postulated that unpaired d electrons are necessary to hold the chemisorbed molecules in a weakly bound state, from which they can then be transferred into a strongly bound state.

The existence of such a transition state lowers the activation energy in general. For reactive molecules such as CO and O_2, such transition states are not absolutely necessary and they are therefore adsorbed by most metals.

Next we shall consider the binding of chemical species to the metal surface in more detail, starting with simple thermodynamic considerations. Adsorption is an exothermic process in which strong binding forces arise between the adsorbed molecules and the surface atoms of the catalyst. At the same time the degree of freedom of the molecules decreases when they leave the gas phase and are adsorbed on the catalyst. Therefore, the entropy S is negative. For a thermodynamically feasible adsorption process, the Gibb's free energy should be negative:

$$\Delta G = \Delta H - T \Delta S \qquad (4\text{-}22)$$

Since it can be expected that the reaction entropy values will not vary greatly from reactant to reactant, the adsorption enthalpy ΔH will depend, as a first approximation, mainly on the strength of chemical bonding between the gas molecules and the catalyst. Two fundamental types of chemisorption processes can be distinguished [T35]:

– Molecular or associative chemisorption, in which all bonds of the adsorbate molecule are retained
– Dissociative chemisorption, in which the bonds of the adsorbate molecule are cleaved and molecular fragments are adsorbed on the catalyst surface

Molecular chemisorption occurs with molecules having multiple bonds or free electron pairs. For example, on platinum surfaces, ethylene gives up two π electrons

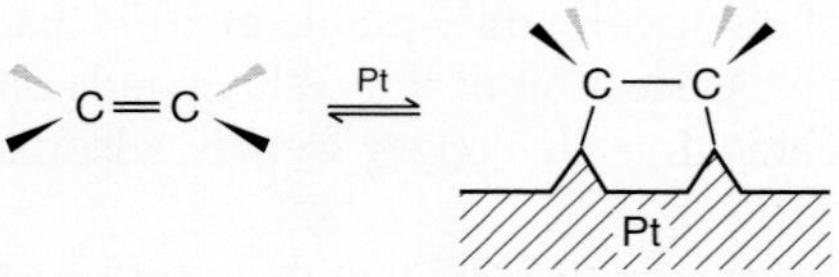

Fig. 4-10. Molecular chemisorption of ethylene on a Pt surface

of its double bond and forms two σ bonds with Pt atoms. The resulting sp³ hybridization results in a tetrahedral arrangement of bonds (Fig. 4-10).

Further examples of molecular chemisorption are:

$$H_2S + M \longrightarrow \underset{M}{\underset{|}{\underset{S}{H\diagup \diagdown H}}} \tag{4-23}$$

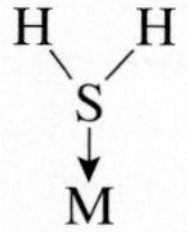

$$(CH_3)_2S + Ni \longrightarrow \underset{Ni}{\underset{|}{\underset{S}{CH_3\diagup \diagdown CH_3}}} \tag{4-24}$$

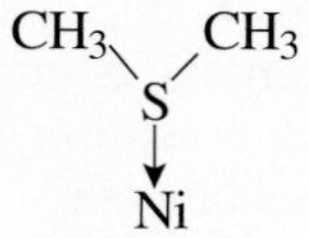

Dissociative chemisorption occurs mainly with molecules containing single bonds, for example, the adsorption of H_2 on nickel, in which the hydrogen is adsorbed in atomic form on the surface. The potential diagram (Fig. 4-11) [10, T43] consists of two intersecting curves, the flatter of which (curve 1) corresponds to the physisorption of molecular hydrogen. Via the state of physisorption with only low heat of adsorption, molecular hydrogen passes through the point of intersection A with the potential curve of atomic hydrogen (curve 2). At this point dissociation begins, initially reaching a state in which the H–H bond is weakened and the new Ni–H bond is forming.

The chemisorbed hydrogen has the lowest potential energy and the shortest distance to the catalyst surface (point B). For the reaction according to

$$1/2\,H_2\,(g) + Ni \longrightarrow H\text{–}Ni$$

the binding strength is given by the reaction enthalpy of -46 kJ/mol. Note that two Ni–H bonds are formed from the single chemical bond in the H_2 molecule. Dissociative chemisorption always increases the number of chemical bonds, and this ensures that the total process is exothermic.

The entropy change for the chemisorption of H_2 on nickel is ca. -68 J (mol H)$^{-1}$ K^{-1} [T22]. Thus the Gibb's free energy of reaction at 300 K can be calculated according to Equation 4-22 as:

$$\Delta G = -46 + (300 \times 0.068) = -25.6 \text{ kJ/mol}$$

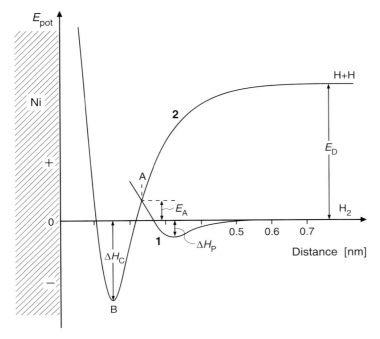

Fig. 4-11. Potential energy and interatomic distances in the adsorption of hydrogen on nickel
Curve 1: physisorption (0.32 nm, $\Delta H_P = -4$ kJ/mol)
Curve 2: chemisorption (0.16 nm, $\Delta H_C = -46$ kJ/mol)
E_D = dissociation energy of H_2 (218 kJ/mol)
E_A = activation energy for adsorption

The probability of reaction is thus extremely high. The diagram also shows that the idea that the H_2 molecule dissociates and is then chemisorbed on the Ni surface is purely hypothetical, and that in fact physisorption precedes chemisorption. The total process can be described schematically as shown in Figure 4-12.

For both types of chemisorption there are numerous examples that exhibit parallells to organometallic chemistry and therefore homogeneous catalysis.

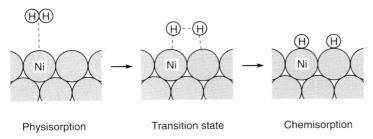

Fig. 4-12. Dissociative adsorption of hydrogen on nickel surfaces

In the chemisorption of alkenes, other surface complexes can occur, for example, π complexes with a donor–acceptor bond and dissociatively bound complexes:

$$\begin{array}{cc} CH_2{=}CH_2 & H\ HC{=}CH_2 \\ \downarrow & |\ \ | \\ M{-}M{-}M & {-}M{-}M{-} \end{array}$$

Metal π complex   Dissociatively chemisorbed ethylene

Dissociative chemisorption occurs preferably with alkenes in which the allylic methyl group is highly activated (e.g., propene). Hydrogen abstraction gives an allyl radical, which can be bound as follows:

$$\begin{array}{cc} CH_2{-}CH{=}CH_2 & \quad\ \ CH \\ | & H_2C\ \ |\ \ CH_2 \\ {-}M{-} & \ \ \ {-}M{-} \end{array}$$

Chemisorbed   Chemisorbed
σ-allyl radical   π-allyl radical

Other species can occur on certain metal oxides:

$CH_3{-}\overset{+}{C}H{-}CH_3$ Carbenium ion on zeolites or ZrO_2

$\underset{*}{CH}{=}CH{-}CH_3$ Propen-1-yl on Al_2O_3

$CH_2{=}\underset{*}{C}{-}CH_3$ Propen-2-yl on Al_2O_3

The molecular chemisorption of ethylene is observed below room temperature, but at higher temperatures the alkene can be cleaved with formation of ethylidyne complexes of the type:

$$\begin{array}{cc} CH_3 & H \\ | & | \\ C & \\ /|\backslash & | \\ M\ M\ M & M \end{array}$$

Another example of dissociative chemisorption is the heterolytic cleavage of hydrogen on metal oxide surfaces. The reaction of hydrogen with a zinc oxide surface produces a zinc–hydride bond and a proton bound to an oxygen center (Eq. 4-25) [T39].

$$H_2 + Zn{-}O{-}Zn{-}O \longrightarrow \overset{H^-}{Zn}{-}O{-}\overset{H^+}{Zn}{-}O \tag{4-25}$$

It is assumed that this reaction is an important step in the catalytic hydrogenation of CO to methanol:

$$CO + 2\,H_2 \longrightarrow CH_3OH$$

Heterolytic chemisorption can also take place on Brønsted acids, as has been shown for MgO (Eq. 4-26) [35].

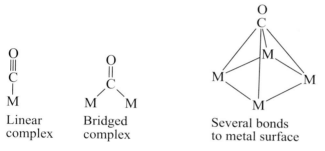

(4-26)

Numerous different adsorption complexes have been observed for CO, which can form linear, bridging, and multicenter bonds with metal atoms:

$$\underset{\text{Linear complex}}{\overset{O}{\underset{|}{\overset{|||}{C}}}-M} \qquad \underset{\text{Bridged complex}}{\overset{O}{\underset{M\quad M}{\overset{||}{C}}}} \qquad \underset{\text{Several bonds to metal surface}}{\text{structure}}$$

The stoichiometry depends on the adsorbing metal and the degree of coverage of the surface by the adsorbate, smooth transitions between the structures being possible (Eq. 4-27).

$$O{=}C{=}M \rightleftharpoons O{=}C\underset{M}{\overset{M}{\diagup}} \rightleftharpoons \underset{M\ M}{\overset{C{=}O}{|\ |}} \qquad (4\text{-}27)$$

The chemisorption of CO is molecular on some transition metals and dissociative on others, depending on the electronic structure of the metal (Table 4-4).

Table 4-4. Chemisorption of CO on transition metals [T20]

Dissociative chemisorption of CO	Boundary region	Molecular chemisorption of CO
Fe 3d^6	Ni 3d^8	Cu 3d^{10}
Mo 4d^4 ⇐	Ru 4d^6 ⇒	50–60 kJ/mol
Ti, Mn, Cr	Re 5d^5	Pd 4d^8
≈ 400 kJ/mol		140–170 kJ/mol
		Pt 5d^8

It should be emphasized once again that the surface state of a solid does not necessarily correspond to the conditions within the solid. For example, it was found that not all copper(II) oxide preparations adsorb CO from the gas phase, and that Cu^{2+} does not react with CO. The active oxides have Cu^+ ions on the surface which can bind CO as a ligand.

Thus predictions of possible bond formation can not be made solely on basis of chemical relationships; geometrical effects must also be considered. For example, CO is adsorbed molecularly on smooth Ni surfaces but dissociatively at steps. The probability of dissociative chemisorption is generally higher at surface defects such as steps and edges.

Nitrogen, which is isoelectronic with CO, is also chemisorbed on metal surfaces. Orbital theory can be used to explain the metal–nitrogen binding strength. Electron density flows to the metal from the bonding π orbitals of the nitrogen molecule, and backdonation occurs from the metal into the antibonding π* N_2 orbitals, weakening the N–N bond. This is of importance in ammonia synthesis. It is assumed that the nitrogen is first molecularly chemisorbed and that the subsequent dissociative chemisorption of the nitrogen molecule is the decisive step of the catalytic cycle.

Oxygen-containing compounds such as alcohols also undergo dissociative chemisorption, an example being the adsorption of gaseous methanol on molybdenum oxide catalysts (Eq. 4-28). Such metal oxides, and in particular mixed metal oxides, act as redox catalysts, as we shall see in Section 4.3.3.

$$-O-\underset{\underset{\diagdown}{\diagup}}{Mo}-O- + CH_3OH \longrightarrow -O-\underset{\underset{\diagdown}{\diagup}}{Mo}-O- \quad (4\text{-}28)$$

(with O on top of left Mo, and H_3CO, OH on right Mo)

The nature of the ligands on metal surfaces is often deduced by comparing their IR spectra with those of comparable inorganic or organometallic complexes [28]. Terminal and bridging CO complexes have been detected on metal surfaces by IR spectroscopy. For the adsorption of CO on Pd surfaces, several readily assignable bands were found:

Terminal
v_{CO} 2060 cm^{-1}

Bridged
1960, 1920 cm^{-1}

Support materials can also strongly influence the spectra of adsorbed CO. For Pt catalysts it was found that the ratio of bridging to terminal ligands was much higher on an SiO_2 support than on Al_2O_3. Thus the strength of adsorption also depends on the nature of the support. For example, an SiO_2 support has only a minor influence on Ni catalysts, whereas Al_2O_3 and TiO_2 have major effects. The CO chemisorption complexes found on supported Ni catalysts are listed in Table 4-5 [T37].

The IR spectra of many hydrocarbon ligands on metal surfaces also resemble those of discrete organometallic species, as shown by the example of ethylene complexes (Table 4-6). Weakening of the double bond is evident both in the supported catalyst and in the isolated complex (for comparison: gaseous ethylene has an IR band at 1640 cm^{-1}).

In the case of nitrogen, coordination to metal surfaces was observed by IR spectroscopy before dinitrogen complexes had been synthesized and characterized.

Table 4-5. IR bands of surface CO complexes on supported Ni catalysts

Bands [cm^{-1}]	Intensity	Structure
1915	strong	μ-CO bridging two Ni
2035	strong	linear Ni–CO
1963	medium	tri-CO on two Ni
2057	medium	linear Ni–CO
2082	weak	linear Ni–CO

Table 4-6. IR bands of ethylene complexes

	Supported catalyst Pd/SiO$_2$	Comparison π complex Pd(C$_2$H$_4$)
ν_{CH_2} [cm^{-1}]	2980	2952
$\nu_{C=C}$ [cm^{-1}]	1510	1502

Method: matrix isolation
Pd in C_2H_4 / Xe = 1 : 100 matrix at 15 K, ultrahigh vacuum

IR spectroscopy was also helpful in elucidating the mechanism of the decomposition of formic acid, a well-known model reaction in heterogeneous catalysis. On metal surfaces formic acid decomposes to hydrogen and carbon dioxide (Eq. 4-29), and it was shown that the reaction proceeds via chemisorbed metal formates.

$$\text{HCOOH} \xrightarrow{A} \underset{\ominus \; \oplus}{\overset{\overset{H}{|}}{\underset{O \; \; \; O}{C}}} \xrightarrow{B} H_2 + CO_2 \qquad \begin{array}{c}H\\|\end{array}$$

(4-29)

Figure 4-13 shows the relative activity of various metal catalysts for the decomposition of formic acid [T20]. The y-axis gives the temperature required to achieve a particular catalytic activity: the lower the temperature, the higher the activity of the catalyst. On the x-axis the heats of formation of the corresponding metal formates are plotted. This so-called volcano plot shows a very good correlation between the strength of adsorption of the formic acid as a metal formate and the heat of formation of the individual compounds.

How can this typical shape of the curve be explained? Left of the maximum are the metals with too weak adsorption (e. g., Ag, Au), and those to the right (Ni, Co, Fe,

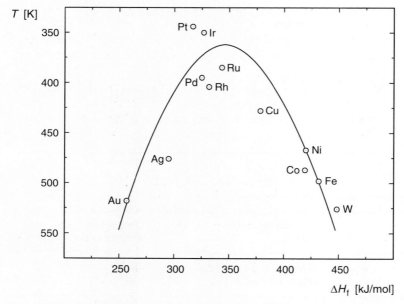

Fig. 4-13. Relative activity of metals for the decomposition of formic acid as a function of the heat of formation of the metal formates (volcano plot)

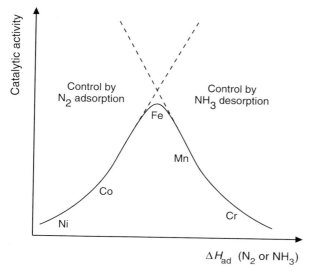

Fig. 4-14. Volcano plot for ammonia synthesis

W) are also poor catalysts since the adsorption complexes are too stable. The most effective catalysts, in the middle, have the appropriate medium binding strength.

Instead of the heat of formation, other thermodynamic quantities can be used, such as the heats of adsorption and desorption. Numerous examples of such volcano plots can be found in the literature.

Another example shows the importance of adsorption strength in ammonia synthesis. In this case the activity of the first transition metal row was measured (Fig. 4-14).

Metals on the left bind N_2 too strongly, and those on the right, too weakly. Exactly the right binding strength was found for iron, the classical ammonia catalyst.

The influence of the electronic structure of the metal and hence its position in the periodic table is also demonstrated by other reactions (Table 4-7).

Table 4-7. Relative reaction rates on transition metal catalysts [T22]

Row	Reaction	Metals and relative reaction rates							
1	Hydrogenation of ethylene (metal catalysts)				Cr 0.95		Fe 15	Co 100	Ni 36	Cu 1.2
2	Hydrodesulfurization of dibenzo-thiophene (metal sulfide catalysts)	Nb 0.5	Mo 2	Tc 13	Ru 100	Rh 26	Pd 3		
3	Hydrogenolysis of CH_3NH_2 to methane (metal catalysts)				Re 0.008	Os 0.9	Ir 100	Pt 11	Au 0.5

Maximum reaction rates are found for metals with six to eight d electrons, a fact which can be explained in terms of electronic effects. The reactants must be rapidly adsorbed on the surface, and the chemisorptive bonding must be strong to attain high adsorbate concentrations. However, these bonds must subsequently be broken so that reaction with other reactants can occur; that is, a compromise is necessary. A general trend, apparent in Table 4-7, is that the bonding strength of chemisorption decreases along a transition metal row. At the beginning of a row, the chemisorption bonds are so strong that they can not be subsequently broken. At the end of a row, the chemisorption bonds are too weak, so that high degrees of coverage of the catalyst surface, and therefore high reaction rates, can not be attained.

To explain the catalytic activity, thermodynamic correlations, often involving the heats of adsorption of the reactants or of related simpler molecules, are widely used. However, the heats of adsorption are often not known, in which case the following quantities are used:

— Heat of desorption
— Heat of formation of intermediates
— IR frequencies of metal–adsorbate bonds, etc.

Such correspondence of the surface chemistry with the physicochemical data of the solid is not always found, since, as we have seen, the surface of a solid rarely corresponds to its interior.

Figure 4-15 describes the adsorption of hydrogen. The chemisorption enthalpy increases from group 4 to 6, then decreases. In groups 8–10 it remains almost constant [T20].

An anomaly occurs at manganese, which is attributed to the half-filled d shell.

Interestingly the heat of formation of ZrH_2 of -163 kJ/mol corresponds exactly to the heat of adsorption of H_2 on Zr. Similar dependences of the adsorption enthalpies as a function of the position of the metal in the periodic table have also been found for N_2, CO, and CO_2.

Finally, let us discuss some industrial reactions with the aid of the concept introduced above. The hydrogenation of unsaturated hydrocarbons is one of the most important catalytic reactions in organic chemistry. In particular the hydrogenation of ethylene was long studied as a model reaction for testing the activity of metal catalysts [22]. Today's models for the catalytic hydrogenation of unsaturated compounds are largely based on the general theory of catalytic hydrogenation developed by Balandin and by Horiuti and Polanyi [30].

In general transition metal catalysts are used, and they can be roughly ordered according to their catalytic activity as follows:

Ru, Rh, Pd, Os, Ir, Pt > Fe, Co, Ni > Ta, W, Cr ≈ Cu

For the hydrogenation of olefinic double bonds, both the alkene and hydrogen must be activated. Various mechanisms have been proposed for alkene hydrogenation, one of which we will discuss in more detail here (Scheme 4-1) [32].

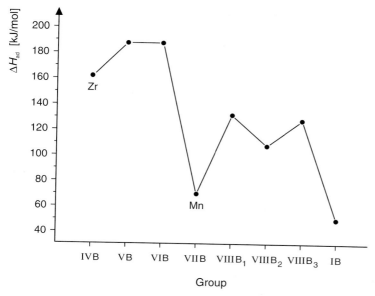

Fig. 4-15. Mean chemisorption enthalies of hydrogen as a function of the position of the elements in the periodic table [T20]

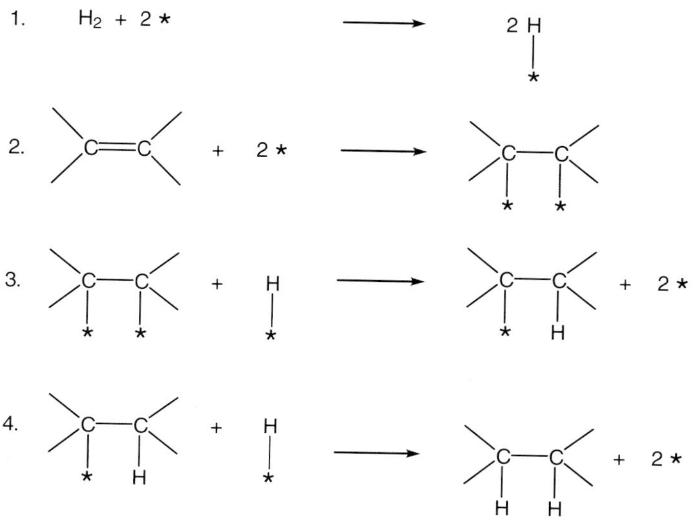

1. $H_2 + 2\,\star \longrightarrow 2\,\underset{\star}{H}$

2. $\mathrm{C{=}C} + 2\,\star \longrightarrow \underset{\star\;\;\star}{\mathrm{C{-}C}}$

3. $\underset{\star\;\;\star}{\mathrm{C{-}C}} + \underset{\star}{H} \longrightarrow \underset{\star\;\;H}{\mathrm{C{-}C}} + 2\,\star$

4. $\underset{\star\;\;H}{\mathrm{C{-}C}} + \underset{\star}{H} \longrightarrow \underset{H\;\;H}{\mathrm{C{-}C}} + 2\,\star$

Scheme 4-1. Mechanism for the hydrogenation of an alkene [32]

The chemisorbed alkene reacts stepwise with atomically adsorbed hydrogen. In step 3, a hemihydrogenated product is formed; in the case of ethylene this is an ethyl radical, which in the final step is hydrogenated to ethane, desorption of which frees the active center for further reaction.

Analytical methods such as adsorption measurements with H_2, H_2/D_2 exchange reactions, and IR spectroscopy have shown that H_2 can be adsorbed on the surface in different forms, as shown here for the example of Pt:

$$\begin{array}{ccc} H_w & H_s & H_w \\ | & / \quad \backslash & | \\ Pt & Pt & Pt \end{array} \quad \begin{array}{c} H_s \\ / \quad \backslash \\ Pt \end{array}$$

H_w = weakly bound species
H_s = strongly bound species

The singly bonded H atoms perpendicular to the surface can be distinguished from the H atoms more strongly bonded between two Pt centers by IR spectroscopy. In addition, molecular adsorption of hydrogen at a surface site also occurs. An analogy to hydride complexes can be seen, for which mono- and dihydrido species are also known, as demonstrated by the example of the Ir complexes [IrHCl$_2$(PR$_3$)$_3$] and [IrH$_2$Cl(PR$_3$)$_3$].

We have already dealt with such complexes as important intermediates in homogeneous catalysis. Hydrogen is homolytically cleaved at the metal center in an oxidative addition reaction to give a dihydrido complex, which can transfer hydrogen stepwise to the coordinated olefin. Hence the similarity between heterogeneous and homogeneous catalysis is not surprising, and industrial reactions can often be catalyzed both heterogeneously and homogeneously by the same metal. For example, [RhCl(PPh$_3$)$_3$] and Rh/activated carbon are both active hydrogenation catalysts.

Palladium catalysts are very important in selective hydrogenation reactions. In the industrial production of alkenes, acetylenes and other compounds must be removed prior to work up. Let us consider the proposed mechanism in more detail. According to Scheme 4-2, the hemihydrogenated intermediate is the vinyl radical, which can react further to give ethylene or ethane. The selectivity of the reaction depends strongly on the catalyst and the reaction conditions. Selectivity generally decreases with increasing hydrogen pressure and decreasing temperature. Palladium usually exhibits complete selectivity for the hydrogenation of acetylene and related compounds. The high selectivity can be attributed to the fact that it adsorbs H atoms dissociatively on the surface in relatively low concentration.

Another standard reaction in which the dissociative adsorption of hydrogen is the rate-determining step is H_2/D_2 exchange in hydrocarbons. The following activity series was found:

Rh > Pd > Pt > Ni > Fe > W > Cr

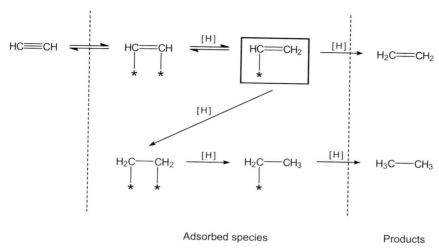

Scheme 4-2. Mechanism for the hydrogenation of acetylene [T20]

This is one of the best methods for determing the nature and reactivity of adsorbed intermediates on the catalyst surface. Model reactants include CH_4, for which the role of adsorbed CH_2 groups was proved, and ethane, for which alkyl/alkene interconversion on the metal surface was investigated.

Exercises for Section 4.3.1

Exercise 4.10

The activation of methane, ethylene, and propylene on metal surfaces is described as follows:

a) $CH_4 + 2\,M \longrightarrow H{-}M + CH_3{-}M$

b) $C_2H_4 + 2\,M \longrightarrow \underset{\underset{M}{|}}{CH_2} - \underset{\underset{M}{|}}{CH_2}$

c) $CH_3{-}CH{=}CH_2 + M \longrightarrow \underset{\underset{M}{|}}{CH_2}{=}CH{=}CH_2$

Compare the three processes.

Exercise 4.11

Modern investigations of the Fischer–Tropsch synthesis by X-ray and UV photo-electron spectroscopy have shown the presence of surface carbides and oxygen atoms in the adsorption of CO on various metals.

- W and Mo dissociate CO below 170 K
- Fe and Ni dissociate CO between 300 and 420 K, whereby Ni reacts faster, forming thermally unstable carbides
- Platinum group metals bind CO mainly nondissociatively

The following mechanism has been proposed: adsorbed hydrogen removes the oxygen as water, which is desorbed, and converts the C fragments to CH and CH_2 groups, which then polymerize:

$$H \quad CH_2 \quad CH_2 \longrightarrow CH_3 \quad CH_2 \longrightarrow CH_3-CH_2 \xrightarrow{H_2} H + RH$$

Which catalytic properties are to be expected for the above-mentioned metals?

Exercise 4.12

At high temperature finely divided titanium reacts with N_2 to give a stable nitride. The rate-determining step in ammonia synthesis with iron catalysts is cleavage of the $N \equiv N$ bond. Why is titanium inactive and iron active as a catalyst for ammonia synthesis?

Exercise 4.13

The effectiveness of Pt in catalyzing the reaction $2 H^+_{(aq)} + 2 e^- \rightarrow H_{2(g)}$ is lowered by the presence of CO. Give an explanation.

Exercise 4.14

The activation energy of a catalytic reaction is 110 kJ/mol. On using catalyst pellets, an activation energy of only 50 kJ/mol is measured. Give an explanation for this finding.

Exercise 4.15

For the adsorption of CO on W surfaces, two values of the activation energy for desorption are given in the literature: 120 and 300 kJ/mol. What could the reason for this be?

Exercise 4.16

The following examples of surface complexes of molecules are depicted in a publication:

$$
\underset{a}{\overset{H_3N-}{\underset{Al^{3+}}{|}}} \quad \underset{b}{\overset{O}{\underset{Pt}{\overset{|||}{C}}}} \quad \underset{c}{\overset{O}{\underset{Pt}{\overset{||}{C}}}\overset{}{\underset{Pt}{}}} \quad \underset{d}{\overset{H}{\underset{Pt}{|}}\overset{H}{\underset{Pt}{|}}} \quad \underset{e}{\overset{CH_3}{\underset{Pt}{\overset{|}{CH_2}}}\overset{H}{\underset{Pt}{|}}} \quad \underset{f}{\overset{H^-}{\underset{Zn^{2+}}{|}}\overset{H^+}{\underset{O^{2-}}{|}}} \quad \underset{g}{\overset{H_2C=CH_2}{\underset{Pt}{|}}} \quad \underset{h}{\overset{H_2C-CH_2}{\underset{Pt}{\diagup}\underset{Pt}{\diagdown}}}
$$

a   b   c   d   e   f   g   h

Discuss these examples.

Exercise 4.17

What is the preferred mode of bonding of H_2S on a catalyst surface?

4.3.2 Steric Effects [T22, T37]

Apart from energetic and electronic effects, steric (geometric) effects also play an important role in chemisorption and heterogeneous catalysis [32]. The porosity and the surface of solids must therefore also be taken into account. A steric factor means that a molecule has to be adsorbed on the catalyst in such a manner that it fits properly on the surface atoms. Only then can it be readily activated.

As early as 1929 Balandin introduced the multiplet theory, which is based on purely structural and geometric considerations, into the field of catalysis. If we assume that the molecule to be adsorbed is large and therefore is not adsorbed at a single active center (single-point adsorption), but at two or more centers (mulipoint adsorption), then it becomes clear that the steric conditions and topology of the surface are of crucial importance for the activation of the reactants. Balandin referred to the principle of "geometric correspondence" between the reactant molecules and the surface atoms of the catalyst.

In enzymatic catalysis this "key/keyhole" mechanism is so pronounced that the reactant molecule must fit exactly to the geometry of the catalyst for reaction to occur. Such reactions, which normally occur with 100% selectivity, are of course not found in heterogeneous catalysis.

Extension of the model then led to the concept of active centers on the catalyst surface, presumably attributable to free valences or electron defects (see Section 4.3.3). Therefore, methods for characterizing catalyst surfaces are of great importance, and they play a key role in understanding catalysis.

It is tempting to use summed parameters such as lattice type and interatomic distances in the lattice to explain particular reactions [T40]. However, this is rarely successful, and one should not expect too much of this concept.

One of the first predictions made on the basis of steric effects was that the ease of chemisorption of diatomic molecules should strongly depend on the lattice dimensions of the metallic catalysts. The reasoning was that for large interatomic distances, diatomic molecules would have to dissociate to be completely chemisorbed, while for closely packed lattices, repulsion effects would hinder chemisorption. This is exemplified by our first example, the dehydrogenation of cyclohexane.

It was shown that only elements with interatomic distances between 0.248 and 0.277 nm catalyze the dehydrogenation of cyclohexane. Table 4-8 lists the lattice distances and lattice types for several metallic catalysts.

Table 4-8. Structure and lattice spacings (distance to next-nearest neighbor in nm) of metals [T40]

Lattice type		
Body-centered cubic (bcc)	Face-centered cubic (fcc)	Hexagonal close packing (hcp)
Ta 0.286	Ce 0.366	Mg 0.320
W 0.272	Ag 0.288	Zr 0.312
Mo 0.272	Au 0.288	Cd 0.298
V 0.260	Al 0.286	Ti 0.292
α-Cr 0.246	Pt* 0.276	Os* 0.270
α-Fe 0.248	Pd* 0.274	Zn* 0.266
	Ir* 0.270	Ru* 0.266
	Rh* 0.268	β-Co* 0.252
	Cu* 0.256	Be 0.224
	α-Co* 0.252	
	Ni* 0.248	

* Metals that catalyze the dehydrogenation of cyclohexane

It can be seen that only metals with close packed structures, that is the highest surface-atom density, catalyze this reaction:

fcc, from Pt to Ni
hcp, from Os to Co

Apparently, many of the best metal catalysts have the fcc lattice structure. A prerequisite for fundamental investigations of heterogeneous catalysis is that the surface structure of the metal be exactly known and that no impurities are present. Single crystals are preferred for such investigations. Since metals are crystalline, the atoms at the surface form regular two-dimensional arrangements.

A widely used system for describing the lattice planes of a crystalline structure are the Miller indices. These indicate which and how many crystallographic axes of a unit cell are intercepted by a lattice plane. The indices give the relative axis sections a, b, and c in reciprocal whole-number form (Fig. 4-16).

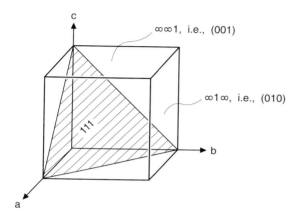

Fig. 4-16. Lattice planes in a cubic lattice with Miller indices

The surface of a cube intersects only one axis and therefore has the designation (100). The surface of a prism intersects two axes of the cubic system, i.e., (110). An octahedron surface is designated (111) since it intersects all three axes at equal distances.

Many catalysts have the fcc structure. The arrangement of the atoms in the above-mentioned surfaces is depicted in Figure 4-17. Also shown is the number of neighboring atoms and free valences of the surface atoms for the example of the nickel lattice [T33]. The highest number of free valences, namely five, occurs for the prismatic faces.

Single crystals a few centimeters in size can be grown for many transition metals and cut so that a specific surface is exposed. Such single-crystal surfaces have the advantage that they can be precisely characterized by modern methods of surface analysis, and molecules adsorbed on the surface, such as CO, N_2O, O_2, and hydrocarbons, can be detected and their bonding modes determined.

Single-crystal surfaces are of course of no importance as practical catalysts, but they provide interesting information about the processes that can take place on real polycrystalline surfaces [18]. Industrial catalysts consist of numerous small crystallites that are randomly oriented and whose surfaces present many crystallographic planes to the reactants. In addition they exhibit steps and lattice defects. The dispersity of a catalyst (particles/cm^3) and the surface of a catalyst are closely interrelated.

Figure 4-18 shows schematically the stepped surface of a catalyst with lattice defects, protruding atoms, which may also be adsorbed species, and kinks.

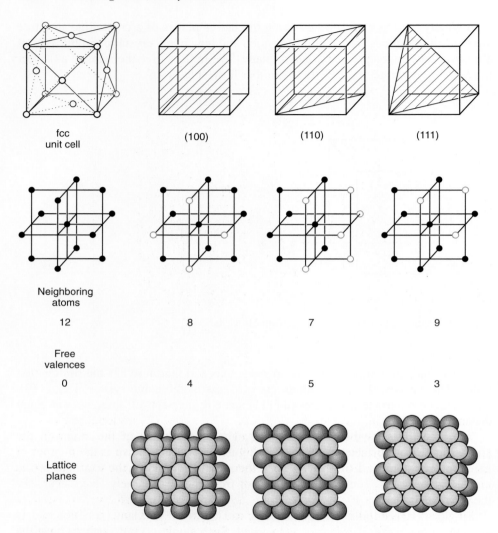

Fig. 4-17. Neighboring atoms and free valences of nickel surfaces in the face-centered cubic (fcc) lattice

In addition to terraces of surfaces with high density such as (111) and (110), there are also steps of monoatomic height. For example, the (557) surface of platinum consists of terraces of (111) surfaces linked by monoatomic (001) surfaces. Such stepped surfaces can also be characterized by the methods of surface science. Interestingly, these stepped surfaces are remarkably stable under various reaction conditions [32].

Especially crystallites of greater than 10 nm in diameter can exhibit high-index crystal surfaces, and the kinks can even be seen in scanning electron micrographs.

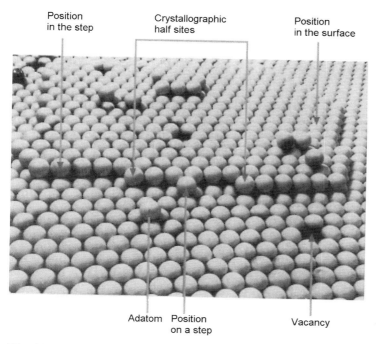

Fig. 4-18. Model of a single-crystal surface (BASF, Ludwigshafen, Germany)

The species actually responsible for the catalytic activity are atoms or groups of atoms (active centers) in the catalyst surface whose chemisorption properties depend strongly on the degree of dispersion of the solid. Thus the catalyst turnover number TON is used as a measure of atomic regions of the catalyst surface (see also Section 1.2) [26].

Since steric factors are not important in all heterogeneously catalyzed reactions, a distinction can be made between structure-sensitive reactions, which react to changes in the surface structure, and structure-insensitive reactions. Numerous examples exist for both types of reaction, and we shall distinguish between them on the basis of the parameters that influence them (Table 4-9).

Table 4-10 lists industrial examples of both types of reaction.

Complete separation according to reaction type is, however, not possible, as will be shown in the following examples. Whereas the hydrogenation of ethylene on nickel catalysts is structure-sensitive, it proceeds on platinum crystals, foils, and supported catalysts with almost constant rate and activation energy. The reaction on rhodium is also structure-insensitive. At normal pressure, the Pt(111) and Rh(111) surfaces are both covered by a monolayer of strongly chemisorbed ethylidyne C_2H_3.

Hydrodesulfurization is structure-insensitive over Mo catalysts but structure-sensitive on Re catalysts. Ethylene oxidation on Ag catalysts is classified as struc-

Table 4-9. Classification of metal-catalyzed reactions [T24]

Reactions	Effects and their influences				
	Structure	Alloy formation	Cat. poisoning	Type of metal	Multiplicity of the active centers
Structure insensitive	–	low	moderate	moderate	1 or 2 atoms
Structure sensitive	moderate	large	large	very large	multiple centers

Table 4-10. Steric effects in chemical reactions

Structure-sensitive reactions	Structure-insensitive reactions

Hydrogenolysis:
  Ethane (Ni)
  Methylcyclopentane (Pt)
  Cyclohexane (Pt)
Hydrogenation:
  Benzene (Ni)
  Ethylene (Ni)
Isomerization:
  Isobutane, hexane (Pt)
Cyclization:
  Hexane, heptane (Pt)
  Ammonia synthesis (Fe)
Methanization
	Ring opening: cyclopropane (Pt)
	Hydrogenation: benzene (Pt)
	  ketones
	Dehydrogenation:
	  cyclohexane (Pt)
	CO oxidation
	Oxidation of ethylene to ethylene oxide (Ag)

ture-insensitive. Probably the oxygen modifies the surface such that each surface reacts the same.

Let us consider the hydrogenation of ethylene on Ni catalysts in more detail. The adsorption of ethylene on nickel is associative, especially in the presence of hydrogen. Spectroscopic investigations have shown that the ethylene double bond opens, forming two σ bonds to neighboring Ni atoms and giving the ethane structure.

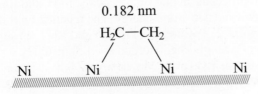

The bond should, however, not be too strong, so that further reaction is possible. For the nickel surfaces with low Miller indices, two Ni–Ni bond lengths were found: 0.25 and 0.35 nm. The results of LEED investigations are summarized in Table 4-11 [T19, T23].

Table 4-11. Adsorption and hydrogenation of ethylene on nickel surfaces [T23]

Ni–Ni distance	Surfaces	Ni–C–C angle	Binding	Catalytic effect
0.25 nm	(111)	105°	stable, strong	low
0.35 nm	(100), (110)	123°	weaker	high

The following explanation can be given for the experimental findings. For the bond length of 0.25 nm, an Ni–C–C bond angle of 105° can be calculated for two-point adsorption. Since this is close to the tetrahedral angle of 109°, stable chemisorption of ethylene on the (111) face can be assumed. For the longer Ni–Ni distance of 0.35 nm, the geometrical situation is less favorable, and chemisorption is therefore weaker. The ethylene molecule is strained and thus can more readily be hydrogenated.

These considerations can also be applied to other metals. Thus the (100) planes of metals with larger atomic spacings than nickel (e.g., Pd, Pt, and Fe) should exhibit weaker chemisorption, and the same should also be true of metals with shorter interatomic distances such as tantalum. Figure 4-19 shows the rate of ethylene hydrogenation as function of metal–metal distance (volcano plot).

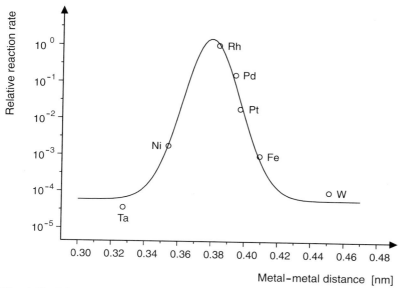

Fig. 4-19. Ethylene hydrogenation as a function of the metal–metal distance in the lattice [T35]

Since two-point adsorption is no longer possible if the metal–metal distance is too large, the optimal catalyst for ethylene hydrogenation should have a certain medium interatomic distance. This is the case for rhodium with 0.375 nm, but since energetic aspects (adsorption enthalpies) must also be taken into account, it can not be said that this is solely the result of steric effects.

It is found in many reactions that a particular surface is favored. For example the (111) surface is particularly active in fcc and hcp metals. A strong dependence is found for ammonia synthesis on iron catalysts (Table 4-12) [T35]. Ammonia synthesis is one of the most structurally sensitive reactions. The opposite order was found for the decomposition of ammonia on copper, i. e., (111) > (100). In the decomposition of formic acid, the (111) surface is three times more active than (110) or (100).

Table 4-12. Activity of surfaces in ammonia synthesis [T35]

Surface	Relative activity
110	1
100	21
111	440

In the hydrogenation of benzene and the dehydrogenation of cyclohexane, six-point adsorption of the molecule on the catalyst has been found. The double bonds open, and chemisorption occurs by formation of σ bonds. Here, too, the (111) surface is clearly favored.

In the next example we will consider the catalytic oxidation of CO [4]. LEED studies have shown that in CO adsorption, the Pd surface is covered by an ordered monolayer. The CO molecules are undissociatively adsorbed by at least two surface sites. The IR spectra of CO adsorbed on Pd (111) surfaces showed bands whose position depends on the degree of coverage of the surface. At low degrees of coverage up to $\theta = 0.18$, a weak $\nu(CO)$ band is observed at 1823 cm^{-1}. It was concluded that the CO molecule is bound to three Pd atoms, which are present in excess, weakening the C–O bond:

up to $\theta = 0.18$ $\theta = 0.5$ and above
$\nu_{CO} = 1823$ cm^{-1} $\nu_{CO} = 1920–1946$ cm^{-1}, increasing intensity

At higher degrees of coverage, fewer free Pd atoms are present on the surface, and a bridging structure with a stronger C–O bond is formed. This structure is retained up to the formation of a monomolecular layer. Thus the CO molecules are bound not by single metal atoms but by an ensemble of several metal atoms.

In the oxidation of CO, the reaction partner oxygen must also be considered. There are two possibilities for the adsorption of two gases on solid surfaces:

- Cooperative adsorption: the two partners form an common ordered surface structure
- Competitive adsorption: the two gases hinder one another in adsorption and form two independent surface structures

Our example is a case of competitive adsorption. If the catalyst is first exposed to O_2, then atomic adsorption of oxygen on the surface occurs. If CO is then introduced at room temperature, the reaction proceeds rapidly by the Eley–Rideal mechanism (Eq. 4-30).

$$O^* + CO \longrightarrow CO_2 \quad \text{fast} \tag{4-30}$$

If CO is first adsorbed and then oxygen is introduced then no reaction occurs (Eq. 4-31).

$$CO^* + \tfrac{1}{2} O_2 \nrightarrow CO_2 \tag{4-31}$$

Finally, if a Pd surface partially covered with CO is allowed to react with O_2, then the latter is adsorbed at the free sites, and ordered surface structures are formed. Only at the boundary layers of the two adsorbed reactants is reaction then possible, and this proceeds slowly according to the Langmuir–Hinshelwood mechanism (Eq. 4-32).

$$CO^* + O^* \longrightarrow CO_2 \quad \text{slow} \tag{4-32}$$

Hence the oxidation process depends on the fastest reaction (Eq 4-30), and this is also the case when mixtures of CO and O_2 react. At low temperatures CO blocks the surface and the reaction is slow. With increasing temperature, above ca. 100 °C, CO is partially desorbed, and O_2 is chemisorbed on the surface. The reaction rate passes through a maximum around 200 °C, after which it falls again. The reaction is structure-insensitive over a wide range, as has been shown on various Pd surfaces [32].

A similar course of reaction was found on Pt surfaces [T36]. Again, CO undergoes molecular adsorption, and the degree of coverage decreases rapidly with increasing temperature (Fig. 4-20a). This is shown by the residence times on the surface:

Room temperature ∞
150 °C ca. 1 s
400 °C ca. 10^{-4} s

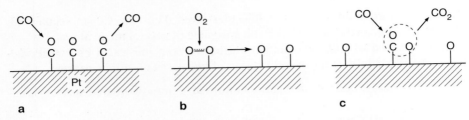

Fig. 4-20. Oxidation of CO on platinum surfaces

The O_2 is initially adsorbed in molecular form as a peroxide-like compound, which rapidly dissociates with release of energy (Fig. 4-20b). Since the oxygen atoms require several free centers for adsorption, saturation coverage with oxygen is rapidly reached.

The low degree of coverage allows adsorption of CO between the oxygen atoms, and the reaction proceeds by the Langmuir–Hinshelwood mechanism. The CO_2 product is only weakly bound on the surface and is rapidly desorbed into the surrounding gas phase (Fig. 4-20c). As in the case of Pd catalysts, no reaction is observed between chemisorbed CO and O_2 from the gas phase.

Platinum catalysts are of major importance for the activation of hydrogen and in reactions of hydrocarbons (e. g., hydrogenation, dehydrogenation, hydrogenolysis). In many cases steps and kinks on the surface have a major influence on the catalytic activity.

Modern surface analysis methods such as LEED allow the number of step and kink atoms to be determined. For example, 2.5×10^{14} step atoms per square centimeter were found for a ($\bar{5}57$) surface of platinum, and 2.3×10^{14} step atoms and 7×10^{13} kink atoms per square centimeter for a ($\bar{6}79$) surface [32].

It was found that oxygen and hydrogen are not adsorbed on smooth (100) and (111) surfaces of Pt but on surfaces with an ordered step structure. The (111) surface is also inactive in the dehydrogenation of cyclohexane. A paticularly strong dependence of the activity on the density of step and kink atoms was observed in the hydrogenolysis of cyclohexane to *n*-hexane. Here the Pt atoms at kinks are an order of magnitude more active than the step atoms. On the basis of strength of coordination and catalytic activity, three types of Pt atoms can be distinguished:

- Largely coordinatively saturated surface atoms: low activity
- Step atoms: more active, catalyze the cleavage of C–H and H–H bonds
- Highly coordinatively unsaturated kink atoms: preferably catalyze the cleavage of C–C bonds

The step and kink atoms resonsible for C–H and C–C bond cleavage do not become covered by carbon and therefore are not subject to deactivation by surface coking. It is assumed that any carbon layer forming here is immediately removed by hydrogenation.

Similar investigations have been carried out on ethylene and CO. Here, too, the reactivity of steps was found to be much higher than that of smooth surfaces. The following chemisorption complexes of CO on Pt were detected by IR spectroscopy:

v(CO) on steps: 2066 cm^{-1}, low coverage, weaker CO bond
v(CO) on terraces: 2090 cm^{-1}, high coverage

Another good example for the function of stepped surfaces is the adsorption and decomposition of acetonitrile on Ni surfaces (Fig. 4-21) [11].

It was shown that on smooth (111) surfaces, the binding of acetonitrile is weak and reversible. At 90 °C the molecules are desorbed, with only 1–2% undergoing cleavage with loss of hydrogen to leave C and N fragments on the surface. On (110) surfaces, which have a higher density of steps, 90% of the molecules are decomposed at 110 °C. This experimental finding is explained by the fact that the CN group is perpendicular to the surface in both cases. On smooth surfaces there is no interaction of the CH$_3$ group with the catalyst surface (Fig. 4-21 a), and the molecule remains largely intact. In contrast, molecules that are adsorbed on or next to steps can be readily decomposed (Fig. 4-21 b, c).

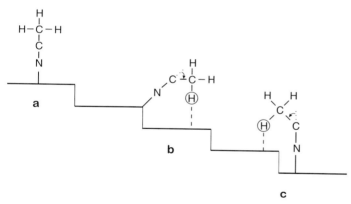

Fig. 4-21. Adsorption and cleavage of acetonitrile on Ni surfaces [11]

As the above examples have shown, the atomic surface structure of a catalyst can have a considerable influence on the catalyst activity and the selectivity of heterogeneously catalyzed reactions. The surface structure of a catalyst metal particle is characterized by the nature of the surface and the ratio of surface, step, and corner atoms. While the characterization of a single particle is relatively simple, it is practically impossible for a real catalyst or supported catalyst due to the distibution of particle sizes and shapes [20].

Measurement of the metal dispersion (ratio of metal surface to total metal content) allows qualitative assignments to be made. In the case of noble metals it is generally determined by adsorption measurements with H$_2$ or CO. Corner atoms dominate for small highly dispersed metal particles, the maximum number of step atoms occurs at medium dispersity, and, as would be expected, terrace atoms are predominant at low degrees of dispersion.

The following were measured for a uniformly dispersed supported Pd catalyst:

Pd particle size	4–10 nm
Pd specific surface area	9.5 m^2/g
Dispersion	21%
BET specific surface area	500 m^2/g

At the usual commercial metal concentrations of 0.1–1%, the dispersion can range from 40 to nearly 100%, with particle sizes of 1–4 nm [T41].

In general, catalyst activity increases with increasing size of the catalyst surface. However, since many reaction rates are strongly dependent on the surface structure, a linear correlation between catalyst activity and surface area can not be expected. In some reactions the selectivity of the catalyst decreases with increasing surface area.

The surface of the support is also important. Catalytic transformations such as hydrogenation, hydrodesulfurization, and hydrodenitrogenation are favored by large support surface areas, whereas selective oxidations such as olefin epoxidation do not require a support surface to suppress problematic side reactions.

Modern methods of surface characterization allow relationships to be found between catalyst structure and catalyst behavior, even for highly complex industrial catalysts. A goal of catalyst research is to use such methods to optimize catalyst production.

Exercises for Section 4.3.2

Exercise 4.18

a) Assign the lattice planes (100), (110), and (111) to the following surfaces:
   – Cube surface
   – Octahedron surface
   – Prism surface

b) What significance do these surfaces have in heterogeneous catalysis?

Exercise 4.19

A three-dimensional right-angled lattice is formed by a unit cell with sides of length a, b, and c. The following figure shows a view with the a- and b-axes. What are the Miller indices of the three planes?

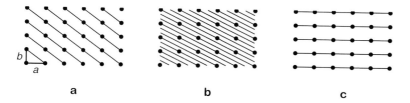

a b c

Exercise 4.20

It is reported in the literature that in the aromatization of hydrocarbons the reaction rate on Pt(111) is an order of magnitude higher than on Pt(100).

a) What is the meaning of the numbers in parentheses?
b) Discuss the findings.

Exercise 4.21

The hydrogenation of ethylene on Pt crystallites, films, foils, and supported catalysts proceeds with practically the same activation energy of 45 kJ/mol. Explain this finding.

Exercise 4.22

In automobile catalytic converters, why is the Pt/Rh catalyst present as a fine dispersion on a ceramic surface rather than as a foil?

4.3.3 Electronic Factors [29, L33]

The concept of electronic factors in catalysis deals with the relationship between the electronic structure of solids, which depends on their physical properties, and the reactivity of adsorbed intermediates.

The key question was: how does the catalytic activity of a solid depend on its geometrical and electronic properties?

In the 1960s extensive searches for electronic effects were undertaken, but although much data and many understandings were obtained, a generally valid catalyst concept could not be developed. Nevertheless, the concept is useful for explaining many experimental findings and in classifying catalysts. For solids, two classes of catalysts can be distinguished:

Redox Catalysts [T34]

This group of catalysts comprises solids exhibiting electrical conductivity, that is, having mobile electrons (metals and semiconductors). Many reactions proceed by the redox mechanism, for example:

- Hydrogenation of alkenes, aromatics, and other compounds with double bonds
- Hydrogenation of CO and CO_2 to methane
- Ammonia synthesis
- Synthesis of hydrocarbons and alcohols from synthesis gas
- Oxidation of hydrocarbons, SO_2, NH_3, etc.
- Dehydrogenation of organic compounds
- Decomposition of formic acid
- Polymerization of hydrocarbons

These are all homolytic processes in which chemical bonds are broken with the aid of the catalyst (Eq. 4-33). This leads to formation of radicals, followed by electron transfer between the reaction partners.

$$Cat.^{\bullet} + A:R \longrightarrow A:Cat. + R^{\bullet} \qquad (4\text{-}33)$$

Typical redox catalysts are metals, semiconductors (e.g., metal oxides in various oxidation states), and special metal complexes. Metals that form an oxide layer on the surface under oxidizing conditions can also be regarded as semiconductors.

Acid/Base Catalysts (Ionic Catalysts)

These catalysts have no mobile charge carriers and thus behave as isolators. With increasing temperature the isolator property is partially lost. Ionic catalysts do not cleave electron pairs in the reactants. Charge is carried by ions, mainly protons. Such heterolytic reactions with the catalyst can be formulated as shown in Equation 4-34).

$$A:Cat. \longrightarrow A^{+} + :Cat.^{-} \quad \text{or}$$
$$A:Cat. \longrightarrow A:^{-} + Cat.^{+} \qquad (4\text{-}34)$$

Heterolytic cleavage is energetically less favorable than homolytic cleavage. Such catalytic processes include:

- Hydrolysis
- Hydration and dehydration
- Polymerization and polycondensation
- Cracking reactions
- Alkylation

- Isomerization
- Disproportionation

These reactions require ionic intermediates and are catalyzed by acidic or basic solids like Al_2O_3 or CaO and especially mixed oxides such as Al_2O_3/SiO_2 and MgO/SiO_2.

Electronic effects can also successfully explain the phenomena of catalyst promotion and catalyst poisoning. Solid-state catalysts can be classified according to their electrical conductivity and electron-transfer properties as shown in Table 4-13.

Table 4-13. Classification of solid-state catalysts

	Conductors	Semiconductors	Insulators
Conductivity range, Ω^{-1} cm^{-1}	$10^6 - 10^4$	$10^3 - 10^{-9}$, increases with increasing temperature	$10^{-9} - 10^{-20}$
Electron transfer	electron exchange metal/adsorbate	electron transfer at high temperatures	–
Examples	numerous metals, mostly transition metals and alloys	metalloids (Si, Ge, etc.); nonstoichiometric oxides and sulfides (ZnO, Cu_2O, NiO, ZnS, Ni_2S_3, etc.)	stoichiometric oxides (Al_2O_3, SiO_2, B_2O_3, MgO, SiO_2/MgO, SiO_2/Al_2O_3, etc.), salts, solid acids

Having discussed the electronic properties of the catalyst, let us now turn our attention to electron transfer between substrate and catalyst. The following classification is relative to the substrate:

- Acceptor reactions: electrons flow from catalyst to substrate; the adsorbate acts as an acceptor (examples: starting materials with high electron affinity; reactions in which oxygen is mobilized)
- Donor reactions: electrons flow from substrate to catalyst (examples: substrates that readily release electrons, i.e., reducing agents with low ionization energies; reactions in which H_2 or CO is mobilized

This classification is shown schematically in Figure 4-22.

4.3.3.1 Metals [T27, T35]

For metals and metal alloys in particular, relationships have been sought between collective properties and catalytic behavior. The metallic state was generally described by the simple band model or the Pauling valence structure theory.

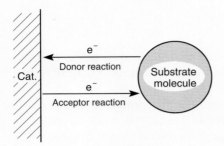

Fig. 4-22. Electron transfer between catalyst and substrate

In metals the valence shell is formed by the s or d band. The main-group elements with their s bands are typical electron donors and form strong bonds with electron acceptors such as sulfur or oxygen; stable sulfides and oxides are formed. These metals are therefore not suitable as catalysts. In contrast the transition metals with their d bands are excellent catalysts. It is noteworthy that both hydrogenations and oxidations can be carried out with d-block elements.

Let us now describe the electronic structure of the transition metals with the aid of the band model. According to this model the metal is a collective source of electrons and electron holes (Fig. 4-23). In a row of the periodic table, the metals on the left have fewer d electrons to fill the bands. There are two regions of energetic states, namely, the valence band and the conduction band with mobile electrons or positive holes. The potential energy of the electrons is characterized by the Fermi level, which corresponds to the electrochemical potential of the electrons and electron holes.

The position of the Fermi level also indicates the number density of electrons in the band model. The energy required to transport an electron from the edge of the Fermi level into vacuum corresponds to the work function ϕ_0 (Fig 4-24a). For the d-block metals, the work function is around 4 eV and therefore in the UV range.

A certain number of free levels or d-holes are available for bonding with adsorbates. The lower the Fermi level, the stronger the adsorption. How do donors and acceptors function in the band model? In the surface layer, the free electrons or holes allow molecules to be bound to the surface, whereby the strength of binding depends on the position of the Fermi level. An acceptor (e.g., O_2) removes electron density from the conduction band of the metal, as a result of which the Fermi level drops to E_F and the work function $\phi_A > \phi_0$ (Fig. 4-24b). A donor (e.g., H_2, CO, C_2H_4) donates electrons to the conduction band of the metal, and the work function becomes corresponding lower: $\phi_A < \phi_0$ (Fig. 4-24c).

Metals normally have a narrow d band. The catalytic properties are strongly influenced by the occupational density of the electrons in this band. In many cases a direct relationship has been found between the catalytic activity of transition metals and the electronic properties of the unfilled d bands. This is shown by the general trend of the rate of adsorption along the transition metal rows. For atomic species

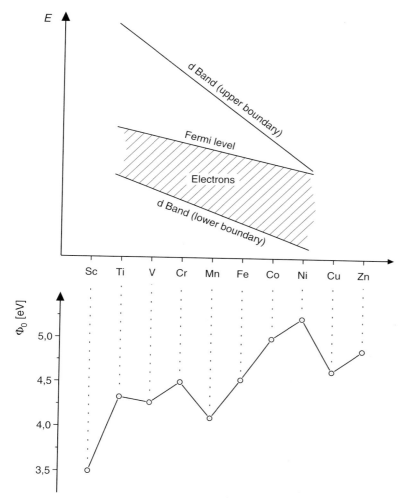

Fig. 4-23. Electron density of the 3 d band and work function ϕ_0 of the transition metals of the fourth period

strong binding is observed on the left-hand side of a row. For molecular species it was found that the rate of dissociative adsorption on the noble metals increases from right to left as a function of the d-band occupation.

In the following example we shall examine the hydrogenation of CO on various metal catalysts. A clear dependence of reaction rate on d-band filling is observed (Fig. 4-25). Thus the familiar volcano plots can also be explained by an electronic factor [38].

Besides the electron occupation of the d bands, another description can be used for obtaining correlations, namely, the valence bond theory of metals. The bonding

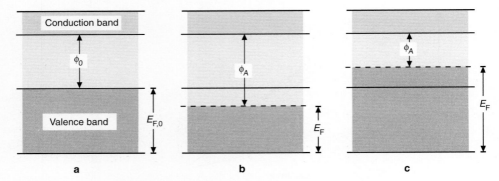

Fig. 4-24. Acceptor and donor function according to the band model
a) No adsorption; b) Acceptor; c) Donor
$E_{F,0}$ = Fermi level; E_F = Fermi energy

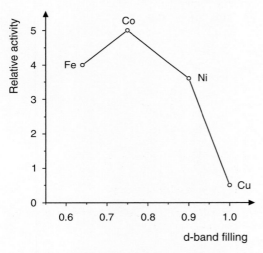

Fig. 4-25. Hydrogenation of CO to methane on various metal catalysts

in a transition metal is partially due to unpaired electrons in bonding d orbitals. The contribution of these d electrons to the valence bonding was termed "percentage d character" of the metallic bonding by Pauling, who made a distinction between three types of d orbitals in transition metals:

– Bonding d orbitals involved in covalent dsp hybrid bonds
– Metallic (free) d orbitals
– Atomic d orbitals

The percentage d character δ can be calculated by using Equation 4-35.

$$\delta = \frac{\text{number of bonding d electrons} \times 100}{\text{bonding electrons} + \text{metallic orbitals}} \qquad (4\text{-}35)$$

For nickel a δ value of 40% has been calculated, and the highest values are found for Ru and Rh (50%).

Relationships have indeed been found between the percentage d character and the catalytic activity, as we shall see for the hydrogenation of ethylene [T20]. However, the results (Fig. 4-26) can not be attributed exclusively to electronic effects. According to Pauling, the d character of the metallic bonding increases with decreasing lattice constants. Therefore, it is possible that geometric factors play the crucial role.

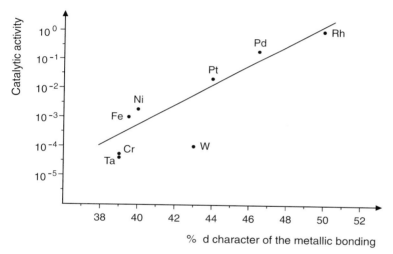

Fig. 4-26. Dependence of the catalytic activity of transition metals in the hydrogenation of ethylene on the percentual d character of the metallic bonding

The degree to which the d band is filled with electrons has considerable influence on the chemisorption capability of metals. Alloying an active metal with another active or even inactive metal can increase or decrease the activity. This is shown by the example of the metals of groups 8–10 of the periodic table, which are particularly active in hydrogenation and dehydrogenation.

Alloying of these metals with the hydrogenation-inactive group 11 metals Cu, Ag, and Au leads to d-band filling of the base metal and lowers the hydrogenation activity. On investigating such bimetallic alloys, enrichment of copper on the surface was found, indicating that phase separation occurred during production of the alloy. This was determined by H_2 adsorption measurements. Hydrogenation of ethylene was investigated on Cu/Ni, Cu/Pt, and Cu/Pd alloys. Increasing the copper content raised the Fermi level and thus led to a lower reaction rate. In contrast, the activity of Ni is increased on alloying with Fe.

Similar effects are observed in other donor reactions, namely, the decomposition of formic acid and the decomposition of methanol. The opposite effect occurred in the decomposition of hydrogen peroxide. Here raising the Fermi level by adding copper accelerates the reaction. It is believed that an acceptor reaction is the rate-determining step, probably formation of O^- or OH^- ions.

In the case of Cu/Ni alloys it was found that the surface is primarily covered with copper over a wide range of compositions (18–95% Ni). This can be shown by adsorption measurements with hydrogen (Fig. 4-27).

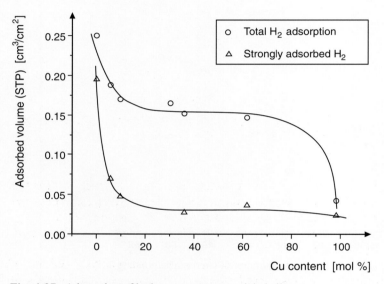

Fig. 4-27. Adsorption of hydrogen on copper–nickel alloys

These Ni/Cu alloys exhibit special selectivity effects, which has been demonstrated in the hydrogenolysis of ethane to methane (Eq. 4-36) and the dehydrogenation of cyclohexane to benzene (Eq. 4-37).

$$C_2H_6 + H_2 \longrightarrow 2\ CH_4 \tag{4-36}$$

$$C_6H_{12} \longrightarrow C_6H_6 + 3\ H_2 \tag{4-37}$$

As shown in Figure 4-28 the rate of hydrogenolysis of ethane decreases by three orders of magnitude on addition of 5% Cu. It was found that at least two neighboring Ni sites are required to take up carbon fragments during the cleavage reaction. Increasing the content of copper, which is enriched on the surface, drastically reduces the number of mutually adjacent Ni sites.

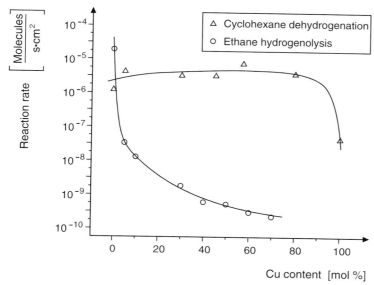

Fig. 4-28. Specific activity of copper–nickel alloys for the dehydrogenation of cyclohexane and the hydrogenolysis of ethane to methane at 316 °C

In contrast, the rate of cyclohexane dehydrogenation increases slightly for small contents of Cu in the alloy, then remains constant over a wide range, and only decreases at high Cu contents. Such effects are also noticeable for other alloys in cyclohexane dehydrogenation. For example, Pd/Ni, Pd/Ru, and Pd/Pt alloys have higher activities than Pd alone.

This effect has some industrial relevance. Thus the hydrogenolysis activity of supported Ru/Os reforming catalysts can be reduced by adding small amounts of copper, so that more alkenes are formed. These high surface area catalysts (ca. 300 m^2/g) contain the metal in the form of mixed crystals, often less than 5 nm in diameter ("bimetallic clusters"). Here, too, the Cu is found exclusively on the surface of the noble metal Ru.

Alloying with active or inactive metals can both accelerate desired reactions and suppress undesired reactions. For example, the addition of Sn to Pd gives selective catalysts for the removal of acetylene from ethylene streams. Similar effects are also observed for Zn, Pb, Ag, and Au [T41]. On alloying Pd with inactive Au, the rate of the reaction between H$_2$ and O$_2$ is increased by a factor of 50. As an additive to Pt, Au increases the rate of isomerization of *n*-hexane tenfold. These effects are explained in terms of a "widening" of the metal–metal bond by another metal.

The range of variation of the catalytic properties of the noble metals is demonstrated by some industrial examples (Table 4-14) [T23].

In more recent work, bimetallic Pd catalysts were investigated in the hydrogenation of saturated and unsaturated aldehydes, and fundamental mechanisms were de-

Table 4-14. Modification of the catalytic properties of the platinum group metals by addition of other metals [T23]

Base metal	Additive	Reaction	Effect of additive
Pt	5–20% Rh	ammonia oxidation	increased NO yield, lower Pt losses
Ag	Au	ethylene oxidation	higher selectivity of ethylene oxide formation
Ag	10% Au	cumene oxidation	increased rate of formation of cumene hydroperoxide
Pt	Ge, Sn, In, Ga	dehydrogenation and hydrocracking of alkanes	increased lifetime due to lower carbon deposition
Pt	Sn + Re	dehydrocyclization and aromatization of alkanes	increased catalyst activity and stability
Pt	Pb, Cu	dehydrocyclization and aromatization of alkanes	effectivity of aromatization
Pt, Pd, Ir	Au	oxidative dehydrogenation of alkanes, n-butene to butadiene, methanol to formaldehyde	improved selectivity
Ir	Au (Ag, Cu)	hydroforming of alkanes and cycloalkanes	high aromatics yield above 500 °C
Pd	Sn, Zn, Pb	selective hydrogenation of alkynes to alkenes	

termined [T32]. The following activity series was found for the hydrogenation of crotonaldehyde with metals of groups 8–10:

$$Ir, Co < Rh < Ni < Pt < Pd$$

The opposite sequence applies in the hydrogenation of n-butyraldehyde to butanol. Both reactions are one-center processes in which the rate-determining step is the formation of a hemihydrogenated intermediate. The following mechanisms have been given for both reactions:

1) Hydrogenation of the alkene in a nucleophilic ligand addition reaction:

$$\underset{\underset{M}{\delta-H}}{\overset{}{C=C}} \xrightarrow{+H_2} \underset{\underset{M}{H\,C\,H}}{\overset{H\,C}{}} \longrightarrow \underset{\underset{M}{\delta-\,H}}{H-C-C-H} \qquad (4\text{-}38)$$

Electron-donor second metals should increase the nucleophilicity

2) Hydrogenation of the carbonyl group in an electrophilic ligand addition reaction:

$$\underset{M}{\overset{\delta + H}{O=C}} \xrightarrow{+H_2} \underset{H\ \ M}{\overset{-\overset{|}{C}-\overline{O}-H}{\underset{H}{\diagup}}} \longrightarrow \underset{\delta + H\diagdown M}{\overset{-\overset{|}{C}-\overline{O}-H}{\underset{H}{|}}} \qquad (4\text{-}39)$$

Electron-acceptor second metals should increase the electrophilicity

The hydrogenation of the double bond in crotonaldehyde to form butyraldehyde proceeds smoothly with pure Pd/Al$_2$O$_3$ supported catalysts (Eq. 4-38). No significant influence of the alloying elements Fe, Sn, and Pb was found. However, these alloying elements accelerate the hydrogenation of butyraldehyde according to Equation (4-39). It was concluded that these elements act as electron acceptors and thus favor the electrophilic ligand addition reaction of hydrogen.

However, this is in disagreement with other results of test reactions reported in the literature, which found an electron-donor function for Fe, Sn, and Pb, partly by IR spectroscopy. This is a further example of the inconsistency of catalyst concepts, which are better regarded as working hypotheses.

According to the current state of knowledge, the band model of metals has several shortcomings. As a simple physical model, it fails to take into account the various types of bonding and surface states. For example, chemisorption processes, which can not cause a change in conductivity, are not considered. Problems occur in particular in explaining the behavior of alloys. The electronic interactions between metal and adsorbate may be masked by steric effects, and experimental results are often not readily interpretable.

For these reasons we shall look at the suitability of metal catalysts in a more empirical manner, giving a few general rules [T40]:

1) Metals are used as catalysts for hydrogenation, isomerization, and oxidation.
2) For reactions involving hydrogen (alone or in combination with hydrocarbons), the following activity series holds:

   Ru, Rh, Pd, Os, Ir, Pt > Fe, Co, Ni, > Ta, W, Cr ≈ Cu

3) Pd is an excellent catalyst that is often active and selective. Pd enables selective hydrogenation of double bonds to be carried out in the presence of other functional groups.
4) Activities sometimes correlate with the percentage d character of the metallic bonding, but there are many exceptions.
5) Activities sometimes correlate with the lattice parameters of the metal.
6) The following metals are particularly stable towards oxygen and sulfur:

   Rh   Pd   Ag
   Ir   Pt   Pd

7) The activity of metals decreases in the order:

(W–Mo) > Rh > Ni > Co > Fe

Numerous relative activity series for particular rections can be found in the literature (Table 4-15). They differ widely and are often contradictory. Therefore, care must be taken in transferring them to other reactions and reaction conditions.

Table 4-15. Relative catalytic activity of metals [T33, T40, T41]

Hydrogenation of olefins	Rh > Ru > Pd > Pt > Ir ≈ Ni > Co > Fe > Re ⩾ Cu
Hydrogenation of ethylene	Rh,Ru > Pd > Pt > Ni > Co,Ir > Fe > Cu
Hydrogenolysis	Rh ⩾ Ni ⩾ Co ⩾ Fe > Pd > Pt
Hydrogenation of acetylenes	Pd > Pt > Ni,Rh > Fe,Cu,Co,Ir,Ru > Os
Hydrogenation of aromatics	Pt > Rh > Ru > Ni > Pd > Co > Fe
Dehydrogenation	Rh > Pt > Pd > Ni > Co ⩾ Fe
Double bond isomerization of alkenes	Fe ≈ Ni ≈ Rh > Pd > Ru > Os > Pt > Ir ≈ Cu
Hydration	Pt > Rh > Pd ≫ Ni ⩾ W ⩾ Fe

4.3.3.2 Semiconductors [16, 35, T27]

Semiconductors are a group of nonmetallic solids whose electron structure is better understood than that of the metals. The band model, already discussed above, is useful for explaining the semiconductor character and catalytic properties of this class of substances.

Two energy bands are present in these crystalline solids: the lowerenergy, electron-containing valence band and the considerably higher lying conduction band. The valence band contains all the electrons of the chemical bonds and the ionic charges in the substance; it has no conductivity. The conduction band contains allowed electronic states, which, however, are all unoccupied.

The electronic properties of the solid depend on the size of the forbidden zone between the two bands. For semiconductors a distinction is made between i- (intrinsic), n-, and p-type semiconductors [T27]. In the i-type semiconductors electrons result from the splitting of homopolar bonds in the solid under the action of heat or light (photoconductivity) (Fig. 4-29).

These excited electrons can jump over the forbidden zone and occupy free states in the conduction band. At the same time a gap arises in the valence band, known as a positive hole. The size of the forbidden zone that must be overcome can be determined. One measure for this is the wavelength at which optical absorption begins. The corresponding energy ε is sufficient to raise an electron from the uppermost level of the valence band into the lowest level of the conduction band. Table 4-16 gives examples of crystals with the Si structure that are regarded as semiconductors.

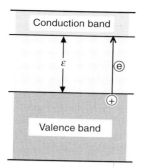

Fig. 4-29. Intrinsic semiconductor with excitation energy

Table 4-16. Excitation energies of semiconductors

Substance	Excitation energy ε, eV (Fig. 4-29)
C (diamond)	5.2
Si	1.09
Ge	0.6
Sn (gray)	0.08

The fraction of electrons of the valence band that are raised to the conduction band by thermal energy corresponds to the Boltzmann factor $\exp(-\varepsilon/2kT)$. The i-type semiconductors play only a minor role in catalysis; the n- and p-type semiconductors are far more important. Nonstoichiometric oxides and sulfides are of industrial importance. The conductivity of these materials is low but can be considerably increased by doping with foreign atoms.

Assume that some of the building blocks of the crystal are replaced by foreign atoms that are electron donors, that is, atoms that readily release electrons on heating. These electrons are located in the forbidden zone, just below the conduction band, and therefore require only a small ionization energy E_i to reach the conduction band (Fig. 4-30a). The positive charge then remains localized on the donor atoms, and we have pure electron conductivity (n conductivity, n = negative).

It is also possible to incorporate electron acceptors in the crystal lattice. They readily take up an electron from the valence band (Fig. 4-30b). On heating, an electron from the valence band enters the acceptor level and remains there, so that a positive hole is generated in the valence band. Thus we now have pure p-type conductivity (p = positive), and the ionization energy E_i in this case is also low.

In semiconductors the Fermi level lies in the forbidden zone. It is the electrochemical potential intermediate between the highest filled and the lowest empty band. The Fermi level can easily be measured, and it is much higher in n-type semiconductors than in p-type semiconductors. Figure 4-30 also shows the corresponding work functions ϕ_A.

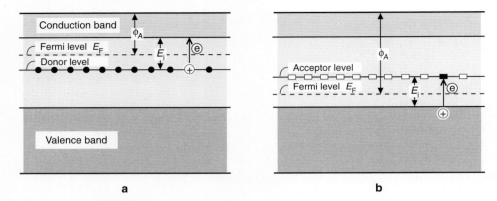

Fig. 4-30. Semiconductors and how they function
a) n-type semiconductor; b) p-type semiconductor

What connection is there between the structure of semiconductors and their properties? As already mentioned nonstoichiometric semiconductor oxides play an important role. On heating, their crystal lattices tend to release or take up oxygen. For an n-type semiconductor such as ZnO, the release of oxygen is described by Equations 4-40 and 4-41.

$$2\,Zn^{2+} + O^{2-} \longrightarrow [2\,Zn^{2+} + \tfrac{1}{2}O_2 + 2\,e] \longrightarrow 2\,\boxed{Zn^+} + \tfrac{1}{2}O_2 \tag{4-40}$$

$$2\,Zn^{2+} + 2\,O^{2-} \longrightarrow [2\,Zn^{2+} + O_2 + 4\,e] \longrightarrow 2\,\boxed{Zn} + O_2 \tag{4-41}$$

The semiconductor capability of ZnO in this case is due to the Zn^+ ions and Zn atoms formed by reaction with oxide ions. The above two reactions can be result from raising the temperature or by reaction with reducing gases such as H_2, CO, and hydrocarbons at room temperature. The Zn ions and atoms occupy interlattice sites and act as electron donors. An equivalent number of quasifree electrons gives electrical neutrality. The formula for the nonstoichiometric compound can be written $Zn_{1+x}O$.

If oxygen is chemisorbed on the ZnO, the conductivity is lowered because the oxygen acts as an electron acceptor (Eq. 4-42).

$$Zn^+ + O_2 \rightleftharpoons Zn^{2+} + O_2^- \tag{4-42}$$

Chemisorbed hydrogen acts as an electron donor and increases the conductivity according to the reaction:

$$Zn^+ + O^{2-} + \tfrac{1}{2}H_2 \rightleftharpoons Zn^{2+} + OH^- \tag{4-43}$$

For a p-type semiconductor like NiO, the take up of oxygen by the lattice is desribed by Equation 4-44.

$$4\,Ni^{2+} + O_2 \longrightarrow 4\,Ni^{3+} + 2\,O^{2-} \tag{4-44}$$

The incorporation of an O_2 molecule in the lattice in the form of O^{2-} ions leads to formation of four Ni^{3+} ions, each of which gives rise to a positive hole, whose mobility in the lattice is responsible for the observed conductivity.

The p-type or defect semiconductor has the formula $Ni_{1-x}O$. Metals that form such p-type oxides are those that exist in several oxidation states. The oxides contain the lower oxidation state form (e. g., Ni^{2+}, Co^{2+}, Cu^+), which can then enter the higher oxidation state (Ni^{3+}, Co^{3+}, Cu^{2+}). The n-type oxides, in contrast, are those that exist in only one oxidation state or in which the highest state is present (e. g., ZnO, TiO_2, V_2O_5, MoO_3, Fe_2O_3).

The conductivity of both n- and p-type oxides is generally low. How can the increased conductivity due to doping be explained? In p-type semiconductors the number of positive holes must be increased, and this can be achieved by incorporating another oxide of lower oxidation state in the lattice. Thus replacing Ni^{2+} ions by Li^+ ions in the nickel oxide lattice leads to an excess of O^{2-} ions (to give electrical neutrality) and formation of Ni^{3+} ions. Doping with trivalent ions such as Cr^{3+} leads to the opposite effect.

In contrast, in an n-type semiconductor like ZnO, doping with Ga_2O_3, Cr_2O_3, or Al_2O_3 leads to increased conductivity, while addition of Li_2O lowers it. Only small amounts of foreign atoms are required for doping, normally less than 1%.

The general behavior of nonstoichiometric semiconductor oxides is summarized in Table 4-17. Table 4-18 classifies the most important oxides according to their electronic behavior.

There are several possibilities for measuring the semiconductor properties of a substance. One of these is to determine the conductivity of the solid at various temperatures; this describes the magnitude of the effect and its energy level. Other possibilities are to investigate the effect of photoelectric and photoelectromagnetic effects on the conductivity and the electron work function.

In practice the results of these measurements are the subject of controversy. A solid can contain various impurities (e. g., Zn and Zn^+ in ZnO) and can have both

Table 4-17. Behavior of nonstoichiometric semiconductor oxides

	n-Type	p-Type
Oxides with ions in interlattice sites	ZnO, CdO	UO_2
Oxides with vacant lattice sites	TiO_2, ThO_2, CeO_2	Cu_2O, NiO, FeO
Type of conductivity	electrons	positive holes
Addition of M_2^IO	lowers conductivity	increases conductivity
Addition of $M_2^{III}O_3$	increases conductivity	lowers conductivity
Adsorption of O_2, N_2O	lowers conductivity	increases conductivity
Adsorption of H_2, CO	increases conductivity	lowers conductivity

Table 4-18. Classification of the metal oxides according to their electronic properties

n-Type	p-Type	i-Type (intrinsic semiconductors)	Isolators
Oxides of main group elements			
ZnO, GeO$_2$, CdO, HgO, SnO$_2$, As$_2$O$_5$, Sb$_2$O$_5$, PbO$_2$, Bi$_2$O$_5$; Al$_2$O$_3$ (at high temperatures)	NiO, Cr$_2$O$_3$, MnO, FeO, CoO, Cu$_2$O, Ag$_2$O, PtO	Fe$_3$O$_4$, Co$_3$O$_4$, CuO	BeO, B$_2$O$_3$, MgO, Al$_2$O$_3$, SiO$_2$, P$_2$O$_5$, CaO, SrO, BaO
Oxides of transition metals			
Sc$_2$O$_3$, TiO$_2$, V$_2$O$_5$, Fe$_2$O$_3$, ZrO$_2$, Nb$_2$O$_5$, MoO$_3$, Ta$_2$O$_5$, HfO$_2$, WO$_3$, UO$_3$			

donor and acceptor levels. The measurements can be carried out on the isolated solid or in the presence of reactants. Interpretation of conductivity, ionization energy, and work function data is difficult. Once again, surface effects must be examined separately from effects inside the lattice.

Thus, similar to the case for metals, the applicability of electronic theory to catalysis is limited. The predictions often contradict experimental results; for example:

1) Oxides and sulfides of the transition metals are the most active, most selective, and industrially most important catalysts. According to electronic theory, numerous other semiconductors should have good catalytic properties, which could be influenced by modification. However, apparently chemical factors are predominant in the catalytic activity.
2) Additives that lead to large changes in the conductivity ought to strongly influence the catalytic properties of the material. Semiconductor oxide and sulfide catalysts are, however, considerably less susceptible to poisoning than metal catalysts. Furthermore, the composition of semiconductor mixed crystals can be varied over a wide range without affecting their catalytic properties.

In the case of this concept, too, empirical findings are of greater interest than exact theoretical predictions of catalytic activity. It is particularly useful for explaining many chemisorption effects and for oxidation reactions.

It can be of practical importance to modify the electronic properties of cheap semiconductor catalysts by doping such that their activity corresponds to that of expensive noble metal catalysts. Two industrial examples of such substitutions are the SCR process (waste-gas purification) and the selective oxidation of methanol to formaldehyde.

Chemisorption on Semiconductors

The chemisorption of simple gases on semiconductors can be relatively simply understood in terms of the chemical reaction of the adsorbate with the catalyst. Reducing gases like hydrogen and CO are strongly and irreversibly adsorbed. On heating, only water and carbon dioxide are detectable. On adsorption, H_2 mainly undergoes heterolytic dissociation (Eq. 4-45):

$$M^{2+} + O^{2-} + H_2 \longrightarrow HM^+ + OH^- \tag{4-45}$$

On heating, the hydroxyl ion is decomposed to water and anionic defects, and a corresponding number of cations are reduced to atoms.

On n-type semiconductors, H_2 and CO almost totally cover the surface, whereas chemisorption on p-type semiconductors is less extensive. In this strong chemisorption a free electron or positive hole from the lattice is involved in the chemisorptive bonding. This changes the electrical charge of the adsorption center, which can then transfer its charge to the adsorbed molecule.

The change in the electrical charge density on the surface can hinder the further adsorption of molecules of the same gas. A decrease in the heat of adsorption with increasing degree of coverage is then observed, and hence a deviation from the Langmuir adsorption isotherm occurs.

Chemisorption of CO usually occurs initially on metal cations, after which it reacts with an oxide ion according to Equation 4-46. This reaction can eventually lead to complete reduction of the oxide to the metal.

$$CO \cdots M^{2+} + O^{2-} \longrightarrow M + CO_2 \tag{4-46}$$

When oxygen is adsorbed on an n-type semiconductor, electrons flow from the donor level, and O^- and O^{2-} ions can be observed. The surface of the solid becomes negatively polarized, and the adsorption of further oxygen requires more and more energy. Therefore the adsorption of oxygen on n-type semiconductors is subject to very rapid auto-inhibition. If n-type semiconductors like ZnO have their exact stoichiometric composition then they can not chemisorb oxygen. If they are oxygen deficient, they can chemisorb precisely the amount of oxygen required to fill the anionic defects and reoxidize the zinc atoms.

Metals that favor the adsorption of oxygen have five, seven, eight, or ten d electrons. The order of preference is:

$$Cu^+ \approx Ag^+ > Pt^{2+} > Mn^{2+} > Rh^{2+} > Ir^{2+} > Co^{2+} > Hg^{2+}$$

Therefore the corresponding p-type semiconductors Cu_2O, Ag_2O, MnO, and PtO are highly effective catalysts for the activation of oxygen.

In n-type semiconductors, metal ions having one, two, or five d electrons are advantageous for the adsorption of oxygen. The following series was determined experimentally:

$$V^{5+} > Mo^{6+} > W^{6+} > Cr^{3+} > Nb^{5+} > Ti^{4+} > Mo^{4+}$$

Accordingly, n-type semiconductors like V_2O_5, MoO_3, WO_3, Cr_2O_3, and TiO_2 are effective oxidation catalysts.

We have already encountered the chemisorption of oxygen on p-type oxides (Eq. 4-44). It results in high degrees of coverage and eventually in complete coverage of the surface by O^- or O^{2-} ions. At the same time Ni^{2+} ions are oxidized at the surface (Eq. 4-47). The heat of adsorption remains practically constant while the surface becomes saturated with oxygen.

$$2\ Ni^{2+} + O_2 \longrightarrow 2\ (O^- \cdots Ni^{3+}) \tag{4-47}$$

The course of reaction on a semiconductor oxide may also depend on the sites to which the starting materials are bound and the manner in which they are bound. Consider the adsorption of hydrogen. It has been shown that hydrogen is heterolytically cleaved on a ZnO surface, so that simultaneous formation of a donor and an acceptor takes place. Active hydrides are bound to the ZnO surface:

$$\begin{array}{cc} H^- & H^+ \\ | & | \\ Zn^{2+}\!\!-\!\!&\!\!O^{2-} \end{array}$$

Cr_2O_3 can heterolytically cleave H_2 in two ways:

$$Cr^{3+}\!-\!O^{2-} + H_2 \longrightarrow \begin{array}{cc} H^- & H^+ \\ | & | \\ Cr^{3+}\!\!-\!\!&\!\!O^{2-} \end{array} \tag{4-48}$$

$$2\ Cr^{2+}\!-\!O + H_2 \longrightarrow 2\ \overset{\overset{\displaystyle H^-}{|}}{Cr^{3+}}\!-\!O \tag{4-49}$$

The adsorption of various gases on special TiO_2 surfaces with surface defects (oxygen holes) has been studied in detail. The following results were obtained [17, 35]:

- H_2 is dissociatively bound on Ti
- O_2 is dissociatively bound and fills O_2 holes
- CO is bound molecularly on Ti atoms with O_2 holes
- CO_2 reacts with O^{2-} ions to form surface carbonate; this is not influenced by the O_2 holes

All of these findings are important for understanding reaction mechanisms on semiconductors catalysts.

Reactions on Semiconductor Oxides [T20, T40]

The knowledge obtained about chemisorption on semiconductor oxides makes possible a better understanding of the behavior of these materials as oxidation catalysts. An oxidation reaction consists of several steps:

1) Formation of an electron bond between the starting material to be oxidized (e. g., a hydrocarbon) and the catalyst; chemisorption of the starting material.
2) Chemisorption of oxygen.
3) Transfer of electrons from the molecule to be oxidized (the donor) to the acceptor (O_2) by the catalyst.
4) Interaction between the resulting ion, radical, or radical ion of the starting material and the oxygen ion with formation of an intermediate (or the oxidation product).
5) Possible rearrangement of the intermediate.
6) Desorption of the oxidation product.

Hence the oxidation catalyst must be capable of forming bonds with the reactants and transferring electrons between them. Oxides of the *p*-type, with their tendency to adsorb oxygen up to complete saturation of the surface, are more active than *n*-type oxides. Unfortunately, activity and selectivity mostly do not run parallel, and the *p*-type semiconductors are less selective than than the *n*-type semiconductors. The *p*-type semiconductors can often cause complete oxidation of hydrocarbons to CO_2 and H_2O, while the *n*-type semiconductor oxides often allow controlled oxidation of the same hydrocarbons to be performed.

The ratio of adsorbed oxygen to hydrocarbon on *p*-type semiconductor oxides is generally high and is difficult to control even at low partial pressures of oxygen. The result is often complete combustion of the hydrocarbon. In contrast the amount of adsorbed oxygen on *n*-type semiconductors is generally small and can readily be controlled by means of the nature and amount of dopant, making selective hydrocarbon oxidation possible.

However, in practice neither *p*- nor *n*-type semiconductors are good catalysts for highly selective oxidations. Experience has also shown that a combination of the two semiconductor types also does not give any outstanding results with respect to activity and selectivity. Nevertheless, many simple oxidation reactions have been investigated with semiconductor catalysts, as we shall see in the following examples.

Variations in the selectivity of oxidations is explained in terms of:

- Electronegativity differences
- Ionization potentials of the metals

– Strength of oxygen bonding on the surface of the oxidation catalyst ("removability of lattice oxygen")

Simple two-step oxidation/reduction mechanisms are often used to explain industrial reactions. The oxidation of a molecule X can proceed by two mechanisms (Scheme 4-3).

A $\quad \frac{1}{2} O_{2(G)} \longrightarrow O^*$
$\quad X^* + O^* \longrightarrow$ products

B $\quad X^* + O_{lattice} \longrightarrow$ products + lattice vacancy
$\quad \frac{1}{2} O_{2(G)} +$ lattice vacancy $\longrightarrow O_{lattice}$

Scheme 4-3. Mechanisms of oxidation [T22]

In case **A**, O_2 is more rapidly adsorbed than the substrate X, and X* then reacts to remove this "excess" oxygen. Oxidation then proceeds through to the final products carbon dioxide and water.

In case **B**, adsorbed molecules of the starting material react with lattice oxygen. The result is selective oxidation, as is observed for patially oxidized molecules such as carbonyl compounds and unsaturated species in particular. Selective catalysts that react according to this Mars–van Krevelen mechanism formally contain a cation with an empty or filled d orbital, for example:

$$Mo^{6+} \quad V^{5+} \quad Sb^{5+} \quad Sn^{4+}$$
$$4d^0 \quad 3d^0 \quad 4d^{10} \quad 4d^{10}$$

Metals in their highest oxidation states readily release lattice oxygen, formally as O^{2-}. The well-known V_2O_5 catalysts have been intensively investigated. It was found that the rate of oxidation depends on the number of V=O bonds on the surface, and this was confirmed for the oxidation of H_2, CO, ethylene, butenes, butadiene, and xylene. The decisive step is the transition of V^V into the lower oxidation state V^{IV} (Eq. 4-50). A similar description can be applied to the oxidation of methanol to formaldehyde on Mo=O bonds.

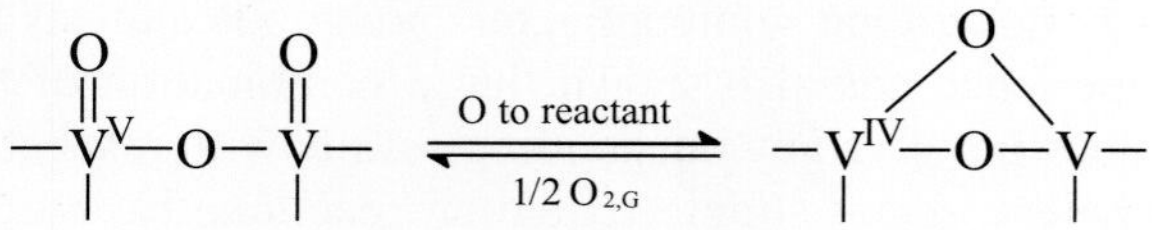

$$-\overset{\overset{O}{\|}}{\underset{|}{V^V}}-O-\overset{\overset{O}{\|}}{\underset{|}{V}}- \underset{1/2\,O_{2,G}}{\overset{O \text{ to reactant}}{\rightleftharpoons}} -\underset{|}{V^{IV}}-O-\underset{|}{\overset{\overset{O}{\diagup \diagdown}}{V}}- \qquad (4\text{-}50)$$

Metal oxides, especially those of the transition metals, can oxidize, dehydrogenate, decarboxylate, decarbonylate, and cleave C–C bonds. Numerous empirical activity and selectivity series can be found in the literature [32, T41].

4.3 Catalyst Concepts in Heterogeneous Catalysis

Activity Series

For the oxidation of H_2 in excess oxygen:

$Co_3O_4 > CuO > MnO > NiO > Fe_2O_3 > ZnO > Cr_2O_3 > V_2O_5 > TiO_2$
  46   55   59   59   63   100   76   76 E_a [kJ/mol]

Oxidation of ammonia:

$Co_3O_4 > MnO_2 > Cr_2O_3 > CuO > NiO > Fe_2O_3 > ZnO > TiO_2$

For both the above series the dissociative adsorption of oxygen is the rate-determining step.

Selectivity series are of greater practical importance.

Oxidation of propene to acrolein:

$MoO_3 > Sb_2O_5 > V_2O_5 > TiO_2 = Fe_2O_3 > SiO_2 > WO_3 \cdot Al_2O_3$

Oxidation of benzene to maleic anhydride:

$V_2O_5 > Cr_2O_3 > MoO_3 > Co_2O_3$

Oxidative dimerization and ring closing of propene to give 1,5-hexadiene and benzene:

$ZnO > Bi_2O_3 > In_2O_3 > SnO_2 > Ga_2O_3 > CdO$

These few examples show that there can be no generally valid selectivity series for oxidation catalysts; each reaction must be investigated individually. In the following section we shall discuss some well-known reactions in more detail.

A well-investigated model reaction is the decomposition of N_2O (Eq. 4-51).

$$2\, N_2O \longrightarrow 2\, N_2 + O_2 \qquad (4\text{-}51)$$

The mechanism is described as follows:

$$N_2O + e \longrightarrow N_2 + O^-_{ads} \qquad (4\text{-}52)$$
$$O^-_{ads} + N_2O \longrightarrow N_2 + O_2 + e \qquad (4\text{-}53)$$

The release of electrons to the catalyst is the rate-determining step; in addition, a good catalyst should readily adsorb oxygen. The transfer of an electron from adsorbed O^- only takes place if the Fermi level of the surface is lower than the ionization potential of adsorbed O^-. This situation is more likely for *p*-type semiconductors. An activity series has been given for this reaction, the criterion being

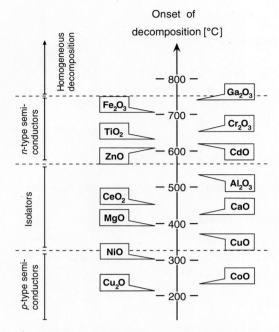

Fig. 4-31. Relative activities of metal oxides in the decomposition of N_2O [5]

the temperature at which decomposition begins. The catalysts can be classified in three groups, as shown in Figure 4-31. As expected, the *p*-type semiconductor oxides are the most active catalysts, followed by the insulator oxides, and finally the *n*-type semiconductor oxides. This ranking has been verified by subsequent work, and relative activities have been determined. The overall trend does not follow any correlative relationship, and the results are presumably influenced by other effects such as dispersity, impurities, and number and symmetry of the active centers. The influence of donors on NiO catalysts can be clearly be seen, as the following series shows [T35]:

$$(NiO + 2\% \, LiO_2) > NiO \gg (NiO + 2\% \, Cr_2O_3)$$

Thus *p*-type doping has a positive effect, and *n*-type doping, a negative one.

A donor reaction was also found to be rate-determining for the *n*-type semiconductor catalyst ZnO, which showed considerably higher activity after doping with Li_2O. The oxidation of CO (Eq. 4-54) has also been thoroughly investigated with the *p*-type semiconductor NiO and the *n*-type semiconductor ZnO as catalyst (Table 4-19).

$$2 \, CO + O_2 \longrightarrow 2 \, CO_2 \tag{4-54}$$

4.3 Catalyst Concepts in Heterogeneous Catalysis

Table 4-19. Oxidation of CO with metal oxide catalysts

Catalyst	E_a [kJ/mol]
1. NiO (p)	63
Doped with Cr_2O_3	80
Doped with Li_2O	50
2. ZnO (n)	118
Doped with Ga_2O_3	84
Doped witht Li_2O	134

As expected the *p*-type semiconductor NiO is the better catalyst. An increased *p*-type conduction due to Li_2O and a decrease due to Cr_2O_3 is understandable. It is assumed that a donor step is rate-determining, namely, the chemisorption of CO.

$$CO + [+] \longrightarrow CO_{ads}^+ \qquad (4\text{-}55)$$

This assumption is supported by the observation that the reaction is first order in CO. The chemisorption of O_2 (Eq. 4-56), an acceptor step, is fast, and the reaction rate is not dependent on the oxygen partial pressure.

$$\tfrac{1}{2} O_2 \longrightarrow O_{ads}^- + [+] \qquad (4\text{-}56)$$

The final step is neutral and follows the Langmuir–Hinshelwood mechanism (Eq. 4-57).

$$CO_{ads}^+ + O_{ads}^- \longrightarrow CO_{2,G} \qquad (4\text{-}57)$$

The influence of donors on ZnO at first appears remarkable. Here one would expect that increasing the *n*-type conductivity by adding trivalent donors would lower the reaction rate. However, this does not happen. This leads to the conclusion that in this case an acceptor reaction is the rate-determining step. Presumably it is the chemisorption of oxygen, since considerable dependence of the reaction rate on the oxygen partial pressure was observed. The example shows how the reaction mechanism can change from catalyst to catalyst.

The next reaction that we will study is the decomposition of ethanol (Eq. 4-58). Depending on the catalyst, dehydrogenation (A) or dehydration (B) can occur. Table 4-20 summarizes the results [T35].

$$C_2H_5OH \xrightarrow{\text{Cat.}} \begin{array}{c} A \\ \longrightarrow \\ B \\ \longrightarrow \end{array} \begin{array}{l} CH_3CHO + H_2 \\ CH_2{=}CH_2 + H_2O \end{array} \qquad (4\text{-}58)$$

Table 4-20. Decomposition of ethanol on semiconductor oxides

Catalyst		Decomposition of ethanol [%]	
		$CH_3CHO + H_2$	$C_2H_4 + H_2O$
$\gamma\text{-}Al_2O_3$		1.5	98.5
Cr_2O_3	increasing	9	91
TiO_2	n-type	37	63
ZrO_2	character	55	45
Fe_2O_3		86	14
ZnO	↓	95	5

The results can easily be explained. The extent of dehydrogenation increases with rising Fermi level and increasing n-type character, while dehydration follows the opposite trend.

The hydrogenation of ethylene on the catalyst ZnO at ca. 100 °C has been thoroughly studied by IR spectroscopy. The catalytic centers on the surface are ZnO pairs. Adsorption measurements have shown that these pairs lie spatially far apart on the surface [T24]. We have already seen that ZnO can cleave hydrogen heterolytically. The hydrogen atom bound to oxygen can be transferred to other oxygen atoms in the lattice (Scheme 4-4).

The starting material ethylene is also initially bound to oxygen by physisorption and then chemisorbed. The reaction of neighboring hydrogen and ethylene ligands

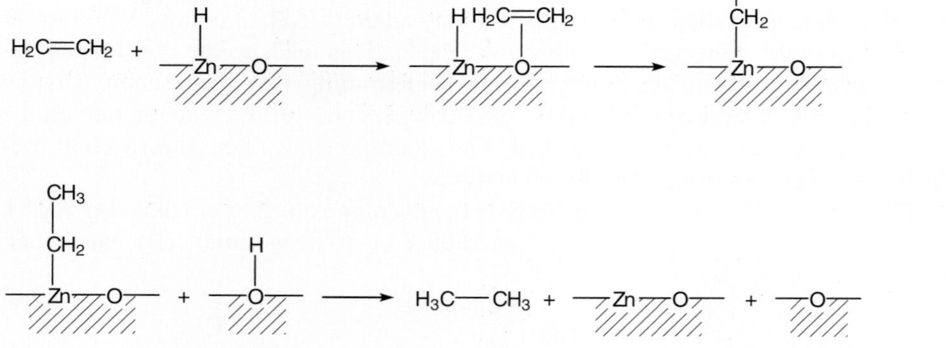

Scheme 4-4. Hydrogenation of ethylene on ZnO

leads to formation of a σ-ethyl complex on Zn. This complex is then hydrogenated to ethane by chemisorbed H, which migrates from oxygen to Zn centers.

Binary oxide catalysts are of major industrial importance. Such compounds are combinations of the oxides of Fe, Co, Ni, Cu, and Zn with those of Cr, Mo, and W. They form mixed oxide phases such as chromites, molybdates, and tungstates. Important industrial processes involving mixed oxides are:

- Oxidation of methanol to formaldehyde: Fe/Mo, Fe/W
- Selective hydrogenation and dehydrogenation: Cu/Cr
- Desulfurization, denitrogenation, and deoxygenation: Co/Mo, Ni/Mo
- Methanol synthesis: Zn/Cr, Zn/Cu
- CO conversion: Fe/Cr

In methanol synthesis Cu^I ions are dispersed in a ZnO matrix. Copper chemisorbs CO, and ZnO sites adsorb hydrogen. The heterolytically cleaved hydrogen reacts with the chemisorbed CO to give CH–OH fragments, which are further hydrogenated to methanol.

The Cr_2O_3/Al_2O_3 catalysts are used for the dehydrogenation of butanes to butenes and butadiene. With the addition of alkali metal oxides, they are used for the aromatization of n-alkanes. In these catalysts Al_2O_3 does not act only as a support, it also forms mixed phases with the chromium oxide. The active centers in these catalysts are Cr^{2+} and Cr^{3+} ions [T41].

Semiconductor oxides are also important support materials. Even if a support is inactive in the reaction under consideration, it can considerably change the reactivity of the catalyst that it supports. As an example, metals such as Ni and Ag are often applied to doped Al_2O_3 by vapor-phase deposition. The resulting catalyst system behaves like a rectifier in that electrons flow from the support through the catalyst metal to the reactants (Eq. 4-59). Hence in this case acceptor reactions are favored.

$$Al_2O_3 \xrightarrow{e} Ni \xrightarrow{e} \text{Reactant} \qquad (4\text{-}59)$$
n-doped

Such systems are of course less well suited to donor reactions, for which a p-doped support with an electron-withdrawing effect would be more favorable. There are many examples of support effects, which are discussed in more detail in Section 4.4.

Major influences have been observed in the reactions listed in Table 4-21.

Table 4-21. Reactions with supported catalysts

Reaction	Catalyst
Formic acid cleavage	Ni/Al_2O_3
	Ag/SiC
Ethylene hydrogenation	Ni/ZnO
Aldehyde hydrogenation	Pd/C

Finally let us summarize the knowledge gained in some general rules [T40]:

1) Transition metal oxides catalyze oxidation and dehydrogenation reactions.
2) Simple oxides with several stable oxidation states are generally the most active catalysts.
3) Alkalis generally stabilize high oxidation states, and acids, low oxidation states.
4) Activity and selectivity often follow opposite trends in catalytic oxidations.
5) Metal oxides with d^0 or d^{10} electronic structures are often selective oxidation catalysts.
6) Activity correlates with
   – the strength of bonding of oxygen to the surface
   – the heat of formation of the metal oxide
   – the number of O atoms in the oxide
   – the availability of lattice oxygen
7) The catalytic activity in the oxidation of H_2, CO, or hydrocarbons correlates with the bonding energy of oxygen on the surface:

$$Co_3O_4 > MnO_2 > NiO > CuO > Cr_2O_3 > Fe_2O_3 > ZnO > V_2O_5 > TiO_2 > Sc_2O_3$$

For a fundamental understanding of catalytic reactions, it is not sufficient to simply consider the global electronic properties of the catalyst. The surface geometry, the orbital structures of catalyst and starting materials, and other effects must also be considered. Refinement of the electronic concept of semiconductor catalysis is therefore essential.

4.3.3.3 Isolators: Acidic and Basic Catalysts [T24, T39, T41]

Catalysts belonging to this group are less common, and their activity for redox reactions is relatively low, at least at low temperatures. The solid oxides of the third period Na_2O, MgO, Al_2O_3, SiO_2, and P_2O_5 are isolators, and they exemplify the transition from basic through amphoteric to acidic character. The oxides of the elements of other periods behave similarly.

Since the catalytic properties can not be explained directly by means of electronic properties, it is appropriate to introduce another catalyst concept. In this case, the acid/base concept is suitable. Well-known catalysts with insulator properties are Al_2O_3, aluminosilicates, SiO_2/MgO, silica gels, phosphates such as $AlPO_4$, and special clays activated by chemical treatment. All these catalysts have acid centers on their surface.

A special class of crystalline aluminosilicates are the highly active and selective zeolites, which are discussed separately in Chapter 6.

In comparison, the basic catalysts play only a minor role. Well-known acidic/basic catalysts are listed in Table 4-22.

Table 4-22. Classification of acid/base catalysts [T41]

Solid acid catalysts	Solid basic catalysts
1. Oxides such as Al_2O_3, SiO_2, TeO_2	1. Oxides, hydroxides, and amides of alkali and alkaline earth metals (also on supports)
2. Mixed oxides such as Al_2O_3/SiO_2, MgO/SiO_2, ZrO_2/SiO_2, heteropolyacids	
3. Mineral acids (H_3PO_4, H_2SO_4) on solid porous supports	2. Anion exchangers
4. Cation exchangers	
5. Salts of O-containing mineral acids; heavy metal phosphates, sulfates, tungstates	3. Alkali and alkaline earth metal salts of weak acids (carbonates, carbides, nitrides, silicates, etc.)
6. Halides of trivalent metals (e.g., $AlCl_3$) on porous supports	
7. Zeolites (H form)	
8. Superacids: ZrO_2 or TiO_2, treated with H_2SO_4	4. Superbases: MgO doped with Na

Surface Acidity

Oxidic catalysts with acidic properties catalyze many industrial reactions, including the dehydration of alcohols, the hydration of olefins, cracking processes, and olefin polymerization. How does the acidity of such solids arise?

For surface acids a distinction is made between protic (Brønsted centers) and nonprotic (Lewis centers). Brønsted centers can release surface protons, while Lewis centers represent surface acceptor sites for electron pairs and thus bind nucleophiles.

Let us consider the role of Brønsted and Lewis centers in catalysis, using the example of aluminum oxide. Aluminum oxide contains bound water, the amount depending on the temperature. Freshly precipitated, water-containing Al_2O_3 is completely hydroxylated on the surface up to a temperature of 100 °C. The OH groups act as weak Brønsted acids. Above 150 °C the OH groups are gradually lost as water. This dehydroxylation liberates some of the Al atoms in the second layer, and these act as Lewis acid centers. At 400 °C the surface of partialy dehydroxylated Al_2O_3 exhibits Lewis acid sites with coordination holes (Al^{3+} ions), Lewis base sites (O^{2-} ions), and Brønsted acid sites (Fig. 4-32).

At 900 °C the fully dehydroxylated Al_2O_3 exhibits only Lewis acid and Lewis base sites.

It has been shown that the Brønsted acid sites are largely responsible for the polymerization of olefins, the cracking of cumene, and the disproportionation of toluene to benzene and xylene. In contrast, a strong influence of the Lewis acid centers was found in the decomposition of isobutane [32].

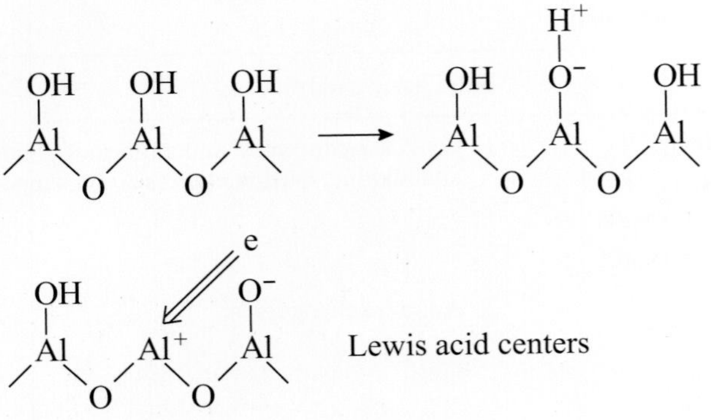

Fig. 4-32. Acid centers in Al_2O_3

The catalytic function of solid acids and bases is fundamentally similar to that of their counterparts in liquid systems. Thus the Brønsted equation is also applicable to heterogeneous catalysis. Since surface-acidic compounds do not dissociate, in contrast to liquid systems, the Brønsted equation in its special form for concentrated acids applies (Eq. 4-60).

$$\lg k = \lg a + \alpha H_0 \tag{4-60}$$

where k is the rate constant of the catalytic reaction, α is a measure of the proton transfer ($\alpha < 1$), a is a constant for a particular class of reaction, and H_0 is the Hammett logarithmic acidity function, a measure of the protonation of the acid.

The acidity function H_0 gives information on the acid centers of a catalyst. It can be determined by means of a series of calibrated bases in the presence of special indicators, and in this way, comparison to sulfuric acid of known strength can be made.

Other methods for determining the surface acidity of a catalyst are also available. For example the sum of Brønsted and Lewis centers can be determined by chemisorption of basic substances such as ammonia, quinoline, and pyridine.

Infrared spectroscopy is a powerful method that allows the direct determination of the Brønsted centers. When pyridine (py) is adsorbed on the catalyst simultaneous determination of both types of center is possible, since it is bound to Brønsted centers in the form of a pyridinium ion through a hydrogen bond (Eq. 4-61), whereas on Lewis acid centers, adsorption occurs by a coordinative acid–base interaction (Eq. 4-62).

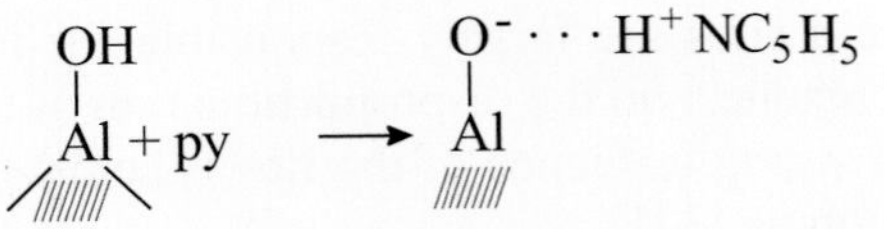

Pyridinium ion
v ca. 1540 cm^{-1} (4-61)
(Brønsted acid center)

$$\text{Al}^{3+} + \text{py} \longrightarrow \text{Al}^{3+}-\text{NC}_5\text{H}_5 \qquad \text{Acid–base complex}$$
$$\nu \text{ ca. } 1465 \text{ cm}^{-1} \qquad (4\text{-}62)$$
$$\text{(Lewis acid center)}$$

In comparison with Al_2O_3, Lewis centers are not so readily formed on the surface of SiO_2 since the OH groups are very strongly bound, so that Brønsted acidity predominates, albeit in a weak form, comparable to acetic acid.

The aluminosilicates have major industrial importance as cracking catalysts. These are derived formally from silicates by partially replacing the Si atoms in the silicate framework by Al atoms [32]. Since each Al center has one nuclear charge less than Si, each Al center has a formal negative charge, which requires additional cations for neutralization. If these are protons, then a very strong, high-polymer acid is formally obtained (Eq. 4-63).

$$(\text{SiO}_2)_n \xrightarrow{\text{AlOOH}} \begin{array}{c} \text{O} \\ | \\ -\text{O}-\text{Si}-\text{O}- \\ | \\ \text{O} \end{array} \begin{array}{c} \text{OH}^+ \\ | \\ \text{Al}^- \\ | \\ \text{O} \end{array} \begin{array}{c} \text{O} \\ | \\ -\text{O}-\text{Si}-\text{O}- \\ | \\ \text{O} \end{array} \qquad (4\text{-}63)$$

The acidity of these catalysts can be determined by titration with alkalis or by poisoning with nitrogen bases such as ammonia and quinoline. Good information about the active centers and the species adsorbed on them can be obtained by ESR spectroscopy.

In contrast to Al_2O_3, aluminosilicates exhibit pronounced Brønsted acidity. This can be explained in terms of dissociatively adsorbed water on the surface (Eq. 4-64).

$$\begin{array}{c} \text{H}-\text{OH} \\ | \\ -\text{Si}-\text{O}-\text{Al}- \\ | \qquad | \end{array} \longrightarrow \begin{array}{c} \text{H}^+ \text{ OH} \\ | \quad \downarrow \\ -\text{Si}-\text{O}-\text{Al}^- - \\ | \qquad | \end{array} \qquad (4\text{-}64)$$

According to Equation 4-64, the Al center can form its fourth bond with a free electron pair of a hydroxide anion. At the same time, the proton can react with a free electron pair of a neighboring O atom, and the formation of a partial bond results in a Brønsted acid center. The Si^{4+} center, which is more electopositive than Al^{3+}, weakens the O–H bond and increases the acidity. Experimentally it was found that maximum acidity occurs at ca. 30 % Al_2O_3. This model also allows the chemisorption of ammonia on Brønsted centers to be explained (Eq. 4-65).

$$\begin{array}{c} \text{H}^+ \text{ OH} \\ | \quad | \\ -\text{O}-\text{Al}^- - \\ | \end{array} + \text{NH}_3 \longrightarrow \begin{array}{c} \text{NH}_4^+ \text{ OH} \\ | \quad | \\ -\text{O}-\text{Al}^- -\text{O}- \\ | \end{array} \qquad (4\text{-}65)$$

It also allows the pronounced increase in the Brønsted acidity resulting from the adsorption of HCl on aluminosilicates to be understood (Eq. 4-66).

$$-\text{O}-\underset{\underset{|}{\text{O}}}{\overset{}{\text{Al}}}-\text{O}- + \text{HCl}_{(G)} \longrightarrow -\text{O}-\underset{\underset{|}{\text{O}}}{\overset{\overset{\text{H}^+}{\downarrow}\ \overset{\text{Cl}}{\downarrow}}{\text{Al}^-}}-\text{O}- \qquad (4\text{-}66)$$

Dehydration of organic molecules can occur on surface acids, for example, the conversion of alcohols to ethers and ketones. A good example is the reaction of ethanol on modified Al_2O_3 catalysts of various acidities (Table 4-23).

Table 4-23. Performance of aluminum oxides in the dehydration of ethanol [7]

Relative acidity* at 175 °C	SiO_2 [%]	Na_2O [%]	Conversion [%]	Selectivities		C(coke) [%]
				Ethene [%]	Ether [%]	
0.021	0.02	0.25	66.1	25.3	70.1	0.1
0.046	0.01	0.06	98.8	99.2	0.2	0.2
0.060	0.13	0.03	85.7	89.2	0.1	0.5

* mmol NH_3/g Al_2O_3

The commercially available aluminas used here contain SiO_2 and Na_2O as the main impurities. Apparently both components influence the conversion and the selectivity with respect to ethylene. The dehydration proceeds by a cyclic mechanism involving the action of an acid and a basic center (Fig. 4-33).

Silica increases the acid content of the surface, and Na_2O influences the basicity. The parameter measured was the relative acidity by ammonia adsorption. The presence of SiO_2 or Na_2O results in an equilibrium between Brønsted acid centers, Lewis acid centers, and Lewis base centers. The catalyst with medium acidity has both the highest activity and the highest selectivity for ethylene. The weakly acidic catalyst with the highest Na_2O content allows greater formation of ether.

Another important factor in industrial reactions is coke formation. As expected, the catalyst with the highest content of SiO_2 (highest acid content) has the most pronounced tendency for coke formation. This is explained by increased formation of carbenium ions, which undergo fast coupling and polymerization reactions that eventually lead to involatile deposits on the surface. This also leads to lower activity and selectivity of the catalyst.

A further finding was that only the moderately active Brønsted acid centers are responsible for dehydration, and that Lewis acid centers such as Al^{3+} are not involved. Evidence for this is that the addition of small amounts of bases such as

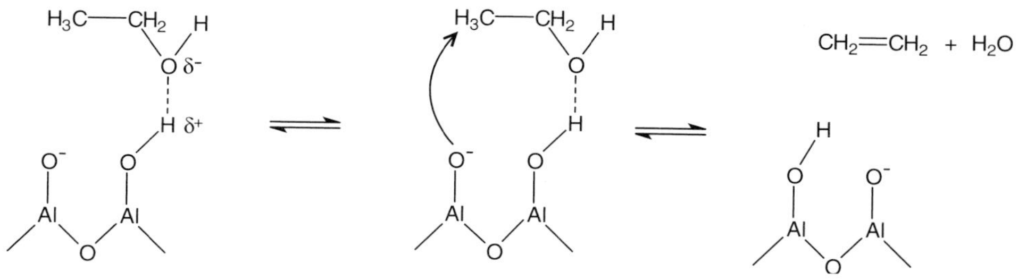

Fig. 4-33. Mechanism of gas-phase dehydration of ethanol on aluminum oxide

NH$_3$ or pyridine does not inhibit the reaction. The formation of ether on Al$_2$O$_3$ is explained by a Langmuir–Hinshelwood mechanism, in which two adjacently adsorbed intermediate alcohol fragments — for example, one bound as an alkoxide – OC$_2$H$_5$ and the other by hydrogen bonding — react with one another.

The chemisorption of olefins on an aluminosilicate catalyst is also believed to proceed by a mechanism similar to that shown in Figure 4-33. As shown in Equation 4-67, the olefin couples with an acid/base pair, that is, a bridging hydroxyl group and a lattice oxygen center on the surface, probably as the result of a direct geometrical correspondence.

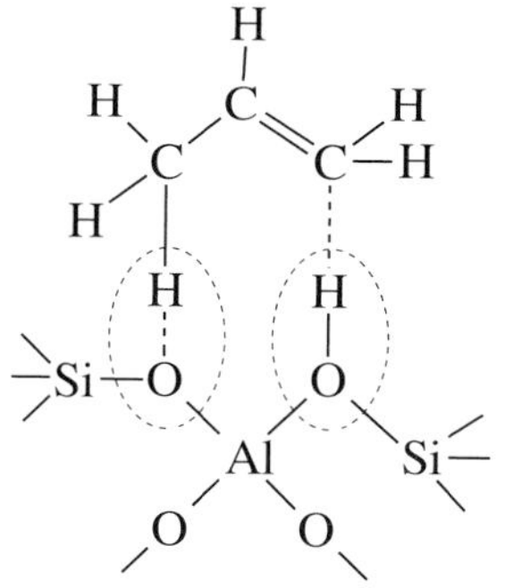

$$(4\text{-}67)$$

Acidic aluminosilicate-based catalysts are of major industrial importance. In terms of product quantity, the most important catalytic process is the cracking of crude oil. The reaction is initiated by the reaction of a Brønsted acidic surface with alkenes in which addition of a proton to the double bond gives chemisorbed carbenium ions (Eq. 4-68).

$$\text{Si}-\overset{\text{H}}{\text{O}}-\text{Al} + \text{CR}_2=\text{CR}_2 \longrightarrow \text{R}_2\text{CH}-\overset{+}{\text{CR}}_2 + \text{Si}-\text{O}^--\text{Al} \qquad (4\text{-}68)$$

Cleavage of long-chain hydrocarbons is accompanied by extensive isomerization, polymerization, and alkylation of the initial products and formation of aromatic hydrocarbons. The same reactions occur in the the homogeneously catalyzed reaction initiated by protons or Lewis acids (BF_3, $AlCl_3$).

It has been shown in many cases that the acid strength of a catalyst of given composition is often comparable to its activity. Thus the polymerization of olefins and the formation of coke depend on the catalyst acidity, for which the following series is given [T40]:

$$SiO_2/Al_2O_3 > SiO_2/MgO > SiO_2 \gg \gamma\text{-}Al_2O_3 > TiO_2 > ZrO_2 > MgAl_2O_4 > UO_2 > CaO \sim MgO$$

$\longleftarrow$ Acidity

Experience has shown that much stronger acids are formed when two oxides whose cations have different coordination numbers or oxidation states are combined. Such catalysts with a broad activity spectrum are listed in Table 4-24. The acid strength and catalytic activity of such solid acids correspond to those of mineral acids. The major advantage of solid acids is their thermal stability, which allows them to be used at much higher temperatures.

Table 4-24. Acid strength of binary mixed oxides [T41]

Components A–B	A [%]	Specific surface area [m^2/g]	Acid strength (Hammett function H_0)
Al_2O_3–SiO_2	94	270	-8.2 ($\approx 90\%$ H_2SO_4)
ZrO_2–SiO_2	88	448	-8.2 to -7.2
Ga_2O_3–SiO_2	92.5	90	-8.2 to -7.2
BeO–SiO_2	85	110	-6.4
MgO–SiO_2	70	450	-6.4
Y_2O_3–SiO_2	92.5	118	-5.6 ($\approx 71\%$ H_2SO_4)
La_2O_3–SiO_2	92.5	80	-5.6 to -3.2

Some interesting results with acid catalysts in selected reactions such as isomerization, polymerization, and cracking reactions confirm the influence of the catalyst acidity (Table 4-25).

Basic Catalysts [16]

Solid basic catalysts are used in only a few industrial processes. The most important group is made up of the compounds of the alkali and alkaline earth metals. Magnesium oxide has been thoroughly investigated [35]. On the surface of alkaline earth metal oxides, water undergoes rapid heterolytic cleavage, covering the surface with hydroxyl groups (Eq. 4-69; cf. Al_2O_3).

Table 4-25. Acidic catalysts for various reactions arranged in order of increasing acidity [T33]

Acid catalyst	Isomerization of n-pentane (Pt + support); Reaction temperature [°C]	Polymerization of propene at 200 °C; Conversion [%]	Cracking of n-heptane (temperature [°C] for 10% conversion)
α-Al$_2$O$_3$	inactive	0	inactive
SiO$_2$	inactive	0	inactive
ZrO$_2$	inactive	0	inactive
TiO$_2$	inactive	0	inactive
Al$_2$O$_3$, small surface area	500	<1	inactive
Al$_2$O$_3$, large surface area	450	0–5	490
Al$_2$O$_3$, chlorinated	430	10–20	475
SiO$_2$–MgO	400	20–30	460
Heteropoly acids	unstable	70–80	unstable
Al$_2$O$_3$, fluorinated	380	>80	420
Aluminosilicate	360	>90	410
Zeolites, exchanged	260	>95	350
Solid phosphoric acid	–	90–95	unstable
AlCl$_3$, HCl/Al$_2$O$_3$	120	100	100

$$\begin{array}{c} H \\ H-O \\ | \\ -M^{2+}-O^{2-}- \\ | \end{array} \longrightarrow \begin{array}{c} \left(\begin{array}{c} H \\ O \end{array}\right)^- \\ | \\ -M^{2+}-\left(\begin{array}{c} H \\ O \end{array}\right)^- \\ | \end{array} \qquad (4\text{-}69)$$

Ion pairs on the surface of MgO can also heterolytically cleave the Brønsted acids HX (Eq. 4-70), acetylene, acetic acid, and alcohols (Eq. 4-71).

$$\{Mg^{2+} \ O^{2-}\} + XH \longrightarrow \{\overset{X^-}{Mg^{2+}} \ \overset{H^+}{O^{2-}}\} \qquad (4\text{-}70)$$

$$-O-Mg-O-Mg- + ROH \longrightarrow -\overset{H}{O}-\overset{OR}{Mg}-\overset{H}{O}-\overset{OR}{Mg}- \qquad (4\text{-}71)$$

The heterolytic cleavage of the alcohol to give RO$^-$ and H$^+$ explains why alkaline earth metal oxides, especially magnesium oxide, are good catalysts for the dehydrogenation of alcohols.

Increasing dehydroxylation resulting from activation at higher temperatures increases the base strength of MgO. Highly dehydroxylated MgO is such a strong base that it deprotonates the weak Brønsted acids NH_3 (pK_a = 36) and propene (pK_a = 35). Heterolytic cleavage of H_2 on MgO has even been demonstrated.

On alkaline earth metal oxides butene is adsorbed as methylallyl anions $(CH_3-CH=\!\!=CH=\!\!=CH_2)^-$. This carbanion is an intermediate in the double bond isomerization of butene. Adsorption was shown to be stronger on CaO (higher basicity) than on MgO. Activation of CaO at 700–900 °C results in maximum Lewis basicity and optimum activity for the isomerization to 2-butene.

Magnesium oxide is a good "solvent" for 3d transition metal ions. For example, Co^{2+} and Ni^{2+} ions are very well dispersed on MgO. The covalent component of the cation–anion bonding lowers the basicity of the oxide, making it "softer". In the hydrogenation of CO, Co/MgO, and Ni/MgO supported catalysts give higher yields of C_2 and C_3 products than those with Al_2O_3 as support.

In the dehydrogenation of alcohols, Co^{2+} ions also increase the selectivity of the reaction. Alkaline earth metal oxides are good catalysts for the dehydrohalogenation of alkyl halides at 100–250 °C. The elimination of hydrogen halide proceeds by a highly selective E2 reaction. The following selectivity series was found for the *trans* elimination:

$SrO \geqslant CaO > MgO > Al_2O_3$

It reflects well the decreasing basicity of the oxides.

Thermally activated MgO, CaO, and BaO can even be used as catalysts for the hydrogenation of alkenes and dienes.

The reaction of benzaldehyde with activated oxides gives the Tischchenko product benzyl benzoate (Eq. 4-72).

$$\text{PhCHO} \xrightarrow{\text{BaO}} \text{PhCO-O-CH}_2\text{Ph} \tag{4-72}$$

The following reactivity series was found for this reaction:

$BaO \gg SrO \gg CaO > MgO$

The most active basic catalysts are alkali metals supported on alumina. Thus the catalyst 5% Na/Al_2O_3 results in complete conversion of 1-butene to 2-butene at 20 °C. Longer chain α-olefins are also readily isomerized by this highly active catalyst.

Reactions for the oxidative coupling of methane are also of much interest. This can be carried out with Li_2CO_3-doped MgO. Good selectivities for ethylene and ethane have been achieved.

In spite of these many examples, only a few base-catalyzed reactions are carried out industrially. Examples are:

- Condensation of acetone to diacetone alcohol with $Ba(OH)_2$ or $Ca(OH)_2$ supported catalysts
- Disproportionation of methylcyclopentene to methylcyclopentadiene and methylcyclopentane with sodium
- Dimerization of propene to 2-methylpentene with supported alkali metal catalysts
- Side-chain alkylation of toluene with Na/Al_2O_3
- Polymerization of butadiene with sodium

Exercises for Section 4.3.3

Exercise 4.23

Classify the following as semiconductor catalysts (S), acid catalysts (A), or isolators (I):

Pd Al_2O_3 ZnO aluminosilicates MgO CoO zeolites

Exercise 4.24

Oxides such as Cu_2O, NiO, and CoO have a high adsorption capacity for CO.

a) Which type of semiconductor are the above-mentioned oxides?
b) How is the reaction with the starting material CO designated?
c) What is the effect of doping the oxides with Li_2O?

Exercise 4.25

Which types of semiconductor are represented by the following oxides:

VO_2 Cu_2O WO_3 MnO_2 Nb_2O_5 CoO

Exercise 4.26

The oxidation of SO_2 can be carried out on chromium(III) oxide catalysts.

$$2\ SO_2 + O_2 \xrightarrow{Cr_2O_3} 2\ SO_3$$

This oxidation catalyst can be both n and p doped by various additives. The activation energy of the reaction increases with n-type doping and decreases on p-type doping.
Discuss these findings.

Exercise 4.27

A donor step is the rate-determining step in a hydrogenation reaction. The following catalysts are available:

a) Ni
b) Ni on Al_2O_3 (n donor)
c) Ni on CoO (p donor)

Which order of catalytic activity can be expected. Give a reason for this.

Exercise 4.28

In the conversion of methane to ethane and ethene, MnO is used as catalyst. Hydrogen abstraction from methane is observed as an intermediate step. On doping the catalyst with Li_2O, the selectivity of the reaction increases considerably.
Explain the course of the reaction.

Exercise 4.29

The selective oxidation of n-butane to maleic anhydride is an industrial process. Butane behaves as a weak base towards metal oxides.
Which properties should a metal oxide catalyst for this reaction have?

Exercise 4.30

a) Explain how a solid can react as an acid, using Al_2O_3 as an example.
b) Arrange the following oxides in order of relative acidity:

γ-Al_2O_3 $MgAl_2O_4$ SiO_2 MgO SiO_2/Al_2O_3

Exercise 4.31

Aluminosilicate surfaces are classified as strong Brønsted acids, whereas silica gel is a weak acid.
Give an explanation for the increased acidity when Al^{3+} is present in the silicon dioxide lattice.

Exercise 4.32

How can the acidity of aluminosilicates be measured?

Exercise 4.33

Explain the cationic polymerization of alkenes on aluminosilicate surfaces with the aid of a reaction equation.

4.4 Interaction of Catalysts with Supports and Additives

4.4.1 Supported Catalysts [T32, T35]

Supported catalysts represent the largest group of heterogeneous catalysts and are of major economic importance, especially in refinery technology and the chemical industry. Supported catalysts are heterogeneous catalysts in which small amounts of catalytically active materials, especially metals, are applied to the surface of porous, mostly inert solids – the so-called supports. The supports can have special forms such as pellets, rings, extrudates, and granules.

Typical catalyst supports are porous solids such as aluminum oxides, silca gel, MgO, TiO_2, ZrO_2, aluminosilicates, zeolites, activated carbon, and ceramics. Table 4-26 lists widely used catalyst supports.

Table 4-26. Important catalyst supports [T40]

Support	Specific surface area, m^2/g	Applications
γ-Al_2O_3	160–250	cracking reactions
α-Al_2O_3	5–10	selective hydrogenation of acetylene; selective oxidation (ethylene oxide)
Aluminosilicates	180–1600	cracking reactions, dehydrations, isomerizations
Silica gel	200–800	NO_x reduction (SCR process)
TiO_2	40–200	TiO_2 on SiO_2: oxidation of o-xylene to phthalic anhydride
Activated carbon	600–1800	vinylation with acetylene, selective hydrogenation with noble metal catalysts
Corundum ceramic	0.5–1	selective oxidation (ethylene oxide, benzene to maleic anhydride, o-xylene to phthalic anhydride)

What are the reasons for the predominant use of supported catalysts in industry?

- **Costs**. The catalytically active components of supported catalysts are often expensive metals. Since this active component is applied in a highly dispersed form, the metal represents only a small fraction of the total catalyst mass. For example, the metals Rh, Re, and Ru are highly effective hydrogenation catalysts for aromatic hydrocarbons. They are sometimes used in mass fractions as low as 0.5 % on Al_2O_3 or activated carbon.
- **Activity.** The high activity leads to fast reaction rates, short reaction times, and maximum throughput.
- **Selectivity** facilitates the following: maximum yield, elimination of side products, and lowering of purification costs; it is the most important target parameter in catalyst development.
- **Regenerability** helps keep process costs low.

Which factors influence these properties? The main factors are the choice of the most suitable support material and the arrangement of the metal atoms in the pore structure of the support. In choosing catalyst supports, numerous physical and chemical aspects and their effects must be taken into account (Table 4-27).

Table 4-27. Selection of catalyst supports [T40]

Physical aspects	Chemical aspects
Specific surface area ($\rightarrow$ activity, distribution of active components)	Specific activity ($\rightarrow$ adaption to heat evolution)
Porosity ($\rightarrow$ mass and heat transport)	
Particle size and shape ($\rightarrow$ pore diffusion, pressure drop)	Interaction with active components ($\rightarrow$ selectivity, bifunctional catalysts)
Mechanical stability ($\rightarrow$ abrasion, durability)	
Thermal stability ($\rightarrow$ catalyst lifetime, regenerability)	
Bulk density (active component content per unit reactor volume)	Catalyst deactivation ($\rightarrow$ stabilization against sintering, poisoning)
Dilution of overactive phases ($\rightarrow$ heat evolution, avoidance of hot spots)	
Separability (filterability of powder catalysts)	No interaction with reactants or solvents

The main function of the catalyst support is to increase the surface area of the active component. Catalytic activity generally increases with increasing catalyst surface area, but a linear relationship can not be expected since the reaction rate is often strongly dependent on the structure of the catalyst surface. However, in many reactions, the selectivity decreases when the catalytic surface is enlarged. As a general rule, catalysts for the activation of hydrogen (hydrogenation, hydrodesulfurization, hydrodenitrogenation) require high support surface areas, while selective oxidations (e. g., olefin epoxidation) need small support surface areas to suppress problematic side reactions.

The choice of the appropriate catalyst support for a particular active component is important because in many reactions the support can significantly influence the reaction rate and the course of the reaction. The nature of the reaction system largely determines the type of catalyst support.

If a support material with a large surface area such as activated carbon is used as support, then the metal is present as discrete crystallites, only a few atomic layers thick, with a very high surface area.

In batch liquid-phase reactions, powder supports are used exclusively, whereas in gas-phase and continuous liquid-phase reactions (trickle columns), supports in pellet or granule form can be employed (see Chap. 9).

The pore structure of the support can also have an influence on the role of the active component, since the course of the reaction is often strongly dependent on the rate of diffusion of the reactants. Furthermore, the size of the support surface can limit the exploitable metal concentration.

Many commercially available catalyst supports, for example, activated carbon and alumina, are offered in various particle sizes, each having a series of different specific surface areas and pore size distributions.

The choice of catalyst support may be restricted by the reaction conditions. Thus the support must be stable under the process conditions and must not interact with the solvent and the starting materials. Depending on the process, supported catalysts can have a low (e. g., 0.3% Pt/Al_2O_3, 15% Ni/Al_2O_3) or a high loading (e. g., 70% Ni/Al_2O_3, Fe/Al_2O_3).

In supported metal catalysts, the support does not only ensure high dispersion of the metal; there are also interactions between metal and support due to various physical and chemical effects:

– Electronic effects: electron transfer up to formation of chemical bonds
– Adhesive forces (van der Waals forces)
– Formation of reduced support species on the metal surface
– Formation of new phases at the boundary surface

Electronic effects and their causes have already been treated in Section 4.3.3; they result from the n- or p-type semiconductor properties of the support material. The interactions can impair the chemisorption capability and effectiveness of a catalyst, as well as restricting the mobility of the disperse phase and delaying its sintering.

In the last few years, the concept of strong metal–support interaction (SMSI) has gained considerable importance [18]. It was introduced in 1978 to explain certain peculiarities in the chemisorption of H_2 and CO on TiO_2-supported platinum group metals. The catalysts were subjected to high-temperature reduction with H_2

(400 °C), after which a strong decrease in the adsorption capacity for H_2, CO, and NO was found. The effect is also exploited in chemical syntheses: platinum group metals on TiO_2 can considerably influence the catalytic activity and product selectivity in the hydrogenation of CO.

In the following we shall discuss some examples of the industrial use of supported catalysts and the above-mentioned metal–support interactions.

Hydrogenation is one the oldest and most widely used applications for supported catalysts. The usual metals are Co, Cu, Ni, Pd, Pt, Re, Rh, Ru, and Ag. There are numerous catalysts for special applications. Most hydrogenation catalysts consist of an extremely fine dispersion of the active metal on activated carbon, Al_2O_3, aluminosilicates, zeolites, kieselguhr, or inert salts such as $BaSO_4$ [22]. An example is the selective hydrogenation of chloronitrobenzene (Eq. 4-73).

$$\text{C}_6\text{H}_4(\text{NO}_2)(\text{Cl}) + 3\,H_2 \xrightarrow{1\% \text{ Pt/C}} \text{C}_6\text{H}_4(\text{NH}_2)(\text{Cl}) + 2\,H_2O \quad (4\text{-}73)$$

99.5% yield

Usually, palladium catalysts are used for the industrial hydrogenation of nitro compounds, but Pd is also an excellent catalyst for the dehydrochlorination reaction, so that aniline is predominantly formed. Therefore, a new, high-selectivity Pt/C catalyst was developed, which gives the desired product o-chloroaniline without affecting the rate of hydrogenation.

In the dehydrogenation of cyclohexanone derivatives (Eq. 4-74), an activated carbon support in which the palladium is uniformly distributed in the support structure is recommended. With increasing ordering of the metal, the catalyst exhibits an increasing metal dispersion and therefore a higher resistance to thermal sintering. Sintering would lead to crystal growth and deactivation of the catalyst.

$$\text{cyclohexanone-R} \xrightarrow{5\% \text{ Pd/C}} \text{phenol-R} + 3\,H_2 \quad (4\text{-}74)$$

The hydrogenolysis of ethane on supported nickel catalysts is a good example for the influence of the degree of dispersion of the metal (Table 4-28). It is known that nickel is more highly dispersed on SiO_2 than on Al_2O_3, and at the same time there is an influence on the crystallite form. A further influence is due to the acid centers of aluminum oxide, which lead to more extensive coke formation, deactivating the nickel catalyst.

The dehydrogenation of cyclohexane to benzene can be explained well in terms of electronic effects (Table 4-29). The benzene selectivity decreases on going from TiO_2 to SiO_2, and this corresponds to the decreasing n character of the support ma-

Table 4-28. Hydrogenolysis of ethane on supported nickel catalysts (10% Ni) [T35]

Support	Reaction rate [mol m^{-2} metal h^{-1} · 10^6]
SiO$_2$	151
Al$_2$O$_3$	57
SiO$_2$/Al$_2$O$_3$	7

Table 4-29. Dehydrogenation of cyclohexane to benzene on supported platinum catalysts at 773 K [T28]

Catalyst	Benzene (%)
Pt/ZnO	—
Pt/TiO$_2$	76.1
Pt/Al$_2$O$_3$	59.8
Pt/MgO	32.3
Pt/SiO$_2$	23.1

terial. Apparently, weak *n*-type semiconductor oxides are the most effective supports for this reaction. In contrast, the strong *n*-type semiconductor ZnO, which has a higher electron concentration than TiO$_2$, gives no reaction.

Extensive investigations have been carried out on the industrially important hydrogenation of CO. Here we shall discuss just a few of the results, some of which are contradictory [T28]. High activities and selectivities for the formation of methanol were found for the catalysts Pd on La$_2$O$_3$, MgO, or ZnO, but high activities and selectivities for the formation of methane with Pd on TiO$_2$ or ZrO$_2$ (Table 4-30). It is no surprise that a high proportion of dimethyl ether is formed with the acidic support Al$_2$O$_3$. However, these investigations did not take degree of dispersion of the metal into consideration.

The hydrogenation of CO can be influenced by means of the support composition and by varying the degree of dispersion of the metal. Thus it is assumed that for metals of Groups 8–10, a low degree of dispersion favors formation of hydrocarbons, and a high degree of dispersion, the formation of oxygen-containing compounds.

Relative activities in CO hydrogenation measured for supported rhodium catalysts are listed in Table 4-31. These experimental findings are supported by H$_2$ chemisorption measurements and active rhodium centers.

In another investigation with supported rhodium catalysts, it was found that the oxidation state of the rhodium influences the type of chemisorption of CO and hence the product distribution according to Equation 4-75.

$$CO/H_2 \xrightarrow{Rh/support} \text{Oxo products} + HC \qquad (4\text{-}75)$$

Table 4-30. Hydrogenation of CO on supported Pd catalysts [T28]

Catalyst	Selectivities (%)			
	CH_3OH	CH_3-O-CH_3	CH_4	C_{2+}
Pd powder	75.0	0	8.8	16.2
Pd/MgO	98.4	1.2	0.3	0.2
Pd/ZnO	99.8	0	0.1	0.2
Pd/Al_2O_3	33.2	62.7	3.3	0.8
Pd/La_2O_3	99.0	0	0.5	0.5
Pd/SiO_2	91.6	0	1.5	0.2
Pd/TiO_2	44.1	8.6	42.1	5.2
Pd/ZrO_2	74.7	0.5	22.3	2.5

Table 4-31. Relative activities of supported Rh catalysts in the hydrogenation of CO [T28]

Support	Relative activity
TiO_2	100
MgO	10
Al_2O_3	5
CeO_2	3
SiO_2	1

Table 4-32. Influence of support materials on the hydrogenation of CO with rhodium catalysts [T22]

Catalyst	Active catalyst	Chemisorption of CO	Products
Rh/SiO_2	Rh(0)	dissociative	CH_x
Rh/ZrO_2	Rh(0), Rh(I)	dissociative/ associative	42% ethanol 12% methanol 32% CH_4
Rh/ZnO Rh/La_2O_3	Rh(I)	associative	94% methanol

Thus dissociative chemisorption of CO leads to hydrocarbons, and associative chemisorption to alcohols as final product (Table 4-32).

In CO hydrogenation with supported copper catalysts (Table 4-33), the results were explained in terms of electronic effects of the support material [13]. The differing CO hydrogenation activity of the catalysts reflects the electronic interaction between the Cu particles on the surface and the support. With *p*-type semiconduc-

Table 4-33. Supported copper catalysts for the hydrogenation of CO [13]

Catalyst	Work function of the support [eV]	TON · 10^3 of CO*	Semiconductor type of support
5% Cu/ZrO$_2$	5.0	0.41	p
5% Cu/Cr$_2$O$_3$	5.8	0.24	p
5% Cu/graphite	4.8	≤0.04	n (metalloid)
5% Cu/ZnO	4.6	≤0.03	n
20% Cu/Al$_2$O$_3$	–	≤0.01	isolator
20% Cu/SiO$_2$	–	≤0.02	isolator
5% Cu/TiO$_2$	3.0	≤0.01	n
5% Cu/MgO	3.5	≤0.01	n
Cu metal	4.55	≤0.02	(metal)

* TON = mol CO/atom surface metal × s; H_2/CO = 3; flow rate 60 mL/min, normal pressure, 275 °C; all catalysts have approximately the same particle size.

tors such as Cr_2O_3 and ZrO_2, which have higher work functions than copper metal, higher activity than with pure copper is observed. This is explained by the fact that in this case, electron density can flow from copper to the support. With the isolators SiO_2 and Al_2O_3, the activity corresponds roughly to that of copper; no electrons can be taken up by the support. In the case of n-type semiconductors such as TiO_2 and MgO, charge transfer from copper to support can not take place, and the catalytic activities are lower than with pure copper.

The next example shows how catalyst bifunctionality can arise from the support material. Platinum metal dehydrogenates napthenes to give aromatic compounds, but it is not able to isomerize or cyclize n-alkanes. This function is adopted by the Al_2O_3 support with its acidic properties. The cooperation of the two catalyst components is shown schematically for the reforming of n-hexane in Scheme 4-5 [T20].

It was shown that neither Pt nor the support material Al_2O_3 can isomerize the alkane starting material. However, acidic Al_2O_3 centers can isomerize n-alkenes, which are then hydrogenated to isoalkanes on Pt. During the activation phase of the catalyst, chlorine is added to achieve the necessary acidity.

The final examples deal with SMSI effects [18]. In the hydrogenation of CO on Pt/TiO$_2$ catalysts, a 100-fold increase in catalyst turnover number was observed after high-temperature reduction. In the high-temperature reduction, the chemisorp-

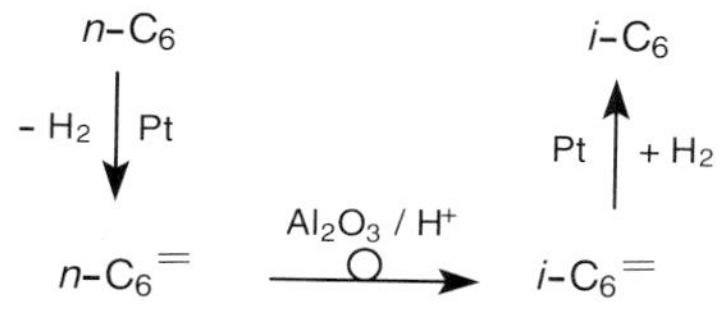

$$n\text{-}C_6 \xrightarrow[\text{Pt}]{-H_2} n\text{-}C_6{=} \xrightarrow{Al_2O_3 / H^+} i\text{-}C_6{=} \xrightarrow[\text{Pt}]{+H_2} i\text{-}C_6$$

Scheme 4-5. Reforming of n-hexane on a Pt/Al$_2$O$_3$ supported catalyst

tion capacity for both starting materials, CO and H_2, was drastically lowered, but no sintering of the metal occurred. It has been shown that partially reduced TiO_x species are distributed over the Pt surface. Interestingly, in spite of the higher catalyst activity, a higher activation energy was measured rather than a lower one [37].

A further example is the model reaction of hydrogenation of acetone to isopropanol (Eq. 4-76).

$$CH_3-\underset{\underset{O}{\|}}{C}-CH_3 + H_2 \longrightarrow CH_3-\underset{\underset{OH}{|}}{CH}-CH_3 \qquad (4\text{-}76)$$

$$\Delta G_{R,0} = -20 \text{ kJ/mol}$$

Kinetic measurements on a Pt catalyst showed no dependence on the size of the crystallites. On an inert SiO_2 support the catalyst turnover number remained virtually constant over the particle size range 2–1000 nm; that is, the reaction is structure-insensitive. With a TiO_2 support, the TON was increased by a factor of 500 following high-temperature reduction (Table 4-34).

Table 4-34. SMSI effect in the hydrogenation of acetone to isopropanol on supported platinum catalysts [37]

Catalyst	TON $\cdot 10^2$ [s^{-1}]	E_a [kJ/mol]
Pt/SiO$_2$	ca. 1.1	67 ± 2,5
Pt/η-Al$_2$O$_3$	2.4	78
Pt/TiO$_2$ (LTR)	ca. 2.8	59 ± 2,9
Pt/TiO$_2$ (HTR)	ca. 565	68 ± 8,3

Reaction conditions: 303 K, 0.1 MPa, H_2/acetone = 3.06; LTR = low-temperature reduction, HTR = high-temperature reduction.

Langmuir–Hinshelwood kinetics involving competetive adsorption of acetone molecules and hydrogen atoms were postulated, and it was assumed that adsorbed acetone dominates. The SMSI effect is explained by the fact that the oxygen atom of the carbonyl group is more effectively activated than in conventional platinum catalysis. It is assumed that the oxygen atom is adsorbed on particularly active Ti^{3+} centers of the partially reduced TiO_x islands on the platinum at the metal/support boundary (Fig. 4-34).

An SMSI effect was also demonstrated in the hydrogenation of crotonaldehyde, and a surprising change in the selectivity of the the reaction was observed. With Pt/SiO$_2$ and Pt/Al$_2$O$_3$ catalysts, only butyraldehyde or butanol, respectively, is obtained as hydrogenation product; with Pt/TiO$_2$ a considerable selectivity of 37% for the unsaturated crotyl alcohol is reported.

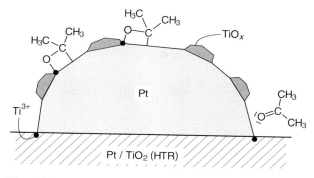

Fig. 4-34. Model for the hydrogenation of acetone (SMSI effect) [37]

Another interesting reaction is the reforming of methane with CO_2 to produce synthesis gas (Eq. 4-77). This endothermic reaction is said to be suitable for storing solar energy in chemical substances [14].

$$CH_4 + CO_2 \rightleftharpoons 2\,CO + 2\,H_2 \tag{4-77}$$

The support is a ceramic honeycomb with a washcoat of Al_2O_3 that contains the rhodium metal catalyst (0.2%). Indications of the mechanism of the reaction are provided by literature data on analogous reactions:

- Al_2O_3 has acidic properties and can catalyze the formation of CH_3^+. At higher temperatures it behaves as an *n*-type semiconductor.
- When CO_2 is adsorbed, it accepts electrons from catalysts with *n*-type semiconductor properties, but releases electrons to catalysts with *p*-type semiconductor properties (Lewis amphoteric behavior).
- When a metal is applied to an *n*-type semiconductor, its electron density increases.
- The SMSI effect influences the binding between support and metal. The bonds can take on ionic character or undergo geometric changes.

Combining these facts leads to the mechanism shown in Scheme 4-6. Methane is the first species to be adsorbed and is partially dehydrogenated. The CO_2 reacts with the methane fragment either from the gas phase or from the adsorbed phase. In the adsorption process, electron flow from the support, through the metal, and to the reactant is assumed for both starting materials. The electrons of CO_2 attack the partially dehydrogenated methane, whereby CO is formed. In the final step hydrogen is split off and desorbed.

Chemical interactions between metal and support are also observed on main group metal oxides such as SiO_2, Al_2O_3, and MgO, which can normally be regarded as chemically highly inert. Strong interactions have also been found between various metals of Groups 8–10 and carbon supports. Palladium and nickel form car-

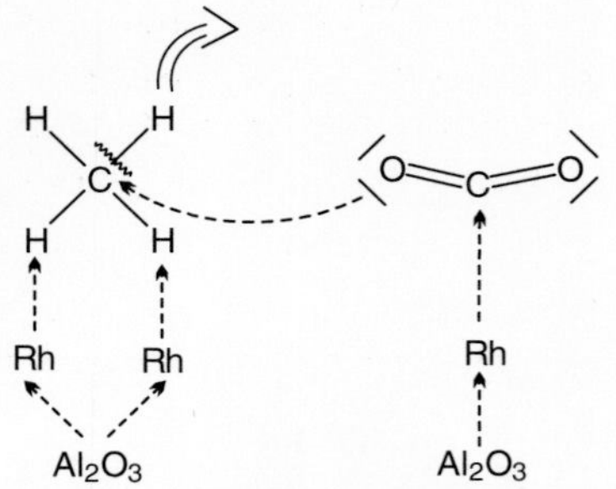

Scheme 4-6. Reforming of methane with CO_2 on supported Rh/Al_2O_3 catalysts [14]

bide phases, and the transformation of carbon and the encapsulation of metal crystallites have been proven [18].

4.4.2 Promoters [39]

Promoters are substances that are themselves not catalytically active but increase the activity of catalysts. The function of these substances, which are added to catalysts in amounts of a few per cent, has not been fully elucidated. There are four types of promoters:

- **Structure promoters** increase the selectivity by influencing the catalyst surface such that the number of possible reactions for the adsorbed molecules decreases and a favored reaction path dominates. They are of major importance since they are directly involved in the solid-state reaction of the catalytically active metal surface.
- **Electronic promoters** become dispersed in the active phase and influence its electronic character and therefore the chemical binding of the adsorbate.
- **Textural promoters** inhibit the growth of catalyst particles to form larger, less active structures during the reaction. Thus they prevent loss of active surface by sintering and increase the thermal stability of the catalyst.
- **Catalyst-poison-resistant promoters** protect the active phase against poisoning by impurities, either present in the starting materials or formed in side reactions.

A catalyst may contain one active component and one or more promoters. The fraction of active components usually exceeds 75%. Since the above four effects tend to overlap in practice, it is sometimes difficult to precisely define the function of a promoter.

Promoters are the subject of great interest in catalyst research due to their remarkable influence on the activity, selectivity, and stability of industrial catalysts.

Many promoters are discovered serendipitously; few are the result of systematic research. This sector of catalyst research is often the scene of surprising discoveries.

Before we discuss some examples of function of promoters, let us examine an overview of promoters for industrial catalysts (Table 4-35).

Table 4-35. Examples of promoters in the chemical industry [T41]

Catalyst (use)	Promoters	Function
Al_2O_3 (support and cat.)	SiO_2, ZrO_2, P	increase thermal stability
	K_2O	poisons coke formation on active centers
	HCl	increases acidity
	MgO	slows sintering of active components
SiO_2/Al_2O_3 (cracking catalyst and matrix)	Pt	increased CO oxidation
Pt/Al_2O_3 (cat. reforming)	Re	lowers hydrogenolysis activity and sintering
MoO_3/Al_2O_3 (hydrotreating, HDS, HDN)	Ni, Co	increased hydrogenolysis of C–S and C–N bonds
	P, B	increased MoO_3 dispersion
Ni/ceramic support (steam reforming)	K	improved coke removal
$Cu/ZnO/Al_2O_3$ (low-temperature conversion)	ZnO	decreased Cu sintering
Fe_3O_4 (NH_3 synthesis)	K_2O	electron donor, favors N_2 dissociation
	Al_2O_3	structure promoter
Ag (EO synthesis)	Alkali metals	increase selectivity, hinder crystal growth, stabilize certain oxidation states

Structure promoters can act in various ways. In the aromatization of alkanes on Pt catalysts, nonselective dissociative reaction paths that lead to gas and coke formation can be suppressed by alloying with tin. This is attributed to the ensemble effect, which is also responsible for the action of alkali and alkaline earth metal hydroxides on Rh catalysts in the synthesis of methanol from CO/H_2 and the hydroformylation of ethylene. It was found that by means of the ensemble effect the promoters block active sites and thus suppress the dissociation of CO. Both reactions require small surface ensembles. As a result, methanol production and insertion of CO into the alkene are both positively influenced.

Promoters can also influence catalytically active phases by stabilizing surface atoms in certain valence states. An example is the effect of chlorine on silver catalysts in the oxidation of ethylene to ethylene oxide. Silver oxide chloride phases were detected on the surface. The selective epoxidation between the electrophilic oxygen and the electron-rich double bond of ethylene is optimized. Cesium promoters stabilize these silver oxide chloride phases of the type

$$\left[\begin{array}{c} O \\ \parallel \\ O^{\diagup} Ag_{\diagdown} Cl \end{array} \right]^{-} Cs^{+}$$

under reaction conditions.

Another example of a structure promoter is Al_2O_3 in ammonia synthesis. It was long assumed that Al_2O_3 hinders the sintering of the iron following reduction of the catalyst, but it is now believed that Al_2O_3 favors the formation of highly catalytically active (111) surfaces of the iron catalyst.

Next, let us take a closer look at electronic effects. Potassium is used as a promoter in many catalytic reactions. The hydrogenation of CO and ammonia synthesis are two well-known examples. The strongly electropositive potassium (or, more often, the oxide K_2O) provides electrons that flow to the metal and then into the chemisorbed molecule. In this way, π backbonding into the π^* orbitals of the adsorbate is considerably strengthened. Figure 4-35 explains this for the example of nitrogen.

The promoter potassium facilitates the dissociation of N_2 and thus increases the rate of formation of NH_3. In investigations of the chemisorption of N_2 on the less active (100) and (110) iron surfaces, it was shown that low concentrations of potassium increase the heat of chemisorption of molecular nitrogen by 16 kJ/mol and increase the rate of N_2 dissociation 300-fold.

This is direct evidence that the rate-determining step in ammonia synthesis is the chemisorption of nitrogen. Commercial iron catalysts contain ca. 1.8 mol% K.

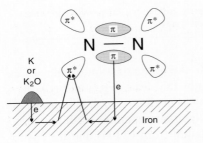

Fig. 4-35. The action of potassium promoters in the dissociative chemisorption of N_2 on iron catalysts

The electron-donor capability of potassium depends strongly on the degree of coverage θ and therefore on the promoter concentration in the catalyst, as has been shown by measurements of the heat of adsorption of potassium on transition metal surfaces. At low degrees of coverage, values of around 250 kJ/mol were measured, which corresponds to complete ionization of the atom. At high degrees of coverage, partial depolarization of the charged potassium species leads to neutralization. At $\theta = 50\%$, the heat of adsorption drops to about 97 kJ/mol, which corresponds to the heat of sublimation of potassium metal. The adsorbed atoms are then clearly no longer ionized. Similar effects were found in the hydrogenation of CO with the transition metals Pt, Ni, and Ru.

Potassium increases the reaction rate and the selectivity for C_{2+} hydrocarbons, as would be expected if dissociation of CO is more facile. Evidence for this is that in the presence of potassium, the CO desorption temperature is 100–200 K higher and the heat of chemisorption increases by 20–50 kJ/mol. Vibrational spectroscopy showed that with increasing degree of coverage by potassium, the CO stretching frequency of 1875 and 2120 cm^{-1} ($\theta = 0$) decreases to 1565 cm^{-1} ($\theta = 0.6$). Thus the influence of potassium lowers the CO bond order from 2 to 1.5, so that CO dissociation can more readily occur.

It was also shown for rhodium catalysts that at low pressures CO is molecularly adsorbed but dissociates in the presence of potassium promoters.

Besides the purely electronic effects that we have discussed up to now, the promoter can also form direct chemical bonds with the adsorbate. An example is the influence of alkali metal cations on the synthesis of methanol with copper catalysts. Sodium and potassium hydroxide can react with CO under relatively mild conditions to form alkali metal formates, which are hydrogenated to methanol by hydrogen dissociatively adsorbed on copper.

Purely chemical promoter effects are also observed with methanation catalysts. The water formed in the reaction is adsorbed on the active centers of the catalyst and thus blocks them. The water can be removed by electron-deficient compounds (Eq. 4-78).

$$MO_{x-1} + H_2O \longrightarrow MO_x + H_2 \qquad (4\text{-}78)$$

The resulting hydrogen is then desorbed. Various reducible transition metal oxides have been tested as promoters, and the following activity series was found:

$$UO_2/U_3O_8 > MoO_2/MoO_3 > WO_2/WO_3 > PrO_3/Pr_4O_{10}$$
$$> Ce_2O_3/CeO_2 > CrO_2/Cr_2O_3$$

Hence the promoter of choice is UO_2, which is added to the catalyst in small amounts.

An interesting promoter effect is exhibited by K_2SO_4 in the oxidation of methanol to formaldehyde on V_2O_5 catalysts. The addition of 10–20% K_2SO_4 drastically increases the reaction rate and raises the selectivity from 85 to 97%. In this case, too, the potassium releases electrons to the oxide, weakening the coordinative V=O bond and accelerating the reaction. Promoters are also developed to strengthen the support or the active component. An important function is influencing the stability of support materials. Oxidic supports can exist in numerous different phases. For Al_2O_3 the preferred phase is γ-Al_2O_3. This oxide has a defect spinel structure, high surface area, a certain degree of acidity, and forms solid solutions with transition metal oxides such as NiO and CoO. Above 900 °C γ-Al_2O_3 is transformed into α-Al_2O_3, which has an hexagonal structure and a smaller surface area. Such high temperatures can occur during catalyst regeneration. Even at lower temperatures a slow phase transition occurs, which shortens the catalyst lifetime. The incorporation of small amounts (1–2%) of SiO_2 or ZrO_2 in γ-Al_2O_3 shifts the γ–α transition to higher temperature and increases the stability of the catalyst.

Promoters are often used to suppress undesired activity of support materials, such as coke formation. Coking is due to cracking reactions on Brønsted acid centers, followed by an acid-catalyzed polymerization to give $(CH_x)_n$ chains, which cover the active centers on the surface and block the pores. Removal of the coke by incineration can lead to loss of activity due to sintering. Acidic centers are best neutralized by bases, preferably alkali metal compounds. Potassium, added as K_2CO_3 during catalyst production is the most effective at minimizing the coking tendency of Al_2O_3 supports.

In the steam reforming of naphtha, potassium promoters accelerate the reaction of carbon with steam. However, this leads to formation of KOH, which sublimes. In this case, potassium aluminosilicate was successfully used as promoter. In the presence of steam and CO_2 it decomposes into K_2CO_3 and KOH to an extent that is just sufficient to remove the coke that is formed. This prolonged catalyst lifetimes to 4–5 years [T35].

Finally, let us take a closer look at the influence of promoters on hydrodesulfurization catalysts. In this important refinery technology reaction, CoMo/Al_2O_3 supported catalysts are used. The schematic reaction sequence is shown in Equation 4-79.

$$R-S \xrightarrow[-H_2S]{+H_2} R^= \xrightarrow{+H_2} R \qquad (4\text{-}79)$$

Hydrogenolysis of the C–S bond is followed by hydrogenation of the resulting alkene. Since the starting materials range from low-boiling compounds to heavy residues, a wide range of technologies is used. However, the fundamental chemistry of the processes is in all cases the same.

Precipitated γ-Al_2O_3-based catalysts with a large surface area (ca. 250 m^2/g) are used. Amounts of about 1% SiO_2 act as texture stabilizer. Cobalt and molybdenum salts are calcined on the support and form oxides such as MoO_3, CoO, Co_3O_4, and $CoAl_2O_4$ in a solid-state reaction. The key precursor of the active component is MoO_3, which is activated by sulfiding to give microcrystalline MoS_2 in which small amounts of Co^{2+} ions are incorporated. Active "CoMoS" centers increase the activation of hydrogen and thus facilitate the cleavage of sulfur. It is assumed that Co acts as a structure promoter and leads to increased dispersion of the sulfided species.

For high-boiling starting materials the catalysts also contain K and P promoters to neutralize acid centers, suppress coking, and to increases the dispersion of the molybdenum component. The last example – even in this strongly simplified form – gives an impression of just how complex the interaction between catalyst components, support materials, and promoters can be, and shows that adapting a catalyst to the requirements of an industrial process is a time-consuming, creative task.

Exercises for Section 4.4

Exercise 4.34

What are the main interactions that can occur between metals and support materials?

Exercise 4.35

What is meant by the term „texture" of a catalyst support?

Exercise 4.36

The chemisorption properties of platinum group metals for CO and H_2 are less pronounced on TiO_2 supports.
The chemisorption of H_2 is reduced on Ni/SiC and SiO_2; formation of Ni–Si alloys is assumed.
Which effect could be responsible for this?

Exercise 4.37

Which catalyst properties can be influenced by promoters?

Exercise 4.38

What influence do potassium promoters have on acidic cracking catalysts?

4.5 Catalyst Deactivation and Regeneration [6]

Catalysts have only a limited lifetime. Some lose their activity after a few minutes, others last for more than ten years. The maintenance of catalyst activity for as long as possible is of major economic importance in industry. A decline in activity during the process can be the result of various physical and chemical factors, for example:

– Blocking of the catalytically active sites
– Loss of catalytically active sites due to chemical, thermal, or mechanical processes

An overview of catalyst deactivation in large-scale industrial processes is given in Table 4-36.

Catalyst deactivation, also known as ageing, is expressed by the decrease in catalyst activity with time. Catalyst activity a is the ratio of the reaction rate at a given time t to the reaction rate at the time that use of the catalyst began ($t = 0$; Eq. 4-80).

$$a(t) = \frac{r(t)}{r(t=0)} \qquad (4\text{-}80)$$

Table 4-36. Causes of deactivation in large-scale industrial processes

Reaction	Reaction conditions	Catalyst	Catalyst lifetime [years]	Deactivation process
Ammonia synthesis $N_2 + 3 H_2 \rightarrow 2 NH_3$	450–550 °C 200–500 bar	$Fe/K_2O/Al_2O_3$	5–10	slow sintering
Methanization $CO + 3 H_2 \rightarrow CH_4 + H_2O$	250–350 °C 30 bar	Ni/Al_2O_3	5–10	slow poisoning by S and As compounds
Methanol synthesis $CO + 2 H_2 \rightarrow CH_3OH$	200–300 °C 50–100 bar	$Cu/Zn/Al_2O_3$	2–8	slow sintering
Hydrodesulfurization of light petroleum	300–400 °C 35–70 bar	$CoS/MoS_2/Al_2O_3$	0.5–1	deposits (decomp. of sulfides)
NH_3 Oxidation $2 NH_3 + 2.5 O_2 \rightarrow 2 NO + 3 H_2O$	800–900 °C 1–10 bar	Pt net	0.1–0.5	loss of platinum, poisoning
Catalytic cracking	500–560 °C 2–3 bar	zeolites	0.000002	rapid coking (continuous regeneration)
Benzene oxidation to maleic anhydride $C_6H_6 + O_2 \rightarrow C_4H_2O_3$	350 °C 1 bar	$V_2O_5/MoO_2/Al_2O_3$	1–2	formation of an inactive vanadium phase

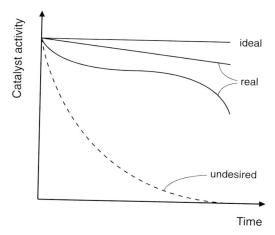

Fig. 4-36. Deactivation behavior of catalysts [8]

The course of the activity of an industrial catalyst with time can be described by means of several basic types (Fig. 4-36).

Not only does the decreasing catalyst activity lead to a loss of productivity, it is also often accompanied by a lowering of the selectivity. Therefore, in industrial processes great efforts are made to avoid catalyst deactivation or to regenerate deactivated catalyst. Catalyst regeneration can be carried out batchwise or preferably continuously while the process is running.

In this chapter we will encounter the most important mechanisms of catalyst deactivation and discuss possible methods of catalyst regeneration [T35].

The four most common causes of catalyst deactivation are:

- Poisoning of the catalyst. Typical catalyst poisons are H_2S, Pb, Hg, S, P.
- Deposits on the catalyst surface block the active centers and change the pore structure (e. g., coking).
- Thermal processes and sintering of the catalyst lead to a loss of active surface area.
- Catalyst losses by evaporation of components (e. g., formation of volatile metal carbonyls with CO).

These processes are shown schematically in Figure 4-37. We will now discuss these effects in more detail and examine some examples from the chemical industry.

Catalyst Poisoning

Catalyst poisoning is a chemical effect. Catalyst poisons form strong adsorptive bonds with the catalyst surface, blocking active centers. Therefore, even very small

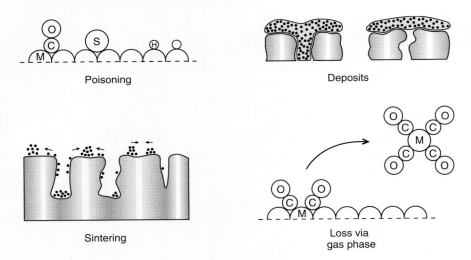

Fig. 4-37. Mechanisms of catalyst deactivation (M = metal) [8]

quantities of catalyst poisons can influence the adsorption of reactants on the catalyst. The term catalyst poison is usually applied to foreign materials in the reaction system. Reaction products that diffuse only slowly away from the catalyst surface and thus disturb the course of the reaction are referred as inhibitors. Table 4-37 lists some catalyst poisons and inhibitors and the way in which they act.

In the investigation of catalyst deactivation by poisoning, the distribution of the active centers, the stoichiometry, and diffusion are of decisive importance. In the following, poisoning of the most important classes of catalysts, i.e., metals, semiconductors, and acidic isolators, is discussed.

Poisoning of Metals

Metal catalysts are highly sensitive to small amounts of certain impurities in the reaction medium. Catalytically active metals make their d orbitals available for adsorption, and this is the key to understanding both their catalytic activity and their sensitivity to poisons.

Poisons for metals can be classified in three groups:

– Nonmetallic ions
– Metal ions
– Unsaturated molecules

Particularly strong catalyst poisons are the ions of elements of groups 15 (N, P, As, Sb, Bi) and 16 (O, S, Se, Te). The poisoning activity depends on the presence of electron lone pairs, which have been shown to form dative bonds with transition me-

Table 4-37. Catalyst poisons and inhibitors in chemical processes [T41]

Process	Catalyst	Catalyst poison, inhibitor	Mode of action
NH_3 synthesis	Fe	S, Se, Te, P, As compounds, halogens	poison: strong chemisorption or formation of compounds
		O_2, H_2O, NO	weak poison: oxidation of Fe surface; reduction possible but causes sintering
		CO_2	inhibitor: reaction with alkaline promoters
		CO	poison and inhibitor: strong chemisorption, reduction to methane; accelerates sintering
		unsaturated hydrocarbons	inhibitor: strong chemisorption, slow reduction
Hydrogenation	Ni, Pt, Pd, Cu	S, Se, Te, P, As compounds, halogens	poison: strong chemisorption
		Hg and Pb compounds	poison: alloy formation
		O_2	poison: surface oxide film
		CO	Ni forms volatile carbonyls
Catalytic cracking	alumino-silicates	amines, H_2O, Ni, Fe, V, (porphyrins)	inhibitor: blocking of active sites
		coke	poison: blocking of active sites
NH_3 oxidation	Pt/Rh	P, As, Sb compounds; Pb, Zn, Cd, Bi	poison: alloy formation, catalyst net becomes brittle
		rust	decomposes NH_3
		alkali metal oxides	poisons: react with Rh_2O_3
SO_2 oxidation	V_2O_5/$K_2S_2O_7$	As compounds	inhibitor → poison; compound formation
Ethylene oxide synthesis	Ag	halogenated hydrocarbons	inhibitor: increase selectivity

tals on chemisorption. If these are involed in bonding to other elements, then the ions are nonpoisons:

Poisons: H_2S, thiophene, NH_3, PH_3, AsH_3
Nonpoisons: SO_4^{2-}, NH_4^+, PO_4^{3-}, AsO_4^{3-}, sulfones

The poisoning effect of metal ions depends on the number of d electrons. Metals with an empty d shell, such as alkali and alkaline earth metals, and those with less than three d electrons are nonpoisons, as shown in the following for the example of platinum:

Poisons: Zn^{2+}, Cd^{2+}, Hg^{2+}, In^{3+}, Tl^+, Sn^{2+}, Pb^{2+}, Cu^+, Cu^{2+}, Fe^{2+}, Mn^{2+}, Ni^{2+}, etc.
Nonpoisons: Na^+, Be^{2+}, Mg^{2+}, Al^{3+}, La^{3+}, Ce^{3+}, Zr^{4+}, Cr^{2+}, Cr^{3+}

Metals readily adsorb unsaturated molecules such as CO and olefins. If they are adsorbed irreversibly in molecular form, then they act as poisons. If dissociation or decomposition occurs, then this can lead to deactivation by coking.

Because of the wide range of chemisorption bond strengths, various effects can occur in the hydrogenation of two unsaturated molecules. Inhibition can range from favored hydrogenation of one component to complete suppression of a reaction in the presence of extremely small quantities of a second unsaturated component. Examples are the poisoning of Pt and Ni hydrogenation catalysts by CO or CN^- and the slower hydrogenation of cyclohexene in the presence of small amounts of benzene.

Halogens and volatile nitrogen compounds generally act as weak catalyst poisons or inhibitors and lead to a reversible or temporary lowering of the catalyst activity.

Catalyst poisoning can be reversible or irreversible, depending on the reaction conditions. For example, sulfur poisoning of nickel catalysts is irreversible at low temperatures, and methanation catalysts can not be regenerated, even by treatment with hydrogen. At higher temperatures sulfur can be removed by hydrogenation and steam, so that nickel catalysts for steam reforming are considerably more resistant to sufur-containing poisons.

Poisoning of metal catalysts can best be avoided by pretreatment of the reactants by:

– Chemical treatment (expensive; can lead to other impurities)
– Catalytic treatment (very effective for organic poisons)
– Use of adsorbers (e. g., ZnO to remove sulfur-containing compounds in natural gas reforming)

The incorporation of promoters can also neutralize catalyst poisons. Thus the sulfur poisoning of nickel catalysts is reduced by the presence of copper chromite since copper ions readily form sulfides.

The appropriate treatment method and the decision whether the catalyst or the process should be modified requires detailed knowledge of the cause of deactivation.

Poisoning of Semiconductor Oxides

Because of the presence of electron-donor or electron-acceptor centers with special surface geometries and the fact that redox reactions are favored, general statements about the poisoning of semiconductor catalysts can hardly be made. Any molecule that is strongly adsorbed on the surface is a potential poison.

Up to now there have been no theoretical models of the poisoning of semiconductor catalysts. They are quite resistant to poisoning, the addition of several per cent of foreign materials being required to give a noticeably lower activity.

Poisoning of Solid Acids

The poisoning of acid centers can easily be explained. Acid centers can be neutralized and thus poisoned by basic compounds such as alkali and alkaline earth compounds and especially organic bases. Alkali and alkaline earth compounds are normally used as promoters and are generally not present in process streams.

In contrast, nitrogen-containing bases are contained in many crude-oil fractions. In a typical starting material, 25–35% of the nitrogen compounds have basic character. The sensitivity of solid acids towards these poisons correlates directly with their basicity. For example, pyridine, quinoline, amines, and indoles are basic, while pyrrole and carbazole are nonbasic. These poisons are best removed by hydrogenation, together with sulfur and most of the heavy metal poisons.

However, in some cases partial catalyst poisoning is desired, for example to lower the catalyst activity or to influence the selectivity. A well-known example is the addition of ppm quantities of H_2S in catalytic reforming with nickel catalysts. Compared to platinum, nickel has a higher hydrogenolysis activity, which leads to formation of gases and coke. Sulfur selectively poisons the most active hydrogenolysis centers and thus drastically influences the selectivity towards the desired isomerization reactions.

Other partially poisoned catalysts have long been used in the laboratory. Supported palladium catalysts, poisoned with lead (Lindlar catalysts), sulfur, or quinoline, are used for the hydrogenation of acetylenic compounds to *cis*-olefins. Another application of this type of catalyst is the removal of traces of phenylacetylene (200–300 ppm) from styrene by selective hydrogenation.

In the Rosemund reaction (Eq. 4-81), acid chlorides are hydrogenated to aldehydes. The catalyst is a supported palladium catalyst (5% Pd/BaSO$_4$) poisoned by sulfur compounds such as quinoline, tiourea, or thiophene to prevent further reduction of the aldehyde.

$$R-\underset{\underset{O}{\|}}{C}-Cl + H_2 \longrightarrow R-\underset{\underset{O}{\|}}{C}-H + HCl \qquad (4\text{-}81)$$

Deposits on the Catalyst Surface

The blocking of catalyst pores by polymeric components, especially coke, is another widely encountered cause of catalyst deactivation. In many reactions of hydrocarbons, side reactions lead to formation of polymers. If these are deposited near the pore openings, catalyst activity and selectivity can be influenced due to impaired mass transport into and out of the pores.

At high temperatures (above 200 °C) these polymers are dehydrogenated to carbon, a process known as coking. Especially catalysts with acidic or hydrogenating/dehydrogenating properties cause coking. Coking on acid centers is observed with zeolite and aluminosilicate catalysts and with acidic supports. The extent of coke formation depends directly on the acidity.

The precursors for coke formation are mainly aromatic and olefinic hydrocarbons, which are either contained in the starting materials or are formed as intermediate products in the process.

In some processes, 5–10% zeolite is added to amorphous cracking catalysts. This increases the activity by several orders of magnitude and drastically reduces coke formation. This is another example of shape-selective catalysis by zeolites, in which the coke-forming intermediates are restricted by the zeolite pores (see Section 6.3.1).

With dehydrogenation catalysts (metals, oxides, sulfides), coke is formed in a different manner than acid cleavage of hydrocarbons. Dehydrogenation steps, followed by hydrogenolysis, lead to formation of carbon fragments C_x. These are highly reactive and are bound in a carbide-like fashion or are present as pseudographite. In the presence of acid support materials, the C fragments migrate from the dehydrogenation centers of the metal to the support, where they are cleaved analogously to acid catalysts (Fig. 4-38).

Dehydrogenative coking mainly occurs in catalytic reforming, hydrodesulfurization, and in cases of metal contamination of the starting materials. In catalytic reforming processes, bimetallic catalysts are successfully used. Addition of Re to Pt greatly increases the stability of the catalyst, as depicted schematically in Figure 4-39. Rhenium inhibits both coking and sintering of the catalyst and thus has a favorable effect on deactivation during the process and on the frequency of regenera-

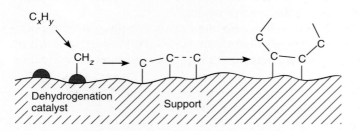

Fig. 4-38. Dehydrogenative coking

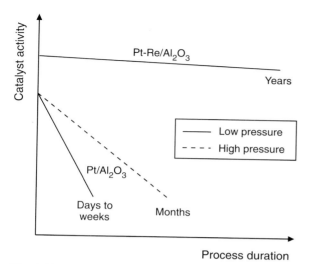

Fig. 4-39. Catalyst deactivation in reforming processes [T35]

tion. By using supported Re/Pt catalysts, the catalyst lifetime can be extended from a few weeks to several years, whereby, however, the H_2 pressure also plays a role. Metal impurities in the starting materials play a role in hydrodesulfurization and hydrodenitrogenation processes. Crude oil fractions contain heavy metal impurities in the form of porphyrins of Fe (up to 150 ppm), Ni (up to 50 ppm) and V (up to 100 ppm). These porphyrins are preferentially adsorbed on Al_2O_3 and aluminosilicates and then decompose to finely divided metals. Nickel is the most active. When the catalyst is regenerated, these metals are oxidized, and the resulting oxides can act as strong oxidizing agents (e. g., V_2O_5). These metals and their oxides have several negative effects: they block active centers, have high catalytic activity, and have a strong coking effect.

Therefore, heavy metals must be removed from crude oil fractions. Various processes are used: chemical or adsorptive removal of the porphyrins, or demetallation by hydrogenation and binding the metals on Al_2O_3. Another effective method is the use of additives. For example, the heavy metals can be alloyed by adding antimony, after which they are deposited on the catalyst in a different form.

Coking of catalysts can be reduced by increasing the hydrogen partial pressure or by partial neutralization of the acid sites with promoters, as we have already seen. Coke that has already formed is removed by periodic regeneration of the catalyst. The deactivated catalyst is purified by controlled combustion of the carbon layer. In fluidized-bed crackers the catalyst circulates continuously between the reactor and the regenerator, in which combustion takes place. The heat of combustion is used to maintain the catalyst at the temperature of the slightly endothermic cracking reaction.

Thermal Processes and Sintering

Thermal influences can often affect the catalyst composition. In many cases one or more metastable phases are formed from the active components or the support materials. Phase changes can limit the catalyst activity or lead to catalyst–substrate interactions. We have already dealt with the transformation of γ-Al_2O_3 into α-Al_2O_3 with its lower surface area. Another example is the phase transformation of TiO_2 from anatase to rutile in V_2O_5/TiO_2/corundum catalysts for the oxidation of o-xylene to phthalic anhydride.

Sintering is a well-known phenomenon in metallurgy and ceramics technology. Sintering processes are also of importance in catalysis, even at low temperatures. Reasons for this are the extremely small crystallites, porous supports, and reactive gases. Catalyst atoms already become mobile at temperatures between one-third and one-half of the melting point.

The rate of sintering increases with increasing temperature, decreasing crystallite size, and increasing contact between the crystallite particles. Other factors are the amount and type of impurities on the crystallite surface and the support composition in supported catalysts.

Increased sintering can also occur if the active catalyst components form volatile compounds with the reactants. An example is the sintering of copper catalysts in the presence of chlorine compounds.

The main effect of sintering is loss of active surface area and the resulting decrease in catalyst activity. However, a change in selectivity can also occur, especially in the case of structure-sensitive reactions. Extensive investigations of sintering have been carried out on highly dispersed metals such as Pt/Al_2O_3.

An informative example is naphtha reforming, in which the influence of regeneration also becomes apparent. The data in Table 4-38 suggest that regeneration of the catalyst by combustion of the coke leads to an increase in crystallite size, since the catalyst activity, measured by H_2 adsorption, decreases steadily with time, in spite of

Table 4-38. Naphtha reforming with 0.6% Pt/Al_2O_3: catalyst deactivation and regeneration [T35]

Catalyst state	Adsorbed hydrogen [cm^3/g cat.]
Fresh	0.242
Coked, 1 d (1% C)	0.054
Regenerated	0.191
Coked, 1 d (1% C)	0.057
Regenerated	0.134
Coked, 5 d (2.5% C)	0.033
Regenerated	0.097

regeneration. Studies of the reforming process with model substances found major changes in selectivity. For example, it was found that with increasing crystallite diameter aromatics formation due to dehydrocyclization decreases, isomerization reactions increase, and the hydrocracking activity remains roughly the same.

Another example is industrial ethylene oxide synthesis. Here, too, it was shown that a decrease in the Ag surface area due to sintering is responsible for the deactivation of the catalyst.

A final example is the selective catalytic reduction of nitrogen oxides on vanadium titanium oxides (SCR process) [20]. The catalyst consists of V_2O_5 on a TiO_2 (anatase) support. Above about 350 °C the less thermally stable TiO_2 sinters, and the anatase surface becomes much smaller. This results in recrystallization of the vanadium pentoxide, which is now present in excess, and growth of threadlike and platelet V_2O_5 crystallites is observed. This results in undesired side reactions such as increased N_2O formation. The thermal stability of the catalyst can be improved by stabilizing the support (addition of sulfate) or by modifying the active components (addition of tungsten oxide).

In general, the effects of sintering can be counteracted by the following measures:

- Addition of stabilizing additives (promoters) to the active components or their dispersion on the surface of the support (e. g., nickel can be stabilized by Cr_2O_3)
- Redispersion of the metals (e. g., chlorine treatment of Pt/Al_2O_3 reforming catalysts: volatile $PtCl_2$ is re-adsorbed on Al_2O_3 and finely distributed)

Catalyst Losses via the Gas Phase

High reaction temperatures in catalytic processes can lead to loss of active components by evaporation. This does not only occur with compounds that are known to be volatile (e. g., P_2O_5 in H_3PO_4, silica gel, $HgCl_2$/activated carbon), but also by reaction of metals to give volatile oxides, chlorides, or carbonyls. In the oxidation of ammonia on Pt/Rh net catalysts (Ostwald nitric acid process), the catalyst reacts with the gas phase to form volatile PtO_2. Furthermore, porous platinum growths are observed on the surface. This can be prevented by addition of rare earth oxides.

In hydrogenation processes with molybdenum-containing catalysts, too high a temperature during regeneration due to the occurrence of hot spots can lead to the formation of MoO_3, which evaporates at temperatures above 800 °C with irreversible loss of activity.

Another example is the use of nickel catalysts in the methanation of synthesis gas. If the temperature of the catalyst bed drops below 150 °C, catalyst is lost by formation of highly toxic nickel tetracarbonyl.

We will now examine some models of catalyst deactivation processes. Given the various causes of catalyst deactivation, it is not surprising that numerous mechanisms and models can be found in the literature.

A relatively simple model of deactivation kinetics is based on Langmuir adsorption. It is assumed that the total number of active surface sites of the catalyst Z_{tot} decreases with increasing lifetime or operating time t', for example, due to poisoning by a component of the starting material. For a monomolecular reaction, the kinetics are then described by Equation 4-82.

$$r = \frac{k'_s Z_{tot} K_1 p_1}{1 + K_1 p_1} = \frac{k_1 p_1}{1 + K_1 p_1} \tag{4-82}$$

The rate of deactivation r_d is defined as the rate of decrease of the activity a with time and can be determined separately (Eq. 4-83).

$$r_d = \frac{da}{dt'} \tag{4-83}$$

In our model r_d can be attributed to the change in the active surface sites. The rate of deactivation depends on the temperature, the activity of the catalyst a, the concentration of the deactivating component c_d, and the activation energy of the process E_d (Eq. 4-84).

$$-r_d = k_{d,0}\, e^{-(E_d/RT)} \cdot f(a, c_d) \tag{4-84}$$

A very high activation energy of 290 kJ/mol is found for the reforming of heptane on Pt/Al$_2$O$_3$ catalysts, for example.

Quantitatively, the rate of loss of activity can often be expressed as a power law of the type given in Equation 4-85.

$$-r_d = -\frac{da}{dt'} = k_d\, a^n\, c_d^m \tag{4-85}$$

If there is no deactivation by poisoning, then $m = 0$. This could then be a process of deactivation by sintering. With the simplifying assumption $n = 1$, Equation 4-85 becomes:

$$-r_d = -\frac{da}{dt'} = k_d\, a \tag{4-86}$$

Integration of Equation 4-86 between $t' = 0$ and t' gives:

$$a(t') = a(t' = 0) \cdot e^{-k_d t'} \tag{4-87}$$

Since by definition $a(t' = 0) = 1$, we obtain the simple exponential equation:

$$a(t') = e^{-k_d t'} \tag{4-88}$$

Such exponential equations can sometimes also describe deactivation by poisoning, provided the concentration of the catalyst poison is constant. Examples are the hydrogenation of ethylene on copper catalysts (poisoning by CO) and the dehydrogenation of alkanes on Cr/Al_2O_3 catalysts.

For sintering processes, the decrease in activity can often be described by a second-order equation. We then obtain:

$$-r_d = k_d a^2 \qquad (4\text{-}89)$$

and after integration

$$a(t') = \frac{1}{1 + k_d t'} \qquad (4\text{-}90)$$

Examples for the application of this hyperbolic law are the dehydrogenation of cyclohexane on Pt/Al_2O_3 catalysts and the hydrogenation of isobutene on Ni catalysts.

After many examples of catalyst deactivation, let us look at the process of catalyst regeneration. The catalyst activity varies with time as shown in Figure 4-40. The activity decreases with increasing operating time in a manner that depends on the reaction conditions and the deactivation kinetics. First, attempts are made to make the deactivation slower by adjusting the operating parameters (e.g., raising the temperature, increasing the pressure in hydrogenation reactions).

The loss of activity can be gradual or very rapid. Examples are the hydrogenative treatment of naphtha, with catalyst lifetimes of several years, and catalytic cracking, in which strong catalyst deactivation occurs after a few minutes. In all cases the deactivation reaches an extent at which the conversion or other process parameters are below specification, and the catalyst must be replaced or regener-

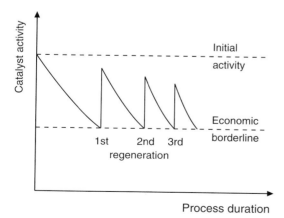

Fig. 4-40. Catalyst regeneration and loss of activity during a process

ated. In practice, the original activity is not attained due to a permanent secondary deactivation. When regeneration steps are no longer economically viable, the catalyst must be completely replaced.

In this chapter we have described the importance of catalyst deactivation in industrial processes. Detailed knowledge of the cause of deactivation is a prerequisite for catalyst modification and process control.

Exercises for Section 4.5

Exercise 4.39

What are the effects of catalyst deactivation?

Exercise 4.40

How can catalyst deactivation be measured experimentally?

Exercise 4.41

Catalysts based on Al_2O_3 are often regenerated by combustion of coke deposits. Which negative effects can occur here?

Exercise 4.42

Why must particular care be taken when nickel catalysts are used in industrial carbonylation reactions?

Exercise 4.43

Platinum metal catalysts are strongly inhibited by halides. Which order of inhibition activity can be expected for the halides?

Exercise 4.44

Zeolites are the preferred catalysts for cracking reactions.
a) How does their deactivation occur?
b) How are cracking catalysts regenerated?

4.6 Characterization of Heterogeneous Catalysts [3, 28]

Both the physical and the chemical structure of a catalyst must be known if relationships between the the material structure of the catalyst and activity, selectivity, and lifetime are to be revealed. The available methods include classical procedures and state-of-the-art techniques for studying the physics and chemistry of surfaces [33].

The physical properties of pore volume, pore distribution, and BET surface area are nowadays routinely monitored in the production and use of industrial catalysts. In contrast, the chemical characterization of catalysts and microstructural investigations, especially of the catalyst surface, are far more laborious and are rarely carried out in industry.

As we have already seen in many examples, the upper atomic layers often have a different composition to that in the catalyst pellet. Promoters, inhibitors, and catalyst poisons can also be deposited. Therefore, in order to understand heterogeneous catalysis, information about the nature and structure of the upper atomic layers is required. Up to about 25 years ago, chemisorption processes were the only methods available for the indirect characterization of the catalyst surface. Only in the 1970s did instruments for the analysis of surfaces become commercially available, opening up new possibilities for fundamental catalyst research [21].

In this section we will encounter some methods for characterizing catalysts and discuss their capabilities and limitations. Most chemical engineers working in catalyst development are not experts in complex industrial analytical chemistry, and only a few major companies and research institutes can afford surface analysis equipment. For these reasons we shall dispense with the details of methods and apparatus and concentrate on practical applications.

4.6.1 Physical Characterization [T41]

An imporant property of catalysts is the distribution of pores across the inner and outer surfaces. The term "texture" generally refers to the pore structure of the particles, including the suface, the pore size distribution, and pore shape. The most widely used method for determining the pore distribution in solids is mercury porosimetry, which allows both mesopores (pore radius 1–25 nm) and macropores (pore radius >25 nm) to be determined. The pore size distribution is determined by measuring the volume of mercury that enters the pores under pressure. The measurement is based on Equation 4-91.

$$p = \frac{2\pi\sigma\cos\alpha}{r_p} \tag{4-91}$$

p = pressure
σ = surface tension of Hg
α = contact angle Hg/solid
r_p = radius of the cylindrical pores

Pressures of 0.1 to 200 MPa allow pore sizes in the range 3.75–7500 nm to be determined. Since the pores are not exactly cylindrical, as assumed in Equation 4-91, the calculated pore sizes and pore size distributions can differ considerably from the true values, which can be determined by electron microscopy.

Mercury porosimetry is advantageously used for characterizing various shaped industrial catalysts in which diffusion processes play a role. The macropore distribution is of major importance for the turnover and lifetime of industrial catalysts and is decisively influenced by the production conditions.

For the actual catalytic reaction, the distribution of meso- and micropores is of greater importance. The specific pore volume, pore size, and pore size distribution of microporous materials are determined by gas adsorption measurements at relatively low pressures (low values of p/p_0 = pressure/saturation pressure). The method is based on the pressure dependence of capillary condensation on the diameter of the pores in which this condensation takes place. To calculate the pore size distribution, the desorption isotherm is also determined. Thus a distinction can be made between true adsorption and capillary condensation. The latter is described by the Kelvin equation (Eq. 4-92).

$$\ln\frac{p}{p_0} = \frac{V}{RT}\frac{2\sigma\cos\theta}{r_p} \tag{4-92}$$

V = molar volume
σ = surface tension of adsorbate
θ = contact angle adsorbate/solid
r_p = radius of the cylindrical pores

Micropores occur in particular in zeolites and activated carbons. They can lead to false values of BET surface areas due to capillary condensation.

The specific surface area (in m^2/g) of a catalyst or a support material can be determined by the proven BET method. The volume of a gas (usually N_2) that gives monomolecular coverage is measured, allowing the total surface area to be calculated. The equilibrium isotherms are of the form shown in Figure 4-41.

In Figure 4-41 the adsorbed volume is plotted against p/p_0 (p = pressure, p_0 = saturation pressure). At low pressures monolayer adsorption obeys the Langmuir equation (Eq. 4-93).

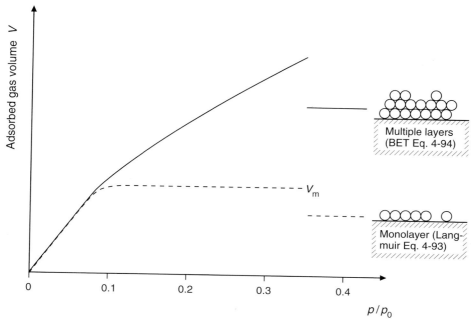

Fig. 4-41. Typical isotherm for physisorption

$$\frac{V}{V_M} = \frac{K(p/p_0)}{[1 + K(p/p_0)]} \qquad (4\text{-}93)$$

V_M = volume of the monolayer
K = constant

However, above about $p/p_0 = 0.1$, multilayer adsorption becomes important. A theoretical model for describing such adsorption processes was developed by Brunauer, Emmett, and Teller, who formulated the familiar BET equation (Eq. 4-94) [T35].

$$\frac{V}{V_M} = \frac{cp}{[p_0 - p][1 + (c-1)p/p_0]} \qquad (4\text{-}94)$$

c includes the heat of adsorption and condensation and is constant for a particular class of compounds (i.e., oxides, metals) with values of over 100.

Equation 4-94 applies up to $p/p_0 = 0.3$, above which capillary condensation begins, initially in the smallest micropores and finally in the mesopores when p/p_0 approaches unity. The determination of V_M by means of Equation 4-94 is impractical, and a much better method was developed by transforming Equation 4-94 into a linear equation (Eq. 4-95).

$$\frac{p}{V(p_0 - p)} = \frac{1}{V_M c} + \frac{(c-1)}{V_M c} (p/p_0) \tag{4-95}$$

Then V_M can easily be determined from the slope m and the ordinate intersection y.

From the difference between the BET surface area and the surface area measured by mercury porisometry, the micropore fraction of materials such as zeolites and activated carbons can be determined.

Table 4-39 lists typical surface areas of catalysts and supports.

Table 4-39. Specific surface areas of catalysts and support materials

Catalyst/support	Specific surface area [m^2/g]
H-zeolite for cracking processes	1000
Activated carbon	200–2000
Silica gel	200–700
Aluminosilicates	200–500
Al$_2$O$_3$	50–350
Ni/Al$_2$O$_3$ (hydrogenation cat.)	250
CoMo/Al$_2$O$_3$ (HDS)	200–300
Fe–Al$_2$O$_3$–K$_2$O (NH$_3$)	10
Bulk catalyst	5–80
V$_2$O$_5$ (partial oxidation)	1
Noble metal/support	0–10
Pt wire (NH$_3$ oxidation)	0.01

Although the specific surface area is one of the most important parameters in heterogeneous catalysis, it must be taken into account that especially in the case of supported catalysts, there is no direct relationship between catalyst activity and the physical surface of the catalyst. Such predictions can only be made with the aid of chemisorption measurements. Here the number of catalytically active surface atoms is determined by chemisorption of appropriate gases such as H_2, O_2, CO, NO, and N_2O at room temperature or above (Fig. 4-42). The choice of gas depends on the metal (Table 4-40).

Table 4-40 lists the advantages and disadvantages of the adsorbates. It can be seen that the adsorbates are not specifically suited to just one type of metal, so that chemisorption on multimetal catalysts has only limited applicability.

Apart from the adsorbates listed in Table 4-40, the so-called surface titration method is used to determine Pt and Pd on supports. The method is based on the dissociative chemisorption of H_2 or O_2, followed by reaction with chemisorbed O_2 or

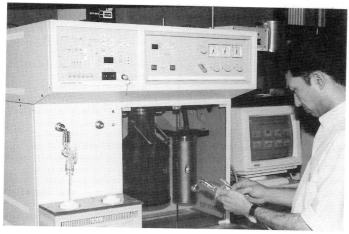

Fig. 4-42. Sorptometer (catalysis laboratory, FH Mannheim)

Table 4-40. Specific chemisorption for the characterization of metal surfaces

Adsorbate	Metal	T [°C]	Advantages	Disadvantages
H_2	Pt, Pd, Ni	0–20 −78 −195	dissociative chemisorption, low adsorption on support, negligible physisorption, simple stoichiometry	possible hydride formation, sensitive to impurities
CO	Pt, Pd	0–25	no solution in volume	physisorption at low temperatures
	Ni, Fe, Co	−78 −195		bridging and linear binding, risk of carbonyl formation or dissociation
O_2	Pt, Ni Ag Cu	25 200 −195	low adsorption on oxide supports	physisorption at low temperatures, oxidation possible
N_2O	Cu, Ag	25	low adsorption on oxide supports, very little oxidation	complicated measurement (volumetric not possible)

H_2, respectively. A problem is the reaction mechanism, which is still the subject of debate in the literature.

Given knowledge of the adsorption stoichiometry, the amount of material adsorbed on a supported catalyst can be used to determine the dispersion of the active component.

The volume of adsorbed gas gives the active surface. The number of active metal atoms is given per gram of catalyst, that is, the degree of dispersion. Normally, a direct proportionality between the measured number of surface atoms and the number of active centers can be assumed. Knowledge of the dispersion allows comparison of the catalyst activity on the basis of reaction rate per unit metal surface.

To a certain extent, chemisorption techniques are also applicable to oxides [33].

4.6.2 Surface Analysis and Chemical Characterization

Of particular importance are the composition, i.e., the distribution of elements in the catalyst, and the detection of phases and surface compounds. Also of interest are differences in composition between catalyst volume and catalyst surface, as well as interactions between active components and support materials and between the active components themselves.

These phenomena are best studied by advanced spectroscopic methods. Since the solid surface plays the decisive role in heterogeneous catalysis, methods for the characterization of surfaces are of major importance in modern catalyst research [28].

Catalyst surfaces, surface compounds, metals dispersed on supports, and adsorbed molecules are investigated by electron spectroscopy, ion spectroscopy, analytical microscopy, and other methods [12, 33].

We will first discuss methods with which the structure of the surface is determined, and then those that determine the chemical composition of the surface (catalyst and substrate). Finding relationships between the structures of material and the catalyst activity requires high-resolution investigation of the microstructure of the catalyst. Since heterogeneous catalysts are often highly nonuniform solids, correct sampling, sample preparation, and choice of the appropriate method are important if meaningful results are to be obtained.

Transmission Electron Microscopy

This method allows the size distribution and shape of metal particles in supported and unsupported catalysts to be characterized down to the level of atomic resolution [27].

Scanning transmission electron microscopy (STEM) uses X-ray backscattering analysis to obtain information on the size, morphology, and chemical composition of the active components on support materials (Fig. 4-43).

Examples of applications:

– Dispersion measurements on Pd/SiO_2 catalysts: good agreement with chemisorption measurements
– Sintering, segregation, and redispersion of metal particles as a result of oxidative treatment
– Study of coking processes
– Detection of surface impurities and surface poisoning

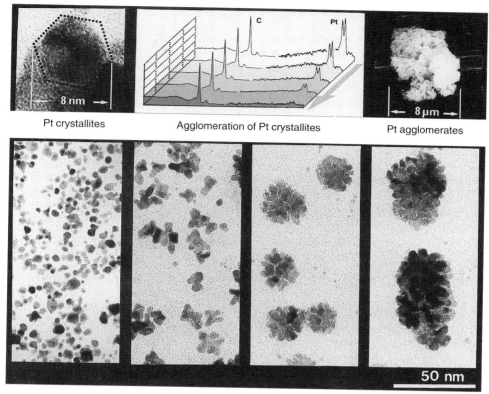

Fig. 4-43. Agglomeration of platinum crystallites in a platinum/graphite catalyst: quantification of the process by XPS (top middle) and visualization by TEM (bottom) (BASF, Ludwigshafen, Germany)

Low-Energy Electron Diffraction (LEED) [9, 25]

Electrons with kinetic energies in the range 2–200 eV are important tools for the investigation of surface properties. Since such electrons interact with the electrons of the solids, they penetrate to a depth of only a few atomic layers, or are emitted from a depth of a few atomic layers in emission processes. Their energy and angular distribution provide information on the properties of the surface region.

In the LEED method, low-energy electrons with high cross sections undergo elastic scattering from atomic cores. Since their wavelengths are comparable to interatomic distances, diffraction effects are observed with single-crystal surfaces. This allows the spatial arrangement of the atoms in the surface region, and hence the periodic structure of the catalyst surface to be determined.

In principle, the LEED method provides the same information about surfaces as is obtained for the interior of solids by X-ray structure analysis.

Examples of applications:

– Changes in the structure of nickel surfaces on chemisorption of oxygen
– Ordered surface structures in the adsorption of CO on Pd surfaces
– Detection of competitive adsorption of CO and O_2 on Pd surfaces
– Surfaces of Au, Ir, Pt, and semiconductors: structures other than those expected from the lattice geometry

IR Spectroscopy [12]

Infrared spectroscopic investigations in special cuvettes can be used to characterize active centers on catalyst surfaces and chemisorbed molecules. Catalysts on strongly absorbing supports, such as activated carbons, can be studied by reflection IR spectroscopy.

Examples of applications:

– Detection of Brønsted acid and Lewis acid surface groups with chemisorbed pyridine
– Ethylene chemisorption on isolated Pd atoms by means of matrix-isolation techniques (metal vapor, 10–30 K, xenon matrix, high vacuum): detection of chemisorption complexes
– Proof of metal oxo compounds (Mo, V) as oxidation catalysts
– IR bands of chemisorbed NO for the characterization of supported metal catalysts, e. g., Mo/Al_2O_3, Pt–Re/Al_2O_3, cracking catalysts

Electron Spectroscopy for Chemical Analysis (ESCA)

This widely used method is particularly suitable for the analysis of surface composition [19]. In an ESCA spetrometer, a sample is exposed to X-ray radiation (Fig. 4-44).

The photoelectrons formed in the probe by ionization are analyzed according to their kinetic energy. From this, the binding energy can be determined and therefore the element identified. The chemical shift of these energy values gives information about the bonding and oxidation state of the elements. ESCA can be used to detect all elements other than hydrogen in the upper 5–10 nm of the catalyst surface.

Figure 4-45 shows the ESCA spectrum of a silver catalyst supported on Al_2O_3. Both the sharp photoelectron lines and the broader Auger electron lines can be seen on a background of scattered electrons. The lines can be assigned by means of tabulated values of the bond energies of the main components of the catalyst [25].

4.6 Characterization of Heterogeneous Catalysts

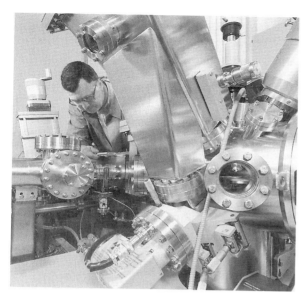

Fig. 4-44. Investigation of a catalyst surface in an ESCA apparatus (BASF, Ludwigshafen, Germany)

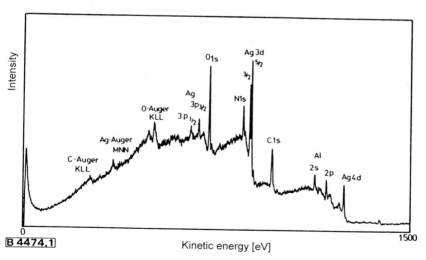

Fig. 4-45. ESCA spectrum of an Ag/Al_2O_3 supported catalyst [25]

Examples of applications:

- Distinction between Al metal and Al_2O_3
- Distinction between aliphatic carbon and acid carbon
- Changes in the oxidation state of tin on pretreatment of a Pd/Sn catalyst
- Effect of pretreatment on the structure and composition of Mo/Al_2O_3 desulfurization catalysts
- Distinction between Fe^0 and Fe^{III} in ammonia synthesis catalysts

Auger Electron Spectroscopy (AES)

In this method the surface of the sample is bombarded with high-energy (1–5 keV) electrons [32]. Similarly to ESCA, photoelectrons are generated. The remaining atom, which now has an electron missing from the K shell, has two possibilities for filling this hole. One possibility is X-ray fluorescence, in which an electron from a higher shell fills the hole, and the energy that is released is emitted as an X-ray quantum. The competing process is Auger emission, in which an electron fills the hole, and the energy released is transferred to a valence electron, which exits the probe as a so-called Auger electron. Similar relationship apply as in ESCA.

The kinetic energy of the Auger electron allows the element to be inferred, and the intensity is a measure of the concentration. Information is obtained for the upper 5 nm. The advantage of AES is that the electron beam can be very tightly focussed. Since the electron beam can be moved across the surface, it is possible to measure a concentration profile along a line or to generate figures. A disadvantage is the relatively low sensitivity and damage to the sample. AES is mainly suited to the determination of surface composition and changes therein.

Examples of applications:

- Alkyne hydrogenation on NiS catalysts: Ni_3S_2 on the nickel surface is active in selective hydrogenation, but NiS is not
- Determination of the Si/Al ratio in zeolite crystals and on their surfaces

Ion Scattering Spectroscopy (ISS)

In this method a surface is bombarded with noble gas ions, and the kinetic energy of the ions is measured after impact with the surface [21, 31]. In simple terms, this can be regarded as playing billiards with the uppermost atoms of the surface. Since the mass and energy of the noble gas ions prior to impact are known, mechanical energy and impulse equations can be used to calculate the mass of the impacted surface atom. In this way a sort of elemental analysis of the uppermost atomic layer is obtained. A disadvantage is that the lines in ISS spectra are relatively broad, and information about bonding is not obtained.

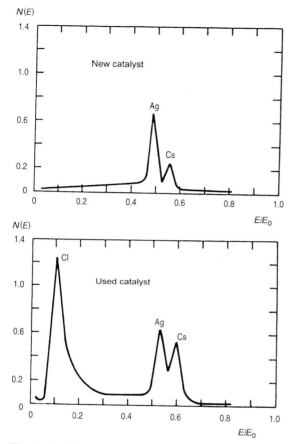

Fig. 4-46. ISS spectra of a new and used ethylene oxide catalyst [26]

Comparing the ISS spectra of an unused and a used silver catalyst for ethylene oxide synthesis shows how the alkali metal promoter on the catalyst behaves during the process (Fig. 4-46). In time, the alkali metal promoter spreads across the surface of the catalyst [25].

Secondary Ion Mass Spectrometry (SIMS)

In this method the surface is also bombarded with noble gas ions (primary ions; 1–10 keV) [9, 31]. In each impact, atoms, groups of atoms, or secondary ions are knocked out of the surface. The charged fragments—both positive and negative ions—are analyzed in a mass spectrometer. SIMS can be used to detect all the elements of the periodic table and their isotopes. By using low-energy ion beams, the depth of penetration can be kept so low that exclusively information about the uppermost atomic layers is obtained (static SIMS). More energetic ion beams can be used to remove the surface layer by layer to obtain a depth-profiled analysis over a range of a few nanometers to several micrometers (dynamic SIMS). This is the most sensitive method of depth-profiled trace analysis of solids. The detection limit is 5×10^{14} atoms/cm^3 or one particle in 10^8, i. e., near the ppb range.

Examples of applications:

- Changes in the Si/Al ratio in the interior of cracking catalysts relative to the particle surface
- Effects of metals in exchanged zeolites
- Proof of the segregation of Pt on the surface of Pt/Re catalysts
- Determination of the surface composition of Cu, Co, and Ni spinels MAl_2O_4
- The action of CO on Ni surfaces: detection of $NiCO^+$ and Ni_2CO^+, i. e., associative chemisorption

Of course, the above-mentioned methods all have advantages and disadvantages. Therefore, it is best to use a combination of methods, e. g., surface analysis, microscopy, and chemisorption measurements.

The main problem of surface analysis methods is that many of them are restricted to special measurement conditions (defined single-crystal surfaces, low temperatures, ultrahigh vacuum). It is questionable whether these results are extrapolable to the behavior of industrial catalysts under process conditions (pressure, high temperatures, impurities). Table 4-41 gives a comparison.

Table 4-41. Comparison of surface physics and industrial heterogeneous catalysis

Surface physics	Industrial heterogeneous catalysis
Ideal, well-defined surface (single crystals)	complex, poorly defined surface
Very pure surface	highly contaminated surface
Pressure ca. 10^{-8} mbar	pressure up to 300 bar
Equilibrium	kinetically controlled

Nevertheless, in the last few years many successes have been achieved that bridge the gap between pure research and applied catalysis. In any case, electron microscopy and surface analysis have made a decisive contribution to the development, optimization and monitoring of catalysts.

Exercises for Section 4.6

Exercise 4.45

The following physicochemical catalyst properties are to be determined:

A) Surface complexes
B) Number and type of active centers
C) Specific surface area and pore radius distribution
D) Element distribution on the pore surface
E) Bonding state of the elements
F) Crystal structure
G) Crystallite size

Which of the following methods are suitable for determining the above properties: ESCA, BET method, reflection IR spectroscopy, SIMS, X-ray structure analysis, scanning electron microscopy, temperature-programmed desorption, ESR.

Exercise 4.46

In a sorptometer, chemisorption measurements are carried out with various gases such as H_2, CO, NO and N_2O at room temperature or at higher temperatures. Which catalyst properties are measured?

☐ Specific pore volume and pore size
☐ Number of active surface atoms per gram of catalyst
☐ True catalyst density
☐ Selectivity of catalysts
☐ Degree of dispersion
☐ Pore size distribution

Exercise 4.47

A sample of γ-Al_2O_3 that was heated to 200 °C, cooled, and then pretreated with pyridine exhibits IR bands at 1540 and 1465 cm^{-1}. Explain this finding.

Exercise 4.48

On a supported nickel catalyst, two strong C–O stretching bands are observed in the IR for chemisorbed CO at 1915 and 2035 cm^{-1}. Interpret the position of the bands. How is the CO bound to the metal?

Exercise 4.49

Carbon dioxide adsorbed on a Rh(111) surface gives the same IR spectrum as adsorbed CO. What can be said about the manner in which the gas is adsorbed?

Exercise 4.50

Ethylene undergoes a reaction on a supported Pd/SiO$_2$ catalyst. The IR band of the adsorbed molecule is observed at 1510 cm^{-1}. Olefins usually have a band at ca. 1640 cm^{-1}. Give an explanation.

Exercise 4.51

What is the LEED method, and what can be measured with it?

5 Catalyst Shapes and Production of Heterogeneous Catalysts

5.1 Catalyst Production [1, T41]

Industrial catalysts are generally shaped bodies of various forms, e.g., rings, spheres, tablets, pellets (Fig. 5-1). Honeycomb catalysts, similar to those in automobile catalytic converters, are also used. The production of heterogeneous catalysts consists of numerous physical and chemical steps. The conditions in each step have a decisive influence on the catalyst properties. Catalysts must therefore be manufactured under precisely defined and carefully controlled conditions [14].

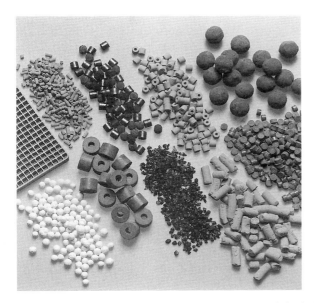

Fig. 5-1. Various shaped catalyst bodies (BASF, Ludwigshafen, Germany)

Since even trace impurities can affect catalyst performance, strict quality specifications apply for the starting materials. Successful catalyst production is still more of an art than a precise science, and much company know-how is required to obtain catalysts with the desired activity, selectivity, and lifetime.

Depending on their structure and method of production, catalysts can be divided into three main groups [8]:

- Bulk catalysts
- Impregnated catalysts
- Shell catalysts

Bulk catalysts are mainly produced when the active components are cheap. Since the preferred method of production is precipitation, they are also known as precipitated catalysts. Precipitation is mainly used for the production of oxidic catalysts and also for the manufacture of pure support materials. One or more components in the form of aqueous solutions are mixed and then coprecipitated as hydroxides or carbonates. An amorphous or crystalline precipitate or a gel is obtained, which is washed thoroughly until salt free. This is then followed by further steps: drying, shaping, calcination, and activation (Scheme 5-1)

The production conditions can influence catalyst properties such as crystallinity, particle size, porosity, and composition.

In the shaping step, the catalyst powder is plastified by kneading and pelletized by extrusion or pressed into tablets after addition of auxiliary materials (Fig. 5-2). The influence of the shaping process on the mechanical strength and durability of the catalyst should not be underestimated. When reactors are filled with catalyst, a dropping height of 6–8 m is usual, and bed heights of up to 10 m are possible.

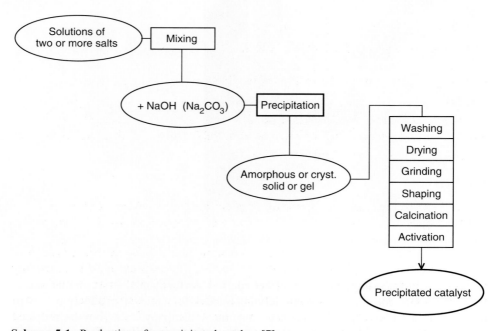

Scheme 5-1. Production of a precipitated catalyst [7]

Fig. 5-2. Production of noble metal catalysts at the company Degussa in Hanau-Wolfgang, Germany

Furthermore, industrial catalysts are subject to high temperatures and also often to changing temperatures.

Typical examples of precipitated catalysts are:

- Iron oxide catalysts for high-temperature CO conversion (Fe_2O_3 with addition of Cr_2O_3)
- Catalysts for the dehydrogenation of ethylbenzene to styrene (Fe_3O_4)

Highly homogeneous catalysts can be obtained by using mixed salts or mixed crystals as starting materials, since in this case the ions are already present in atomically distributed form. Readily decomposible anions such as formate, oxalate, or carbonate are advantageous here.

Examples:

- $Cu(OH)NH_4CrO_4$ as a precursor for copper chromite (Adkins catalyst)
- $Ni_6Al_2(OH)_{16}CO_3 \cdot 4H_2O$ decomposes to give a supported Ni/Al_2O_3 catalyst

One of the best known methods for producing catalysts is the impregnation of porous support materials with solutions of active components [9,10]. Especially catalysts with expensive active components such as noble metals are employed as supported catalysts. A widely used support is Al_2O_3. After impregnation the catalyst particles are dried, and the metal salts are decomposed to the corresponding oxides by heating. The process is shown schematically in Scheme 5-2.

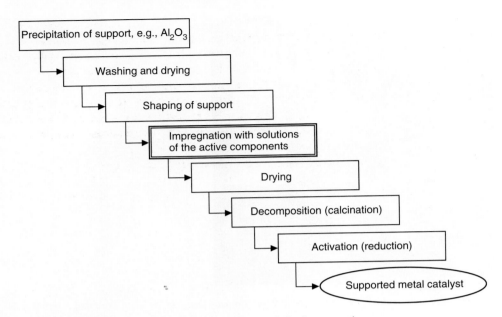

Scheme 5-2. Production of supported metal catalysts by impregnation

In the impregnation process, active components with thermally unstable anions (e.g., nitrates, acetates, carbonates, hydroxides) are used. The support is immersed in a solution of the active component under precisely defined conditions (concentration, mixing, temperature, time). Depending on the production conditions, selective adsorption of the active component occurs on the surface or in the interior of the support. The result is nonuniform distribution.

To achieve the best possible impregnation, the air in the pores of the support is removed by evacuation, or the support is treated with gases such as CO_2 or NH_3 prior to impregnation. After impregnation, the catalyst is dried and calcined.

Calcination is heat treatment in an oxidizing atmosphere at a temperature slightly higher than the intended operating temperature of the catalyst. In calcination numerous processes can occur that alter the catalyst, such as formation of new components by solid-state reactions, transformation of amorphous regions into crystalline regions, and modification of the pore structure and the mechanical properties.

In the case of supported metal catalysts, calcination leads to metal oxides as catalyst precursors, and these must subsequently be reduced to the metals. This reduction can be performed with hydrogen (diluted with nitrogen), CO, or milder reducing agents such as alcohol vapor. In some cases reduction can be carried out in the production reactor prior to process start-up. Here temperature control is a problem.

Impregnated catalysts have many advantages compared to precipitated catalysts. Their pore structure and specific surface area are largely determined by the support. Since support materials are available in all desired ranges of surface area, porosity, shape, size, and mechanical stability, impregnated catalysts can be tailor-made with respect to mass transport properties [9].

In individual cases it is possible to achieve almost molecular distribution of the active components in the pores. As a rule, however, the active substance is distributed in the form of crystallites with a diameter of 2–200 nm. This fine distribution on the support not only ensures a particularly favorable surface to volume ratio and hence makes good use of the active components, some of which are expensive, it also reduces the risk of sintering.

In general, with increasing loading, catalyst activity eventually reaches a limiting value. Therefore, for economic reasons the catalyst loading is 0.05–0.5% for noble metals, and 5–15% for other metals. Examples of industrial impregnated catalysts are:

– Ethylene oxide catalysts in which a solution of a silver salt is applied to Al_2O_3
– Catalysts in the primary reformer of ammonia synthesis, with 10–20% Ni on α-Al_2O_3
– Catalysts for the synthesis of vinyl chloride from acetylene and HCl: $HgCl_2$/ activated carbon; $HgCl_2$ is applied from aqueous solution

Catalysts in which the active component is a finely divided metal are often pyrophoric. The catalyst can be better handled after surface oxidation of the active component (passivation). Reactivation is then carried out in the start-up phase under process conditions.

Shell catalysts consist of an compact inert support, usually in sphere or ring form, and a thin active shell that encloses it [4]. Since the active shell has a thickness of only 0.1–0.3 mm, the diffusion paths for the reactants are short. There are many heterogeneously catalyzed reactions in which it would be advantageous to eliminate the role of pore diffusion. This is particularly important in selective oxidation reactions, in which further reactions of intermediate products can drastically lower the selectivity. An example is acrolein synthesis: two catalysts with the same active mass but different shell thicknesses differed greatly in selectivity at the high conversions desired in industry (Fig. 5-3). Therefore, if acrolein synthesis is to be operated economically, the shell thickness must be optimized.

The best known method for producing shell catalysts is the controlled short-term immersion of strongly adsorbing support materials. A well-known example is the platinum shell catalyst, which can easily be prepared with low loading and a high degree of dispersion. The support is immersed in solution of hexachloroplatinic acid (H_2PtCl_6), and an outer layer of adsorbed $PtCl_4^{2-}$ ions is formed. The adsorption of the hexachloroplatinic acid is so fast that diffusion of the solution into the pores is rate-determining. The treated catalyst particles are then dried without washing and calcined to generate the metal [T35]. Figure 5-4 shows how different impregnation techniques can be used to obtain supported catalysts with special distributions of the metal.

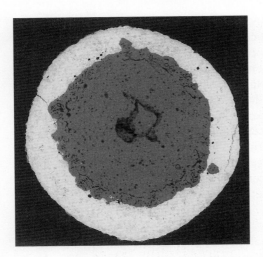

Fig. 5-3. Cross section of a shell catalyst (magnification 18 ×). Influence of the shell thickness on the selectivity of acrolein synthesis (BASF, Ludwigshafen, Germany):

Shell thickness [µm]	150	400
Selectivity [%] at 99% conversion	89	82

The advantages of shell catalysts are short transport or diffusion paths, a pore structure independent of the support, and better heat transport in the catalyst layer. Examples of industrial applications of shell catalysts are:

- Selective oxidation reactions, e.g., production of acrolein from propene and of phthalic anhydride from o-xylene
- Purification of automobile exhaust gases
- Selective oxidation of benzene to maleic anhydride: vanadium molybdenum oxide on fused corundum (catalytically inactive support without pores)
- Autothermal decomposition of liquid hydrocarbons on NiO/α-Al$_2$O$_3$ shell catalysts (high selectivity for lower alkenes [4]

In this chapter we have seen how the different steps of catalyst production can affect the functional properties of catalysts, such as activity and selectivity, and their morphology (Fig. 5-5).

Because of the numerous influencing parameters, prediction of the catalytic properties is not possible. They can only be determnined by measurement of the reaction kinetics. This makes it clear why catalyst production is based on special company know-how and that not all details are publicized.

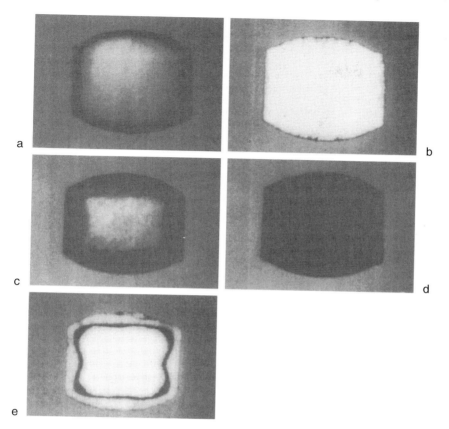

Fig. 5-4. Different metal distributions in pellets of diameter 6 mm consisting of a metal on a support (Degussa, Hanau-Wolfgang, Germany)
a) Shell catalyst with normal shell thickness
b) Shell catalyst with an extremely thin shell
c) Shell catalyst with a thick shell
d) Impregnated catalyst
e) Catalyst with ring distribution

Fig. 5-5. Modern catalyst production plant (BASF, Ludwigshafen, Germany)

5.2 Immobilization of Homogeneous Catalysts

As we have seen in Chapter 3, the industrial use of homogeneous catalysts often leads to problems with catalyst separation and recycling, recovery of the often valuable metal, and short catalyst lifetimes. Therefore, in the last twenty years or so, extensive studies have been carried out on the development of heterogenized homogeneous catalysts, which are intended to combine the advantages of homogeneous catalysts, in particular high selectivity and activity, with those of heterogeneous catalysts (ease of separation and metal recovery). Hence attempts are made to convert organometallic complex catalysts to a form that is insoluble in the reaction medium. This is generally achieved by anchoring a suitable molecule on an organic or inorganic polymer support.

In the following, we will discuss such methods for obtaining immobilized homogeneous catalysts, which are also known as fixed catalysts or hybrid catalysts, and the potential applications of this intersting class of catalysts [3]. To come to the most important point first: the ideal immobilized metal complex for industrial appplications has not yet been found, as is shown by weighing up the advantages and disadvantages of this type of catalyst.

Advantages:

1) Separation and recovery of the catalyst from the product stream is straightforward. This is the main advantage of heterogenization.
2) Mutifunctional catalysts can be obtained in which more than one active component is bound to a carrier.
3) Highly reactive, coordinatively unsaturated species that can not exist in solution can be stabilized by heterogenization.

Disadvantages:

1) The immobilized homogeneous catalysts are not sufficiently stable. The valuable metal is continuously leached and carried away with the product stream.
2) The problems of homogeneous catalysts, such as corrosion, catalyst recovery, and catalyst recycling, have so far not been satisfactorily solved.
3) Lower catalytic activity than homogeneous catalysts because of: poor accessibility of the active sites for the substrate, steric effects of the matrix, incompatibility of solvent and polymer, deactivation of active centers.
4) Inhomogeneity due to different linkages between support matrix and complex.

Particularly intensive investigations have been carried out on catalysts for reactions with CO or alkenes. These reactions, which are typical transition metal catalyzed conversions, provide the best possibility for assessing the properties of heterogenized catalysts. Examples are given in the following overview (Table 5-1). All the examples show that the reaction mechanisms with homogeneous and heterogeneous catalysis are in many respects similar. However, care must be taken in comparing soluble and matrix-bound catalysts, since the matrix can be regarded as a ligand. Thus at least one coordination site of the complex catalyst is no longer available for the catalytic cycle. It is difficult to find the corresponding ligands required for a comparison. For example, a monodentate phosphine ligand like PPh_3 is not directly compa-

Table 5-1. Comparison of homogeneous and heterogenized catalysts in industrial reactions

Reaction	Homogeneous catalyst	Heterogenized catalyst
Hydroformylation of olefins (oxo synthesis)	Co or Rh complex	Co or Rh complex on polymer or SiO_2 support matrix
Oxidation of olefins (Wacker process)	$[PdCl_4]^{2-}$	$PdCl_2$ on support matrix
Carbonylation of methanol to acetic acid	$[Rh(CO)_2I_2]^- + HI$	"$RhCl_3$" on activated carbon or $[RhCl(CO)PR_n]$ on modified polystyrene
Hydrogenation of olefins	$[Rh(PPh_3)_3Cl]$	$[Rh(PPh_3)_nCl]$ on polymer support

rable to a polystyrene matrix with phosphine groups. For meaningful comparisons, the less common multidentate ligands must be used in solution.

There are four basic ways of fixing transition metal complexes on a matrix:

1) Chemical bonding on inorganic or organic supports
2) Production of highly dispersed supported metal catalysts
3) Physisorption on the surface of oxidic supports (supported solid phase catalysts, SSPC)
4) Dissolution in a high-boiling liquid that is adsorbed on a porous support (supported liquid phase catalysts, SLPC)

The immobilization of organometallic complexes on inorganic or organic supports is the most widely used method. Basically the supports act as high molecular mass ligands and are obtained by controlled synthesis. The bonding can be ionic or coordinative. The main aim of the process is to bind the complexes on the solid surface in such a manner that its chemical structure is retained as far as posssible. A common method is the replacement of a ligand by a bond to the surface of the solid matrix. This means that a reactive group must be incorporated in the surface during production of the support.

Numerous polymer syntheses and orgamometallic syntheses are available for the construction of functionalized supports; Equation 5-1 gives just one example.

$$\text{Polymer chain (polystyrene)} \xrightarrow{PCl_3/AlCl_3} \text{[Ar-PCl}_2\text{]} \xrightarrow{2\ RLi} \text{[Ar-PR}_2\text{]}$$

(5-1)

Here triphenylphosphine, the most important ligand in organometallic catalysis, is coupled to the benzene rings of cross-linked polystyrene. An anchored catalyst is then formed by coordination of the phosphine group to the metal center of a rhodium complex (Eq. 5-2).

$$\text{Polymer-}C_6H_4\text{-PPh}_2 \xrightarrow{\text{RhL}_x} \text{Polymer-}C_6H_4\text{-P(Ph)}_2\text{·RhL}_x \qquad (5\text{-}2)$$

The degree of swelling of this copolymer in organic solvents is controlled by means of the amount of divinylbenzene. Hard copolymers of this type take up metal complexes only on the surface. The physical properties of the support can be varied by means of the polymerization method; the metal loading can also be controlled well.

There are many reactions available for applying the organometallic complexes to the surface. Two examples are shown in Equations 5-3 and 5-4.

$$|\text{-COOH} + \text{RuH}_2(\text{PPh}_3)_4 \longrightarrow |\text{-C(O)(O)RuH(PPh}_3)_3 \qquad (5\text{-}3)$$

$$|\text{-CH}_2\text{Cl} + \text{NaMn(CO)}_5 \xrightarrow{-\text{NaCl}} |\text{-CH}_2\text{-Mn(CO)}_5 \qquad (5\text{-}4)$$

Disadvantages of the organic polymer supports are low mechanical durability (e.g., in stirred tank reactors), poor heat-transfer properties, and limited thermal stability (up to max. 150 °C).

There are also several methods available for producing inorganic supports. Here we will discuss a few basic methods. The most important method is the reaction of inorganic supports having surface hydroxyl groups with metal alkyls (Eq. 5-5).

$$\text{Mg}|\text{(OH)}_2 \xrightarrow{\text{Ti(CH}_2\text{C}_6\text{H}_5)_4} \text{Mg}|\text{(O)}_2\text{Ti(CH}_2\text{-C}_6\text{H}_5)_2 \qquad (5\text{-}5)$$

Alkoxides and halides can also be attached to surfaces. Subsequent hydrolysis and dehydration lead to terminal metal oxo structures (Eq. 5-6).

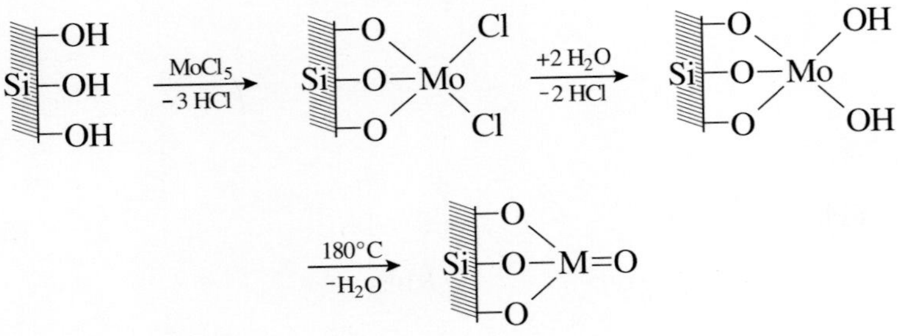

$$\text{Si}(-\text{OH})_3 \xrightarrow[-3\,\text{HCl}]{\text{MoCl}_5} \text{Si}(-\text{O})_2\text{Mo}(\text{Cl})_2 \xrightarrow[-2\,\text{HCl}]{+2\,\text{H}_2\text{O}} \text{Si}(-\text{O})_2\text{Mo}(\text{OH})_2 \xrightarrow[-\text{H}_2\text{O}]{180\,°\text{C}} \text{Si}(-\text{O})_2\text{M}=\text{O}$$

(5-6)

Such immobilized molybdenum oxide catalysts are active in selective oxidation reactions. For example, methanol can be oxidized with air to methyl formate at ca. 500 K with 90–95% selectivity [T22]. The catalyst obtained from γ-Al$_2$O$_3$ and tetrakis(η^3-allyl)dimolybdenum (Eq. 5-7) is considerably more active in ethylene hydrogenation and olefin metathesis than the catalysts prepared by conventional fixation of [Mo(CO)$_6$] followed by calcination.

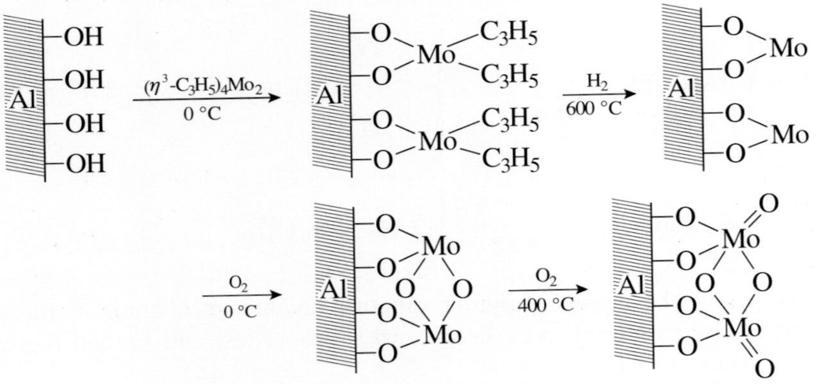

(5-7)

Organofunctional polysiloxanes are a versatile group of catalysts developed by the company Degussa [13]. These are solids with a silicate framework obtained by hydrolysis and polycondensation of organosilicon compounds (Eq. 5-8).

$$\begin{array}{c}
\text{(CH}_2)_3 \quad \text{(CH}_2)_3 \\
| \quad\quad | \\
\text{Si(OR)}_3 \quad \text{Si(OR)}_3
\end{array}
\xrightleftharpoons[+\text{ROH}]{+\text{H}_2\text{O}}
\begin{array}{c}
\text{X} \\
\text{(CH}_2)_3 \quad \text{(CH}_2)_3 \\
| \quad\quad | \\
\text{Si(OH)}_3 \quad \text{Si(OH)}_3
\end{array}$$

$$\xrightleftharpoons[+\text{H}_2\text{O}]{-\text{H}_2\text{O}} \quad \text{surface-bound siloxane} \tag{5-8}$$

X = functional group: sulfane, phosphine, amine

This class of substances is characterized by broad chemical modifiability, a high capacity for functional groups, high temperature and ageing resistance, and insolubility in water and organic solvents. The heterogenized organopolysiloxane catalysts are marketed as abrasion-resistant spheres of various particle sizes. In particular the phosphine complexes of Ru, Pd, Ir, and Pt are interesting catalysts for hydrogenation, hydroformylation, carbonylation, and hydrosilylation.

Highly Dispersed Supported Metal Catalysts [T22]

This method is used to obtain a very fine distribution of metal on a support by decomposition of organometallic compounds (so-called grafted catalysts). For example, by treating TiO_2 with η^3-allyl complexes of rhodium followed by decomposition, highly active hydrogenation and hydrogenolysis catalysts are obtained (Eq. 5-9). Similar catalysts based on polysiloxanes are produced by Degussa; Pd, Rh, and Pt systems are available.

$$Rh(C_3H_5)_3 + \underset{\text{Ti}}{\text{OH} \quad \text{OH}} \xrightarrow{273 \text{ K}} \underset{\text{Ti}}{\overset{C_3H_5}{\underset{O\quad O}{Rh}}} \xrightarrow[H_2]{293 \text{ K}} \underset{\text{Ti}}{\overset{H}{\underset{O\quad O}{Rh}}}$$

$$\xrightarrow[H_2]{473-773 \text{ K}} \underset{\text{Ti}}{(Rh)_n} \tag{5-9}$$

$(Rh)_n$ = small aggregates of 25 or more Rh atoms with particle diameters of ca. 1.4 nm

SSP Catalysts [6,11]

In this group of catalysts, organometallic complexes are anchored on the inner surface of porous supports, mainly by physisorption. These catalysts can be used as catalyst beds through which the reaction medium flows. For example, the complex [Rh(η^3-C$_3$H$_5$)(CO)(PPh$_3$)$_2$] is adsorbed on γ-Al$_2$O$_3$ and used as a hydrogenation catalyst. The fixed complexes often exhibit considerably lower activity and selectivity than in the homogeneous phase, and this limits their range of applications. The SLP catalysts are a better alternative.

SLP Catalysts [11,15]

In this process a solution of the complex in a high-boiling solvent spreads out on the inner surface of a porous support, which generally consists of an inorganic material such as silica gel or chromosorb. The reaction takes place in the liquid film, which the starting materials reach by diffusion. The products are also transported away by diffusion out of the film, which is retained on the support.

The use of SLP catalysts is generally restricted to the synthesis of low-boiling compounds. Oxo synthesis with SLP catalysts has been the subject of much interest. An example is the hydroformylation of propene with [RhH(CO)(PPh$_3$)$_3$] in liquid triphenylphosphine on γ-Al$_2$O$_3$. The starting material and the C$_4$ aldehyde are present in the gas phase. In a pilot plant at DSM, low selectivity was found and diffusion problems were encountered. Further examples are the oxidation of ethylene to acetaldehyde with aqueous solutions of PdCl$_2$ and CuCl$_2$ on kieselguhr, and the oxychlorination of alkenes with a CuCl$_2$/CuCl/KCl/rare earth halide melt on silica gel [T22].

From these examples, most of which are based on laboratory investigations, it becomes clear that heterogenization is not a general method for solving problems in catalysis. It is, however, an interesting addition to the spectrum of catalytic methods.

Finally we shall discuss some examples in which heterogenized catalysts have been successfully used in industrial processes.

Chromium complexes on the basis of chromocene or chromium salts on SiO$_2$ are used for the polymerization of α-olefins and for the production of linear polyethylene in the Phillips process. The structure of the active surface species is unknown.

Heterogenized titanium complexes are used for the polymerization of propylene and give high yields of isotactic polypropylene [T31].

Another example for the use of a multifunctional solid catalyst is the Aldox process for the production of 2-ethylhexanol (Eq. 5-10).

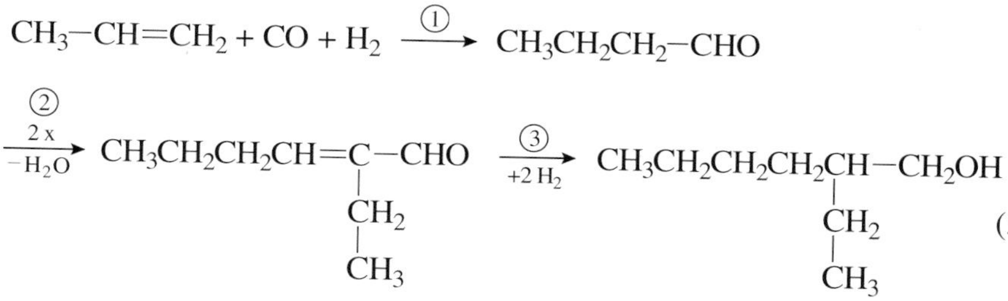

$$CH_3-CH=CH_2 + CO + H_2 \xrightarrow{\text{①}} CH_3CH_2CH_2-CHO$$

$$\xrightarrow[-H_2O]{\text{②} \; 2\times} CH_3CH_2CH_2CH=\underset{\underset{CH_3}{|}}{\underset{CH_2}{|}}C-CHO \xrightarrow[+2H_2]{\text{③}} CH_3CH_2CH_2CH_2\underset{\underset{CH_3}{|}}{\underset{CH_2}{|}}CH-CH_2OH \quad (5\text{-}10)$$

In industry the hydroformylation (reaction 1) is catalyzed by Rh or Co complexes in solution. The aldol condensation (reaction 2) is acid or base catalyzed, and the hydrogenation of the unsaturated aldehyde (reaction 3) is catalyzed by metals such as nickel. On this basis a catalyst with a metal function (Rh) and a base function (amine) was developed (Fig. 5-6), and is active for the formation of 2-ethylhexanol. The rhodium center catalyzes the hydroformylation and the partial hydrogenation of the aldol product, in which the aldehyde group is retained, while the amino group catalzes the aldol condensation [16].

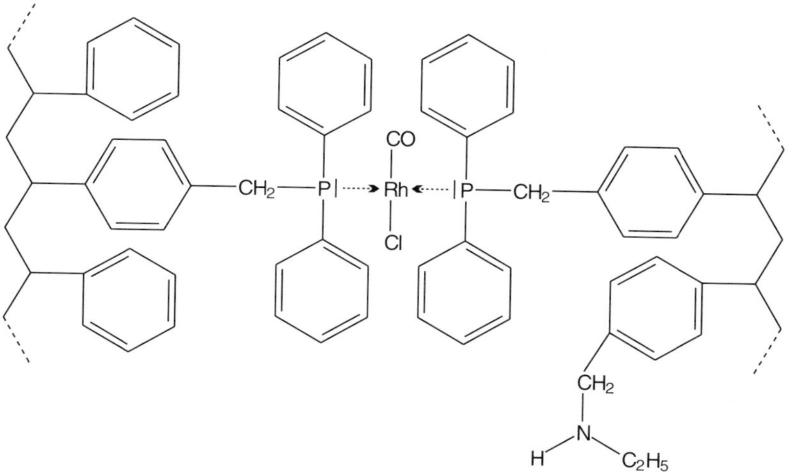

Fig. 5-6. Multifunctional, polymer-fixed solid catalyst for the Aldox process [16]

These examples show that the area of heterogenization of catalysts represents an enormous potential for research. Some of these catalysts show high activities under mild conditions with interesting and sometimes unexpected selectivities. The processes for the production of these catalysts, the investigation of their precise structures, and the elucidation of their reaction mechanisms are still at an early stage.

It would seem that the use of heterogenized catalysts is best suited to small molecules (oxidation, hydrogenation), and that inorganic supports are more promising than organic supports. The field of heterogenization has led to a closer approach between heterogeneous and homogeneous catalysis.

Exercises for Chapter 5

Exercise 5.1

Which are the main physical properties of a catalyst that are influenced by the production conditions?

Exercise 5.2

What are the advantages of impregnated catalysts compared with precipitated catalysts.?

Exercise 5.3

Name porous supports with which impregnated catalysts can be manufactured.

Exercise 5.4

Which two types of support are preferentially used for oxidation catalysts?

Exercise 5.5

For which reactions are supported catalysts impregnated near the surface particularly suitable?

Exercise 5.6

a) Why do monolith and honeycomb catalysts have to be coated before they are loaded with catalyst?
b) What is this initial coating called?

Exercise 5.7

a) What are the advantages of shell catalysts compared to bulk catalysts?
b) What is the preferred support material for shell catalysts?

Exercise 5.8

Why have numerous dinuclear and multinuclear metal complexes (clusters) been tested in the synthesis of gycol from CO/H_2?

Exercise 5.9

What are the advantages of heterogenized metal catalysts compared to conventional heterogeneous catalysts?

Exercise 5.10

A phosphine-modified plastic matrix is treated with iron pentacarbonyl. What reaction can be expected?

$$\text{|---PR}_2 + \text{Fe(CO)}_5 \longrightarrow \text{ ?}$$

Exercise 5.11

What are the disadvantages of organic polymer supports for the production of immobilized homogeneous catalysts?

Exercise 5.12

How are SLP catalysts produced?

6 Shape-Selective Catalysis: Zeolites

6.1 Composition and Structure of Zeolites [5, 6]

Zeolites are water-containing crystalline aluminosilicates of natural or synthetic origin with highly ordered structures. They consist of SiO_4 and AlO_4^- tetrahedra, which are inerlinked through common oxygen atoms to give a three-dimensional network through which long channels run.

In the interior of these channels, which are characteristic of zeolites, are water molecules and mobile alkali metal ions, which can be exchanged with other cations. These compensate for the excess negative charge in the anionic framework resulting from the aluminum content. The interior of the pore system, with its atomic-scale dimensions, is the catalytically active surface of the zeolites. The inner pore structure depends on the composition, the zeolite type, and the cations. The general formula of zeolites is

$$M^I M^{II}_{0,5}[(AlO_2)_x \cdot (SiO_2)_y \cdot (H_2O)_z] \qquad (6\text{-}1)$$

where M^I and M^{II} are preferentially alkali and alkaline earth metals. The indices x and y denote the oxide variables, and z is the number of molecules of water of hydration. The composition is characterized by the Si/Al atomic ratio or by the molar ratio M

$$M = \frac{SiO_2}{Al_2O_3} \qquad (6\text{-}2)$$

and the pore size of zeolites by the type (A, X, Y).

Zeolites are mainly distinguished according to the geometry of the cavities and channels formed by the rigid framework of SiO_4 and AlO_4^- tetrahedra. The tetrahedra are the smallest structural units into which zeolites can be divided. Linking these primary building units together leads to 16 possible secondary building blocks (polygons), the interconnection of which produces hollow three-dimensional structures.

The entrances to the cavities of the zeolites are formed by 6-, 8-, 10-, and 12-ring apertures (small-, medium-, and widepore zeolites). A series of zeolites is composed of polyhedra as tertiary building units. These include truncated octahedra (sodalite or β-cage, Fig. 6-1), composed of 4- and 6-rings, which can be con-

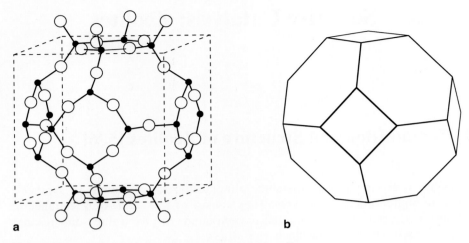

Fig. 6-1. Truncated octahedra as structural units of zeolites
  a) Sodalite cage (β-cage)
  b) Sodalite cage (schematic)

nected in various manners to give the fundamental zeolite structures. The sodalite cage, which consists of 24 tetrahedra, is generally depicted schematically as a polygon, generated by connecting the centers of neighboring tetrahedra with a line. Each vertex of this polyhedron then represents a silicon or aluminum atom, and the midpoint of each edge, an oxygen atom.

The structure of zeolite A, a narrow-pore zeolite, is formed by linking the square faces of the polyhedra via intermediate cubic units (Fig. 6-2). The cavity formed by linking eight truncated octahedra is known as the α-cage (Fig. 6-2). It

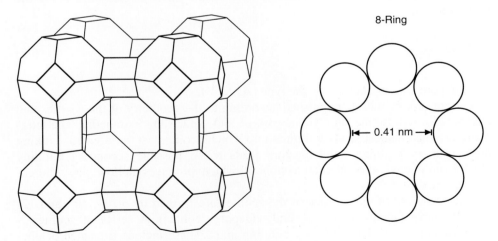

Fig. 6-2. Framework structure of zeolite A with α-cage

6.1 Composition and Structure of Zeolites   227

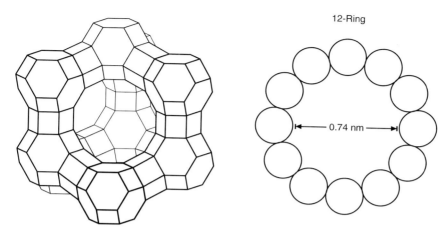

Fig. 6-3. Y zeolite (faujasite)

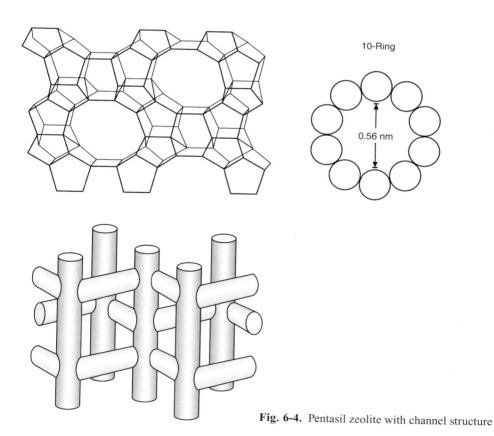

Fig. 6-4. Pentasil zeolite with channel structure

is larger than the β-cage. The wide-pored zeolite Y (faujasite) is formed when the truncated octahedra are linked together by hexagonal prisms. The resulting cavity is larger than the α-cage of zeolite A (Fig. 6-3).

Representatives of the medium-pore zeolites are the so-called pentasils, which belong to the silicon-rich zeolites. In contrast to the structures described above, their polyhedra are composed of 5-rings as secondary building units. These so-called 5–1 units are structurally analogous to methycyclopentane. Linking of the resulting chains gives a two-dimensional pore system in which linear or zig-zag channels are intersected by perpendicular linear channels (Fig. 6-4). An advantage of these zeolites is the uniform channel structure, in contrast to the zeolites A and Y, in which the pore windows provide access to larger cavities. A well-known representive of this class of zeolites is ZSM-5 (from zeolite Socony Mobil no. 5).

Table 6-1 lists the most important synthetic zeolites.

Table 6-1. Characteristics of important zeolites

Type	Pore diameter [nm]	Pore aperture
Zeolite Y (faujasite)	0.74	12-ring
Pentasil zeolite	0.55 × 0.56	10-ring (ellipsoid)
Zeolite A	0.41	8-ring
Sodalite	0.26	4-ring

6.2 Production of Zeolites [T32]

Zeolite syntheses start from alkaline aqueous mixtures of aluminum and silicon compounds. The reactions are sometimes carried out at atmospheric pressure but more often in a high-pressure autoclave. The controlled crystallization of a particular zeolite requires careful control of the concentration and stoichiometry of the reaction partners, the temperature, and the shearing energy of the stirrer. After mixing of the liquid phase and formation of a gel, a transition of the gel phase in to the liquid aqueous phase occurs, whereby crystalline zeolites are formed from the amorphous particles.

The silicon-rich pentasils are mainly synthesized in the presence of organic cations. Their open structures seem to be formed around hydrated cations or other cations such as NR_4^+. In particular, templates such as tetrapropylammonium hydroxide are used, and are of decisive importance for the crystallization of the zeolite structures. The C, H, and N of the tertiary ammonium cation is removed in the subsequent calcination of the microcrystalline product.

6.3 Catalytic Properties of the Zeolites [2–4]

In 1962 the zeolites were introduced by Mobil Oil Corporation as new cracking catalysts in refinery technology. They were characterized by higher activity and selectivity in cracking and hydrocracking. At the end of the 1960s, the concept of shape-selective catalysis with zeolites was introduced to petrochemistry (Selectoforming process), and the zeolites became of increasing importance in catalysis research and applied catalysis [6].

Since then chemists worldwide have prepared numerous "tailor-made" modified zeolites, and the synthetic potential for the production of organic intermediates and high-value fine chemicals is enormous. How can the success of this new class of catalysts in industry and academe be explained? It is due to the outstanding catalytic properties of the zeolites. No other class of catalysts offers so much potential for variation and so many advantages in application. Their advantages over conventional catalysts can be summarized as follows:

- Crystalline and therefore precisely defined arrangement of SiO_4 and AlO_4^- tetrahedra. This results in good reproducibility in production.
- Shape selectivity: only molecules that are smaller than the pore diameter of the zeolite undergo reaction.
- Controlled incorporation of acid centers in the intracrystalline surface is possible during synthesis and/or by subsequent ion exchange.
- Above 300 °C pentasils and zeolite Y have acidities comparable to those of mineral acids.
- Catalytically active metal ions can be uniformly applied to the catalyst by ion exchange or impregnation. Subsequent reduction to the metal is also possible.
- Zeolite catalysts are thermally stable up to 600 °C and can be regenerated by combustion of carbon deposits.
- They are well suited for carrying out reactions above 150 °C, which is of particular interest for reactions whose thermodynamic equilibrium lies on the product side at high temperatures.

Let us first take a closer look at the most important properties of the zeolites:

- Shape selectivity
- Acidity

6.3.1 Shape Selectivity [1]

We have seen that the inner pore system of the zeolites represents a well-defined crystalline surface. The structure of the crystalline surface is predetermined by the composition and type of the zeolite and is clearly defined. Such conditions are otherwise found only with single-crystal surfaces.

The accessibility of the pores for molecules is subject to definite geometric or steric restrictions. The shape selectivity of zeolites is based on the interaction of reactants with the well-defined pore system. A distinction is made between three variants, which can, however, overlap:

- Reactant selectivity
- Product selectivity
- Restricted transition state selectivity

Figure 6-5 shows these schematically with examples of reactions.

Reactant Selectivity

Reactant selectivity means that only starting materials of a certain size and shape can penetrate into the interior of the zeolite pores and undergo reaction at the catalytically active sites. Starting material molecules that are larger than the pore apertures can not react (Fig. 6-5 a). Hence the term "molecular sieve" is justified.

Table 6-2 compares the pore apertures of some zeolites with the kinetic molecular diameters of some starting materials. On the basis of these data, a prelimin-

Table 6-2. Molecular diameters and pore sizes of zeolites [7, T32]

Molecule	Kinetic diameter [nm]	Zeolite, pore size	[nm]
He	0.25	KA	0.3
NH_3	0.26	LiA	0.40
H_2O	0.28	NaA	0.41
N_2, SO_2	0.36	CaA	0.50
Propane	0.43	Erionite	0.38×0.52
n-Hexane	0.49	ZSM-5	$0.54 \times 0.56 / 0.51 \times 0.55$
Isobutane	0.50	ZSM-12	0.57×0.69
Benzene	0.53	CaX	0.69
p-Xylene	0.57	Mordenite	0.67–0.70
CCl_4	0.59	NaX	0.74
Cyclohexane	0.62	AlPO-5	0.80
o-, m-Xylene	0.63	VPI-5	1.20
Mesitylene	0.77		
$(C_4H_9)_3N$	0.81		

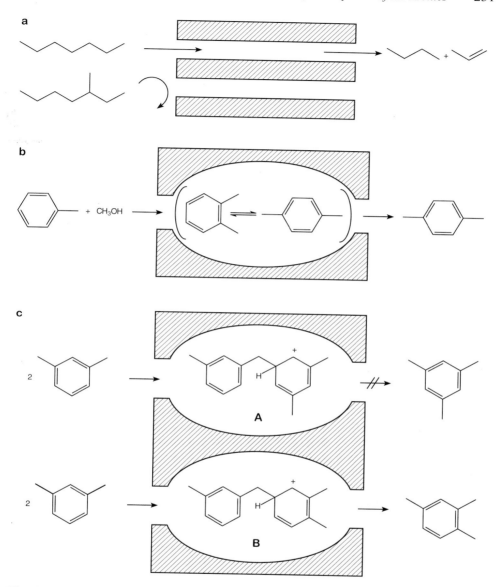

Fig. 6-5. Shape selectivity of zeolites with examples of reactions
    a) Reactant selectivity: cleavage of hydrocarbons
    b) Product selectivity: methylation of toluene
    c) Restricted transition state selectivity: disproportionation of *m*-xylene

ary choice of a suitable zeolite for a particular starting material can be made. However, it should not be forgotten that molecules are not rigid objects and that the kinetic diameter gives only a rough estimate of the molecular size.

The catalytic characterization of zeolites is generally carried out with the aid of test reactions [8]. For example, the constraint index CI (Tab. 6-3) compares the relative rate of cracking of a 1:1 mixture of n-hexane (molecular diameter 0.49 nm) and 3-methylpentane (molecular diameter 0.56 nm).

Table 6-3. Constraint index (CI) for some typical catalysts at 316 °C [T28]

Zeolite	CI
Aluminosilicate, amorphous (at 510 °C)	0.6
HY	0.4
H-Mordenite	0.4
ZSM-4	0.5
ZSM-12	2.3
Offretite	3.7
ZSM-5	8.3
ZSM-11	8.7
Erionite	40

The CI is strongly dependent on the pore size of the zeolite. Small values between zero and two mean little or no shape selectivity (large-pore zeolites), values between two and 12 a medium selectivity (medium-pore zeolites), and values higher than 12 a high shape selectivity (small-pore zeolites).

Thus erionite, with the smallest pore opening of 0.38–0.52 nm, has the highest shape selectivity. It was found that with certain zeolites, the linear alkane n-hexane is cracked 40–100 times faster than the branched isomer 3-methylpentane. This is exploited industrially in the Selectoforming process, in which erionite is added to the reforming catalyst.

Especially ZSM-5 is used for shape-selective reactions. Numerous alkanes with various chain lengths and degrees of branching have been investigated.

The next example shows results for the cracking of heptane isomers over H-ZSM-5 (Table 6-4).

In this case, the critical diameters of the starting material molecules are equal to or slightly larger than the pore openings of the zeolite, but as a result of molecular vibrations under the reaction conditions, are able to enter the zeolite pores, where they react. Here the reactions are largely diffusion controlled.

In particular, the ability of ZSM-5 to cleave unbranched and monomethyl-branched alkanes with retention of more highly branched and cyclic isomers is exploited industrially in the dewaxing process to lower the solidification point of lubricants and in reforming processes to obtain high-octane gasolines (M Forming process) [3].

Table 6-4. Relative rate of cleavage of heptanes on H-ZSM-5 at 325 °C [T35]

C_7 Alkane	r_{rel}
 n-heptane	1.00
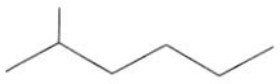 2-methylhexane	0.52
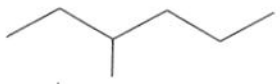 3-methylhexane	0,38
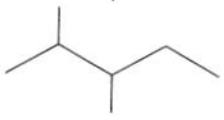 2,2-dimethylpentane	0.09

Another example of reactant selectivity is the dehydration of butanols. On CaA zeolites, the straight-chain alcohol, which fits in the zeolite pores, is much more rapidly dehydrated than isobutanol, which has a larger molecular diameter [T24]. In spite of the considerable molecular sieve effect, 100% selectivity is often not attained because the starting materials can also react to a small extent on the outer surface of the zeolite crystals.

Product Selectivity

Product selectivity arises when, coresponding to the cavity size of a zeolite, only products of a certain size and shape that can exit from the pore system are formed. Well-known examples of product selectivity are the methylation of toluene (Fig. 6-5b) and the disproportion of toluene on ZSM-5.

In both reactions all three isomers o-, m-, and p-xylene are formed. The desired product p-xylene can be obtained with selectivities of over 90%, although the thermodynamic equilibrium corresponds to a p-xylene fraction of only 24%. This is explained by the fact that for the slimmer molecule p-xylene has a rate of diffusion that is faster by a factor of 10^4 than those of the other two isomers. These isomerize relatively rapidly in the zeolite cavity, and the p-xylene diffuses out of the cavity. The selectivity can be further influenced by, for example:

– Increasing the size of the zeolite crystals
– Incorporation of cations or other organic materials in the pore structure
– Closing some of the pore apertures

An industrial application is the Mobil Oil selective toluene disproportionation process (STDP) [T32].

Another example of product selectivity is the alkylation of toluene with ethylene to give ethyltoluene (Table 6-5). The comparison with the conventional Friedel–Crafts catalyst shows the clear advantages of the highly selective zeolite catalyst.

Table 6-5. Product distribution in the ethylation of toluene [T32]

Ethyltoluene	Selectivity (%) Catalyst	
	AlCl$_3$/HCl	ZSM-5
p-	34.0	96.7
m-	55.1	3.3
o-	10.9	0

This form of shape selectivity can also have disadvantages. Large molecules that are unable to leave the pores can be converted to undesired side products or undergo coking, deactivating the catalyst.

Restricted Transition State Selectivity

This third form of shape selectivity depends on the fact that chemical reactions often proceed via intermediates. Owing to the pore system, only those intermediates that have a geometrical fit to the zeolite cavities can be formed during catalysis. This selectivity occurs preferentially when both monomolecular and bimolecular rearrangements are possible. In practice, it is often difficult to distinguish restricted transition state selectivity from product selectivity.

An example is the disproportionation of m-xylene to toluene and trimethylbenzenes in the wide-pored zeolite Y (Fig. 6-5c). In the large zeolite cavity, bulky diphenylmethane carbenium ion transition states can be formed as precursors for methyl group rearrangement, whereby the less bulky carbenium ion **B** is favored. Thus the reaction product consists mainly of the unsymmetrical 1,2,4-trimethylbenzene rather than mesitylene (case **A**). In contrast, in ZSM-5, with its medium sized pores, monomolecular xylene isomerization dominates, and the above-mentioned disproportionation is not observed as a side reaction.

Restricted transition state selectivity is also of importance in the alkylation of benzene with ethylene to give ethylbenzene. High selectivities for ethylbenzene are achieved on H-ZSM-5 owing to suppression of side reactions. These high selectivities were also explained by the fact that the possible bimolecular disproportionation of ethylbenzene is suppressed.

H-ZSM-5 is also used as catalyst in the large-scale MTG (methanol to gasoline) process. The products are hydrocarbons, aromatics in the benzene range, and water. The reaction is based on the dehydration of methanol to dimethyl ether, followed by numerous reactions that proceed via carbenium ion intermediates. The largest molecules observed, e.g., durene (1,2,4,5-tetramethylbenzene), correspond to the high-boiling components of gasoline. The favorable product distribution in this process can be attributed to restricted transition state selectivity.

Restricted transition state selectivity also influences the cracking of alkenes. In the cracking of hexenes with H-ZSM-5, the following order of reactivity is observed:

1-Hexene ⩾ 3-methyl-2-pentene > 3,3-dimethyl-1-butene

The sequence is exactly opposite to that of conventional acid catalysis: The reactants that are best able to form carbenium ions in solution are the least reactive with zeolite catalysis. The restricted transition state selectivity suppresses cracking of the more highly branched hydrocarbons in the cavities [T25].

6.3.2 Acidity of Zeolites [T24, T32]

In the last chapter we have already learnt of the importance of the hydrogen form of the zeolites (H-zeolites). Zeolites in the H form are solid acids whose acid strength can be varied over a wide range by modification of the zeolites (ion exchange, partial dealumination, and isomorphic substitution of the framework Al and Si atoms). Direct replacement of the alkali metal ions by protons by treatment with mineral acids is only possible in exceptional cases (e.g., mordenite and the high-silicon zeolite ZSM-5). The best method is exchange of the alkali metal ions by NH_4^+ ions, followed by heating the resulting ammonium salts to 500–600 °C (deammonization; Eq. 6-3).

$$\text{structure with } NH_4^+ \xrightleftharpoons[+NH_3]{-NH_3} \text{ structure with } H^+ \rightleftharpoons \text{ structure with HO–Si} \tag{6-3}$$

Infrared investigations have shown that the protons are mainly bound as silanol groups but have a strogly acidic character due to the strongly polarizing influence of the coordinatively unsaturated aluminum center. Brønsted acid centers are generally the catalytically active sites of H-zeolites.

Weak to moderately strong acid sites can be generated in zeolites by ion exchange with multivalent cations. Owing to the polarizing effect of the metal cations, water is dissociatively adsorbed, and the equilibrium of Equation 6-4 is established.

$$[M(H_2O)]^{n+} \rightleftharpoons [M(OH)]^{(n-1)+} + H^+ \tag{6-4}$$

The following order of Brønsted acidity is given for cation-exchanged zeolites:

H form ⩾ La form > Mg form > Ca form > Sr form > Ba form

The influence of the exchanged ions is considerable, as shown by the example of cumene dealkylation on faujasite (Table 6-6). Reasons for the large differences in reactivity are the different charges on the ions, and the decreasing ionic radii from Na^+ to H^+ and the associated polarizing power of the ions.

Table 6-6. Effect of the metal ion in faujasite on the dealkylation of cumene [T35]

Cation	Relative activity
Na^+	1.0
Ba^{2+}	2.5
Sr^{2+}	20
Ca^{2+}	50
Mg^{2+}	1.0×10^2
Ni^{2+}	1.1×10^3
La^{3+}	9.0×10^3
H^+	8.5×10^3
SiO_2/Al_2O_3	1.0

The incorporation of transition metal ions into zeolites leads to interesting bifunctional catalysts in which metal and acid centers can act simultaneously.

Another major influence on the acidity of zeolites is the Si/Al ratio. The zeolites can be classified according to increasing Si/Al ratio and the associated acid/base properties (Table 6-7).

Table 6-7. Classification of acidic zeolites according to increasing Si/Al ratio [T24]

Si/Al ratio	Zeolite	Acid/base properties
Low (1–1.5)	A, X	relatively low stability of lattice; low stability in acids; high stability in bases; high concentration of acid groups of medium strength
Medium (2–5)	erionite chabazite chinoptilolite mordenite Y	
High (ca. 10 to ∞)	ZSM-5; dealuminated erionite, mordenite, Y	relatively high stability of the lattice; high stability in acids; low stability in bases; low concentration of acid groups of high strength

Since the ion-exchange capacity corresponds to the Al^{3+} content of the zeolites, those with lower Si/Al ratios have higher concentrations of active centers.

Zeolites with high concentrations of protons are hydrophilic and have high affinities for small molecules that can enter the pores. Zeolites with low H^+ concentrations, such as silicalite, are hydrophobic and can take up organic components (e.g., ethanol) from aqueous solution. The boundary lies at a Si/Al ratio of around 10.

The stability of the crystal lattice also increases with increasing Si/Al ratio. The decomposition temperatures of zeolites are in the range 700–1300 °C. Zeolites of low aluminum content are produced by dealumination with a reagent such as $SiCl_4$, which removes aluminum from the framework with formation of $AlCl_3$. Zeolite Y, which is produced by this method or by hydrothermal treatment with steam at 600–900 °C, is regarded as ultrastable and is employed in cracking catalysts.

The highest proton-donor strengths are exhibited by zeolites with the lowest concentrations of AlO_4^- tetrahedra such as H-ZSM-5 and the ultrastable zeolite HY. These are superacids, which at high temperatures (ca. 500 °C) can even protonate alkanes. It was found that the acid strength depends on the number of Al atoms that are adjacent to a silanol group. Since the Al distribution is nonuniform, a wide range of acid strengths results.

The nonuniform distribution of the proton-active centers in zeolites can be measured by temperature-controlled desorption of adsorbed organic bases. The bases that are adsorbed on the centers of highest activity require the highest temperature for desorption. The IR spectra of adsorbed bases such as ammonia and pyridine give information about the nature of the adsorption centers. For example, the pyridinium ion is indicative of proton-donor sites. NMR and ESR spectroscopy are also useful for elucidating the nature of acid centers.

When an H-zeolite is heated to high temperature, water is driven off and coordinatively unsaturated Al^{3+} ions are formed. These are Lewis acids (Eq. 6-5).

$$2 \begin{array}{c} H \\ | \\ O \quad O \\ Si \diagdown Al \diagdown Si \end{array} \rightleftharpoons \begin{array}{c} O \\ Si \diagdown \boxed{Al} \diagdown Si^+ \end{array} + \begin{array}{c} O \quad O \\ Si \diagdown Al^- \diagdown Si \end{array} + H_2O \quad (6\text{-}5)$$

Brønsted   Lewis acid center
acid center

Bases like pyridine are more strongly bound to such Lewis acid centers than to Brønsted acid centers, as can be shown by IR spectroscopy and temperature-controlled desorption. Figure 6-6 shows the transformation of Brønsted into Lewis acid centers on calcination of an HY zeolite, monitored by IR spectroscopic measurements on the adsorption of pyridine.

As catalysts, zeolites combine the advantages of high density of catalytically active centers with high thermal stability. Practically all reactions that are catalyzed by acids in solution or by acidic ion exchangers are also catalyzed by acid

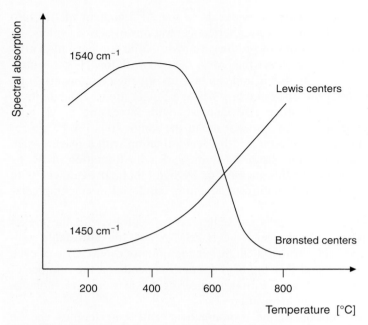

Fig. 6-6. Calcination of an HY zeolite: equilibrium between Brønsted and Lewis acid centers [9]

zeolites. Hundreds of examples are known. However, there are differences to conventional acid–base reactions, as we shall see below.

A simple example is the cracking of alkanes by zeolites with a low density of acid groups. H-ZSM-5 can be regarded as an "ideal solution" of acidic groups, since they are too far apart to influence one another. At low Al contents (up to 4% Al_2O_3) there is a direct relationship between the catalytic activity in the cracking of n-hexane and the concentration of Al^{3+} ions in the zeolite, which can be measured by Al NMR spectroscopy. Since each acidic group has a neighboring AlO_4^- tetrahedron, the activity is proportional to the aluminum concentration [T24].

In contrast, in the cracking of n-octane on H-mordenite, maximum catalytic activity occurs at a SiO_2/Al_2O_3 ratio of about 20. How can this finding be explained? On dealumination, the number of acidic centers decreases, but the acidity of the remaining centers increases up to a degree of dealumination of ca. 50%. The opposite effects of the concentration of acid centers and their acid strength are superimposed, so that maximum reactivity is reached at a certain SiO_2/Al_2O_3 ratio.

The next example is the ethylation of the aromatic compounds benzene and phenol. With normal acid catalysis in solution, ethylene reacts with phenol more rapidly than with benzene, since the more electron-rich ring in phenol more readily undergoes electrophilic attack by the ethyl cation. In zeolites, however, the situation is reversed, and benzene reacts faster than phenol. This has been explained in terms of competitive adsorption. First, the ethylene must be protonated (Eq. 6-6).

$$\underset{\underset{\text{Zeolite}}{|}}{H^+} + H_2C{=}CH_2 \longrightarrow (CH_3CH_2^+)_{ads} \qquad (6\text{-}6)$$

The carbenium ion is "solvated" by the polar, anionic environment of the zeolite pore. The highly reactive carbenium ion can alkylate an aromatic molecule from the surrounding medium. However, if phenol is present in the zeolite pore, then a competing reaction occurs with the less polar olefin at the acid sites. Adsorption of phenol (Eq. 6-7) is favored.

$$ArOH + \underset{\underset{\text{Zeolite}}{|}}{H^+} \longrightarrow (ArOH_2)^+_{ads} \qquad (6\text{-}7)$$

This leads to blocking of the catalytically active centers. However, above 200 °C the influence of phenol adsorption is weaker, some ethylene can be adsorbed, and partial alkylation of phenol is observed [T24].

The composition and therefore the catalytic properties of zeolites can also be influenced by modification of the zeolites. In the following we shall discuss two modification processes in more detail: isomorphic substitution and doping with metals.

6.4 Isomorphic Substitution of Zeolites [T24, T32]

The isomorphic substitution of the tetrahedral centers of the zeolite framework is another possibility for producing new catalysts. A prerequisite is that the ions have a coordination number of four with respect to oxygen and an ionic radius corresponding to the zeolite framework.

The Al centers can be replaced by trivalent atoms such as B, Fe, Cr, Sb, As, and Ga, and the Si centers by tetravalent atoms such as Ge, Ti, Zr and Hf. Silicon enrichment up to a pure SiO_2 pentasil zeolite (silicalite) is also possible [4].

Isomorphic substitution affects zeolite properties such as shape selectivity (influences on the framework parameters), acidity, and the dipersion of introduced components. The following sequence was found for the acidity of ZSM-5 zeolites:

$B \ll Fe < Ga < Al$

Thus the weakly Brønsted acidic boron zeolites allow acid-catalyzed reactions to be carried out with high selectivity. Gallium substitution gives effective, sulfur-resistant catalysts for the synthesis of aromatics from lower alkanes, without the need for noble metal doping [8]. The nonacidic titanium silicalite exhibits very interesting properties in selective oxidation reactions with H_2O_2 [T32].

Recently a completely new class of zeolite-like materials has been synthesized from Al and P compounds, namely the aluminophosphate (AlPO$_4$) molecular sieves. In contrast to the zeolites, the frameworks of the aluminophosphates are electrically neutral, contain no exchangeable ions and are largely catalytically inactive.

In 1988 Davis succeeded in preparing an aluminophosphate with a pore aperture of 1.2 nm: VPI-5 (Virginia Polytechnical Institute no. 5) is the molecular sieve with the largest pore width known up to now [6]. Many possibilities exist for modifying aluminophosphates. Replacing part of the framework P atoms by Si gives the silicoaluminophosphates, which have catalytic properties. Various metals have been introduced into both classes of materials, as shown in the following formula (Eq. 6-8). The catalytic properties of these compounds have barely been explored [7].

$$\text{Silicoaluminophosphate (SAPO)} \quad (6\text{-}8)$$

6.5 Metal-Doped Zeolites [T32]

Zeolites are especially suitable as support materials for active components such as metals and rare earths. With rare earths, the activity of the catalyst and its stability towards steam and heat can be increased. Suitable metals are effective catalysts for hydrogenations and oxidations, whereby the shape selectivity of the carrier is retained. Important factors influencing the reactions of such bifunctional catalysts are the location of the metal, the particle size, and the metal–support interaction.

The bifunctionality of metal-doped zeolite catalysts is explained here for the important example of isomerization and hydrogenation. The metal content facilitates the hydrogenation and dehydrogenation steps, while the acid-catalyzed isomerization step takes place under the restricted conditions of the zeolite cavities (Scheme 6-1).

Bifunctional catalysts are used in many reactions, including hydrocracking, reforming, and dewaxing processes. They usually contain ca. 0.5% Pt, Pd, or Ni. An advantage of nickel-containing hydrocracking catalysts is their lower hydrogenolysis activity compared to conventional catalysts.

A further example is acid-catalyzed disproportionation with [Pt]H-ZSM-5 as catalyst. The metal perfoms the hydrogenative cleavage of more highly aggregated molecules that would otherwise cause coking of the catalyst.

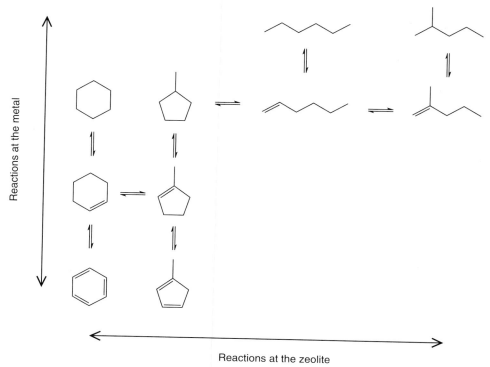

Scheme 6-1. Bifunctionality of metal-doped zeolites: isomerization and hydrogenation

It is understandable that transition metals and transition metal complexes that are used in large-scale industrial processes are also incorporated into zeolites so as to exploit their shape selectivity. Examples are:

- Zeolite X with Rh^{3+} or Ni^{2+}: oligomerization of alkenes
- [Rh]-zeolites: carbonylation reactions (oxo synthesis, methanol carbonylation)
- $[Pd^{II}][Cu^{II}]$-zeolites: oxidation of ethylene to acetaldehyde in the Wacker process
- [Ru]-zeolites: photosensitization of oxygen

Finally, we shall discuss two examples that demonstrate the shape selectivity of bifunctional zeolite catalysts. Thus the diffusivity of *trans*-2-butene in zeolite CaA is 200 times higher than that of *cis*-2-butene. Doping with Pt allows selective hydrogenation of *trans*-2-butene to be carried out [T35]. Also of interest is shape selective hydrogenation on [Pt]ZSM-5, which is compared to hydrogenation on a conventional supported Pt catalyst in Table 6-8. With the zeolite catalyst, hydrogenation of the unbranched alkene is favored.

Table 6-8. Shape-selective hydrogenation [T32]

Alkene	Reaction temperature [°C]	Conversion [%] Catalyst	
		[Pt]ZSM-5	Pt/Al$_2$O$_3$
Hexene	275	90	27
4,4-Dimethyl-1-hexene	275	<1	35
Styrene	400	50	57
2-Methylstyrene	400	<2	58

6.6 Applications of Zeolites [5, 8]

Zeolites have a wide range of applications. They are used as replacements for phosphates in laundry detergents, as adsorbents for purification and separation of materials, and as catalysts. The detergents industry has the largest demand for zeolites (ca. 1.2×10^6 t/a in 1994) and the highest growth rate. Demand for zeolite catalysts is also growing, and in 1994 amounted to ca. 115 000 t/a.

Table 6-9 lists important catalytic processes involving zeolites.

Table 6-9. Important catalytic processes involving zeolites

Process	Starting material	Zeolite	Products
Catalytic cracking	crude oil	faujasite	gasoline, heating oil
Hydrocracking	crude oil + H$_2$	faujasite	kerosene
Dewaxing	middle distillate	ZSM-5, mordenite	lubricants
Benzene alkylation	benzene, ethene	ZSM-5	styrene
Toluene disproportionation	toluene	ZSM-5	xylene, benzene
Xylene isomerization	isomer mixture	ZSM-5	*p*-xylene
MTG	methanol	ZSM-5	gasoline
MTO	methanol	ZSM-5	olefins
Intermediate products	diverse	acidic and bifunctional zeolites	chemical raw materials
SCR process	power station flue gases	mordenite	NO$_x$-free off-gas

Zeolite catalysts are mainly used in refinery technology and petrochemistry [3]:

- **Catalytic Cracking (FCC).** Here heavy heating oil is converted to middle distillate and high-octane gasoline with cerium- and lanthanum-doped Y zeolites. Advantages compared to conventional thermal cracking processes are the better conversion yields and product quality, albeit at the expense of slightly less flexibility with regard to starting materials.
- **Hydrocracking.** In this environmentally friendly process, which operates in a closed system with 100% conversion of heavy crude oil fractions, zeolite is used as a support for a hydrogenating component such as Pd. Bifunctional catalysis is achieved in which the cracking activity of the acidic zeolite is combined with the hydrogenation activity of the palladium.
- **Dewaxing Process.** In this industrial catalytic hydrocracking process, waxy C_{16+} paraffins are cracked and partly converted to aromatics.
- **Methanol to Gasoline Process (MTG) Process.** Methanol, produced from natural gas or coal, can be converted to high-quality, aromatics-rich gasoline in a two-stage fixed- or trickle-bed process with pentasil catalysts. Natural gas based production and cleavage of methanol has been operated since 1985 in New Zealand, where it covers one-third of gasoline demand.
- **Methanol to Olefins Process (MTO) Process.** Methanol can be converted to olefins by using modified pentasil catalysts. This interesting process, which is not yet used on an industrial scale, will presumably be of practical importance some time in the future.

Besides these processes, which lead to a wide product spectrum, controlled large-scale acid-catalyzed organic syntheses can also be carried out:

- **Mobil–Badger Process.** This process is a selective gas-phase alkylation of aromatics on pentasil zeolites. Ethylbenzene is produced from ethylene and benzene in a multistage adiabatic reactor. Compared to the conventional process of homogeneous catalysis with $AlCl_3$ as Friedel–Crafts catalyst, this heterogeneously catalyzed process has several advantages. These include economic and environmental advantages (e.g., up to 95% heat recovery at a reaction temperature of 400 °C), straightforward regenerability, no problems in separating and recovering the catalyst, and freedom from the corrosion and waste-disposal problems encountered with $AlCl_3$ as catalyst.
- **Xylene Isomerization.** This industrial process for obtaining higher contents of *p*-xylene in C_8 aromatics cuts is carried out on pentasil zeolites at 400 °C, generally in the presence of hydrogen. The *para*-selective xylene isomerization and the disproportionation of toluene are among the industrially established processes.

In environmental protection, the use of zeolite catalysts in SCR technology for the denitrogenation of flue gases (e.g., from coal-fired power stations) is the subject of many publications and patent applications (companies: Norton and Degussa). However, up to now the high prices and steam sensitivity of zeolites have prevented industrial realization.

In the last 10–15 years, the use of zeolites in the organic synthesis of intermediate products and fine chemicals has made rapid developments [4]. Especially pentasil zeolites have been used with great success. The syntheses can involve a whole series of steps, some of which are quite complicated. An overview is given in Table 6-10.

Table 6-10. Organic syntheses with zeolite catalysts [8]

Alkylations
Alkylation of arenes, side-chain alkylation, alkylation of heteroaromatics

Halogenation and nitration of arenes, substitution reactions of aliphatics
Ether and ester formation, thiols from alcohols and H_2S, amines from alcohols and NH_3 (mordenite, erionite)

Isomerizations
Isomerization of arenes and aliphatics, double bond isomerizations

Rearrangements
Skeletal rearrangement of alkanes, olefins, and functionalized compounds; pinacolone rearrangement; Wagner-Meerwein rearrangement; epoxide rearrangement; rearrangement of cyclic acetals

Additions and eliminations
Hydration and dehydration, additions to and eliminations from alcohols and acids, additions to N- and S-containing compounds, addition to epoxides

Hydrogenation and dehydrogenation
Dehydrocyclization

Hydroformylation

Oxidations
Oxidation with oxygen and peroxides

Condensations
Aldol condensations; synthesis of N heterocycles, isocyanates, nitriles; O/N exchange in cyclic compounds

---

Apart from simple mechanisms (condensation, hydrogenation, substitutions, alkylations), more complicated reactions can also be performed (Wagner–Meerwein and pinacolone rearrangements, syntheses of heterocycles) [8].

The application potential of zeolite catalysts in organic synthesis is by no means exhausted, and base catalysis remains practically unexplored. Thus the zeolites still have huge potential for future research and development.

Exercises for Chapter 6

Exercise 6.1

a) What are zeolites?
b) What are the three main possibilities for modifying zeolites?

Exercise 6.2

The following figure was found in a textbook:

a

b

Explain these reactions. Which catalyst properties make reactions a and b possible?

Exercise 6.3

Name several advantages that zeolite catalysts have compared to conventional catalysts.

Exercise 6.4

A mixture of olefins is hydrogenated with different catalysts at 275 °C. The following results were obtained:

Catalyst	% Hydrogenation	
	1-Hexene	4,4-Dimethyl-1-hexene
1% Pt/Cs-ZSM-5	90	1
0.5% Pt/Al$_2$O$_3$	27	35

Explain the differing selectivities.

Exercise 6.5

What is shape-selective catalysis?

Exercise 6.6

In the industrial synthesis of gasoline hydrocarbons, two processes compete with one another: Fischer–Tropsch synthesis and methanol cleavage (MTG process). Starting from synthesis gas, the methanol cleavage is two-stage process but still has advantages over the one-step Fischer–Tropsch synthesis. Why is this so?

Exercise 6.7

What are H-zeolites and how are they prepared?

Exercise 6.8

In the large-scale industrial production of methylamines, methanol and NH$_3$ are reacted at 350–500 °C and ca. 20 bar in the presence of Al$_2$O$_3$. A mixture of mono-, di-, and trimethylamine is obtained with an equilibrium content of ca. 62% trimethylamine. However, trimethylamine is of only minor economic importance.

Suggest how the product spectrum could be modified to favor mono- and dimethylamine.

Exercise 6.9

a) In its protonated form ZSM-5 catalyzes the reaction of ethylene with benzene to give ethylbenzene. Suggest a plausible mechanism for this alkylation reaction.
b) It is possible to produce the pure SiO_2 anologue of ZSM-5. Can it be expected that it will be an active catalyst for the alkylation of benzene?

Exercise 6.10

The kinetics of the alkylation of benzene with a rare earth zeolite Y is described by the equation:

$$r = \frac{kc_O c_T}{1 + K_O c_O} \qquad O = \text{olefin}, \quad T = \text{toluene}$$

What can be said about the mechanism of zeolite-catalyzed alkylation?

Exercise 6.11

The hydrothermal treatment (100–300 °C) of aluminophosphate gels in the presence of organic amines and quaternary ammonium bases leads to $AlPO_4$ molecular sieves with open pores and channels.
How do they differ from conventional zeolites?

Exercise 6.12

How can zeolites of low aluminum content be manufactured?

7 Planning, Development, and Testing of Catalysts

7.1 Stages of Catalyst Development [T40]

The development of a catalyst up to industrial application involves three stages:
- The research stage
- Intensive testing in the laboratory and on the pilot-plant scale
- The industrial stage

In the resaerch stage, the first step is to formulate the problem. This involves gathering information about market requirements and estimating the value that a particular catalyst system could have at some time in the future.

Next the concept must be described in chemical terms so that it can be seen whether the project is technically and economically feasible. Estimates must be made whether a profitable yield and selectivity can be achieved, and the raw material supply and future demand for the product must be guaranteed. Only when the results of these estimations are satisfactory can the actual catalyst planning begin.

If several selective catalysts are initially available, then their suitability and lifetime are intensively investigated in a test reactor known as a pilot plant. The final step is then erection and startup of the industrial plant. Before the actual production in the industrial plant begins, detailed tests are carried out so that any teething problems can be identified right at the beginning and eliminated.

The development of an industrial catalyst must also take other parameters into account such as support materials and the type of reactor in which the catalyst will be used. Thus the choice of catalyst depends on many factors, as shown in Scheme 7-1.

With regard to the morphology of the catalyst, a distinction is made between microeffects and macroeffects. Microeffects include the crystallinity, surface properties, and porosity, while examples of macroeffects include particle size and stability. Macroeffects are often not adequately taken into account, although mechanical destruction is one of the most common reasons for changing industrial catalysts.

Although microeffects and macroeffects have their own characteristics, they often act closely together. This is demonstrated by the example of Al_2O_3, in which variation of the crystallite size and controlled phase transition by means of heat treatment have a major effect on the wear resistance and compressive strength of the material. Interactions between the active component and the support material

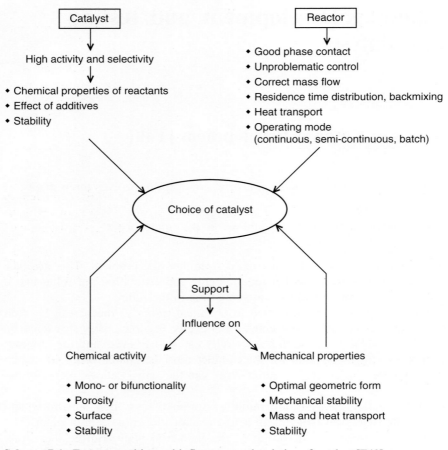

Scheme 7-1. Target quantities and influences on the choice of catalyst [T40]

are discussed in detail in Section 4.4. Chapter 9 deals with the influence of the reactor type on the choice of catalyst.

An example of a successful catalyst development is the production of acrolein by oxidation of propene with air (Eq. 7-1)

$$CH_2=CH-CH_3 + O_2 \xrightarrow[320-420\ °C,\ 1-2\ bar]{Catalyst} CH_2=CH-CHO + H_2O \qquad (7\text{-}1)$$

The first acrolein plant with a bismuth/molybdenum oxide catalyst was brought on stream by Degussa in 1967. Catalyst development concentrated on the optimization of the active phase and the shape of the catalyst. In decades of development work, the selectivity of the catalyst was ever further increased, and the acrolein yield increased from 40 to 80% (Table 7-1).

7.1 Stages of Catalyst Development

Table 7-1. Development of a catalyst for the oxidation of propene to acrolein [7]

	1967	1972	1982	1988
Form	tablet	extrudate	shell catalyst	extrudate
Chemical composition	Bi, Mo, Fe, P, Ni, Co, Sm oxides	Bi, Mo, Fe, P, Ni, Co, W, Si, K oxides	Bi, Mo, Fe, P, Ni, Co, Sm, K, Al, Si oxides	Bi, Mo, Fe, P, Ni, Co, Sm, K, Al, Si oxides
Acrolein yield	40%	70%	76%	>80%

The use of various promoters increased the activity of the catalyst to such an extent that the operating temperature for the formation of acrolein could be lowered from 450–500 to 300–330 °C. In this way the catalyst lifetime was extended to several years.

This example shows just how complex the composition of modern catalysts is, and that the manner in which the catalyst is produced can have a decisive influence on its effectiveness.

7.2 An Example of Catalyst Planning: Conversion of Olefins to Aromatics

In this section we shall discuss an example that is described in detail in the literature [29, T40]. An attempt was made to develop a catalytic process for the production of aromatics from olefins by means of an oxidative dehydroaromatization reaction.

The desired reaction can be formulated as shown in Equation 7-2.

$$2\ CH_3-\underset{\underset{}{|}}{\overset{\overset{R}{|}}{C}}=CH_2 \longrightarrow \text{[1,4-disubstituted benzene with R groups]} \qquad (7\text{-}2)$$

This equation can be regarded as the idea behind the process. The route from this idea to a satisfactorily operating catalytic process is shown in Scheme 7–2. The idea is followed by an initial feasibility study (step II). This involved carrying out simple thermodynamic calculations, which showed that conversion of propene to benzene is at least theoretically possible. The next step is a thorough literature

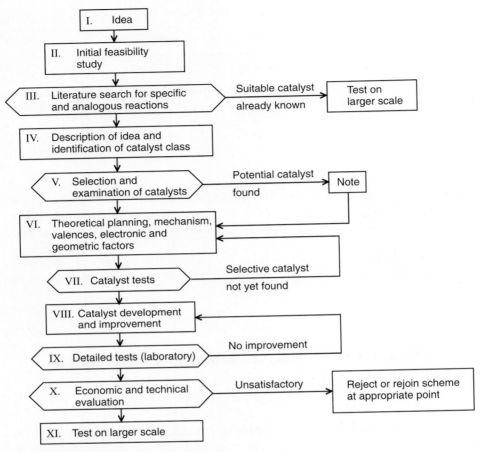

Scheme 7-2. Catalyst planning procedure

search in which one attempts to find out whether this particular reaction or an analogous reaction has already been carried out. In our example, all that was found was the suggestion that the reaction proceeds more selectively in the presence of oxygen.

Step IV is the formulation of the idea and identification of the catalyst class. For this, the probable course of the reaction must be formulated (Eqs. 7-3 to 7-5).

$$2\ CH_2{=}\underset{\underset{\displaystyle R}{|}}{C}{-}CH_3 \longrightarrow CH_2{=}\underset{\underset{\displaystyle R}{|}}{C}{-}CH_2{-}CH_2{-}\underset{\underset{\displaystyle R}{|}}{C}{=}CH_2 + H_2 \qquad (7\text{-}3)$$

$$CH_2{=}\underset{\underset{\displaystyle R}{|}}{C}{-}CH_2{-}CH_2{-}\underset{\underset{\displaystyle R}{|}}{C}{=}CH_2 \longrightarrow R{-}\!\!\left\langle\!\!\bigcirc\!\!\right\rangle\!\!{-}R + H_2 \qquad (7\text{-}4)$$

7.2 An Example of Catalyst Planning: Conversion of Olefins to Aromatics 253

$$R-\langle\rangle-R \longrightarrow R-\langle\rangle-R + H_2 \qquad (7-5)$$

According to Equation (7-3), two olefin molecules form a diene, which undergoes cyclization in the next step (Eq. 7-4). The final conversion of the cyclohexadiene system to an aromatic compound (Eq. 7-5) is, like the other two steps a dehydrogenation reaction. Suitable catalysts for these reactions could be the ionic and the metal oxide catalysts. The possible side reactions of both classes are summarized in Scheme 7-3.

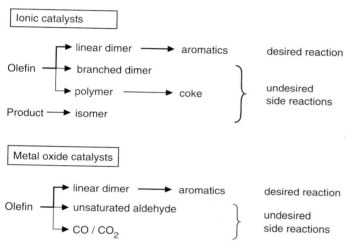

Scheme 7-3. Possible catalyst systems and their disadvantages

The key disadvantage of ionic catalysis is that a carbenium ion is formed as intermediate product and leads to the formation of the undesired branched dimers and high-molecular polymers. Thus only the metal oxides remain as potential catalysts.

Some tests were carried out with some metal oxides already used in other catalytic reactions, including the oxides of Pt, Cr, Mo, Th, and Co. The results, however, were unsatisfactory. Therefore a search was made for a new metal oxide catalyst by using an exact theoretical plan. Thus, we are at the next step of the process, in which the mechanism, oxidation states, and electronic and geometric factors are investigated. Two conclusions were drawn (Scheme 7-4):

1) Under the influence of oxygen, the olefin forms a π-allyl intermediate, which adds to the metal ion.
2) An electron is transferred from this intermediate to the metal center. Since the desired dimer is formed from two molecules that are bound to the same metal ion, the catalyst must be capable of accepting two electrons.

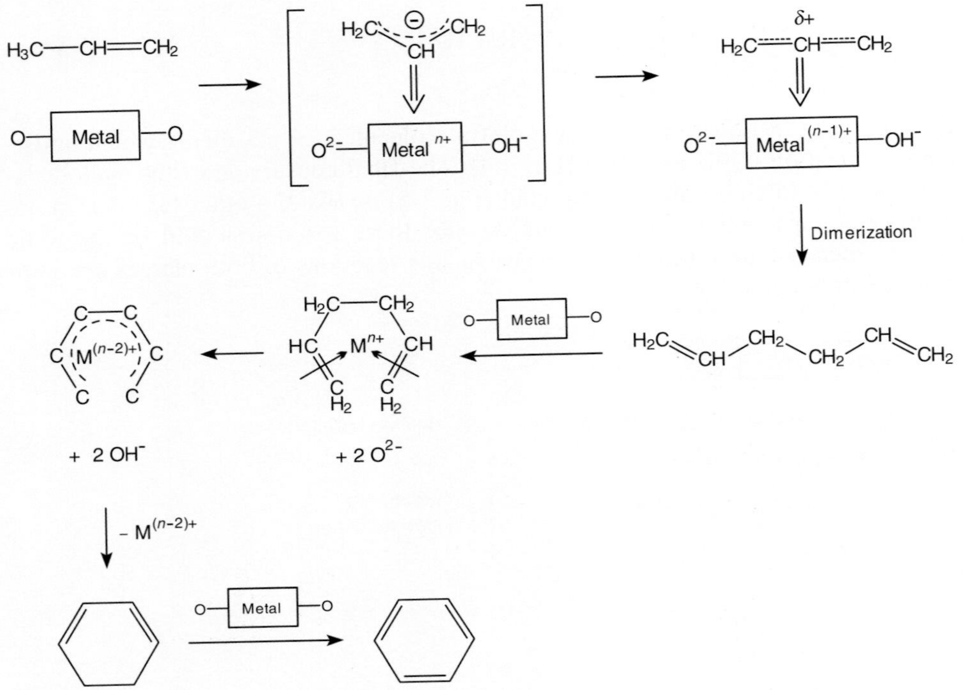

Scheme 7-4. Mechanism of olefin dimerization and cyclization

A search was now made for metal oxides that can adsorb olefins in the oxidized state and whose oxidation states differ by two units. These include thallium, lead, indium, and bismuth. Since the oxides of bismuth and lead are of low thermal stability, attention was focussed on the oxides of thallium and indium.

Let us return to the flow sheet of Scheme 7-2. In step VI we made a preliminary choice of catalyst by using theoretical considerations. However, since experiments are the only sure method for testing the mechanistic hypothesis, the next step is catalyst testing.

These tests showed that thallium is also unsuitable for this reaction because the reduced form of the oxide is lost from the reactor due to its volatility. Hence, only the highly selective indium(I)/indium(III) oxide remained as the catalyst of choice.

Let us briefly examine the entire catalyst planning process once again (Scheme 7-5). The ionic catalysts proved to be unfavorable since they gave large amounts of branched products. Since the proven metal oxide catalysts also had many disadvantages, a completely new catalyst was sought. This search led to metal oxides of Groups 13–15, whereby indium oxide proved to be highly selective. Nevertheless, this oxide also has disadvantages, especially the formation of the side products CO_2 and acrolein.

7.2 An Example of Catalyst Planning: Conversion of Olefins to Aromatics

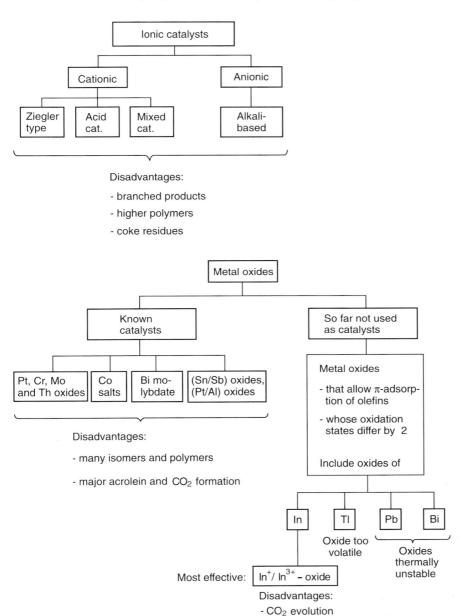

Scheme 7-5. Catalyst planning for the oxidative dehydroaromatization of olefins (Eqs. 7-3 to 7-5)

Let us return to our general Scheme 7-2. The next step is improving the catalyst. In our case this means limiting the oxidation to CO_2 and acrolein (step VIII). Attempts were made to achieve this by manufacturing a catalyst with optimal pore structure and surface properties.

The formation of CO_2 requires the most oxygen of all products. Assuming that the majority of the oxygen is adsorbed on the catalyst, additives that hinder oxygen adsorption should lead to formation of less CO_2. Since the oxygen can form peroxo species, additives such as Ca and Ba, which promote peroxide formation, should be avoided. Since oxygen acts as an electron acceptor, electronegative catalyst additives should counteract the adsorption of oxygen. Such an effect has been observed with bismuth phosphate, which is a more selective catalyst than bismuth oxide.

On the other hand, it can be expected that radical-like allyl ligands will dimerize rather than react with oxygen. Thus, electrons should be removed from the adsorption centers. Dopants that facilitate this electron transfer should have a positive effect. Such an additive is Bi_2O_3, with which the indium oxide was doped. The pore structure of the support material could also have an influence on the over-oxidation. Small pores would promote further oxidation by restricting diffusion. Hence supports with large pores should be best. Many of these suggestions were tested, but, as is often the case in heterogeneous catalysis, conflicting results were obtained. Since the entire process was not very interesting from an economic viewpoint, we will end the discussion here.

To summarize: a suitable catalyst was found by means of mechanistic reasoning, and it was shown that planned research can lead to a satisfactory solution within a relatively short time and with minimum effort.

When the detailed tests are complete, an economic and technical evaluation is carried out (Scheme 7-2, step X). Only when this is satisfactory is a process tested on an industrial scale.

7.3 Selection and Testing of Catalysts in Practice

To shorten the laborious process of purely empirical catalyst selection, which sometimes involves hundreds of tests, today use is made of the various catalyst concepts and statistical methods for test planning [25, T40]. The individual steps of such a procedure are shown in Scheme 7-6. This scheme, with its many steps, clearly shows the efforts involved in finding an optimal catalyst and optimal reaction conditions for the desired reacton.

Scheme 7-6 shows two routes, which differ in the amount of knowledge gained. The more pragmatic procedure **A** dispenses with extensive kinetic measurements and aims at direct optimization of the process, whereas in the detailed

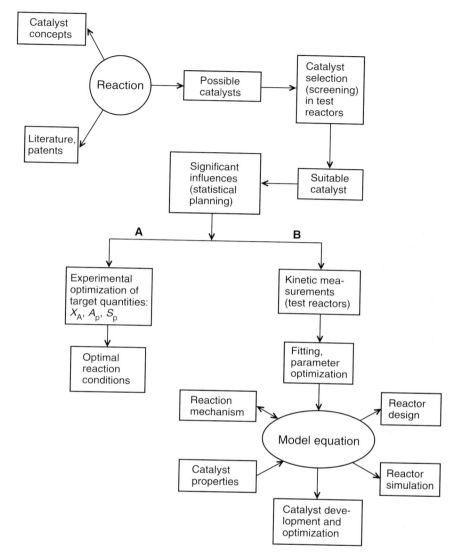

Scheme 7-6. Procedures for choosing a catalyst

procedure **B**, modelling and analysis of the catalytic process provide the foundation for reactor design and simulation. In this chapter we shall discuss both possibilities schematically in order to provide chemical engineers with support in the complex field of catalyst development [27].

7.3.1 Catalyst Screening

A catalytic reaction represents a complex problem that is influenced by numerous factors. In order to find a suitable catalyst or solvent for a particular reaction, screening tests are carried out. This means keeping several reaction conditions constant while only one parameter is varied. The procedure is briefly summarized in Table 7-2 [1].

Table 7-2. Catalyst screening

Measurement method	Advantages	Disadvantages
Conversion under standardized experimental conditions	Rapid predictions due to simple measurement and evaluation procedure	Low reliability due to arbitrary choice of conditions

Screening tests do not allow any absolute predictions about the activity or applicability of a catalyst. Instead they provide measurements that can be compared with one another. Therefore, it is important that parameters, once chosen, are applied to all screening experiments.

Catalyst screening provides a comparison of several catalysts with respect to the desired target parameter. Nonsystematic influences can also be investigated in the course of the screening process, for example:

– Dependence of the reaction on solvent
– Effect of adding reagents and cocatalysts
– Influence of catalyst pretreatment
– Estimation of catalyst lifetime

Let us now examine the screening procedure for the example of a catalytic hydrogenation [13, 30].

In the catalytic hydrogenation of substituted 2-nitrobenzonitrile, the cyano group is also attacked under normal reaction conditions, and several side products are obtained besides the desired product 2-aminobenzonitrile (Eq. 7-6).

$$R\text{-}C_6H_3(NO_2)(CN) + 3\,H_2 \xrightarrow{\text{Cat.}} R\text{-}C_6H_3(NH_2)(CN) + 2\,H_2O \tag{7-6}$$

Side products: amide, diamine, dimer of the nitro compound.

Table 7-3. Hydrogenation of substituted 2-cyanonitro compounds [30]

Catalyst system	Amine yield [%] (reaction time)
$SnCl_2$/HCl in DMF	67
Fe/HCl in methanol	78
Fe/glacial acetic acid, 2-propanol	88
Raney Ni, 2-propanol	91 (24 h)
Pd/$BaSO_4$, dioxane	79 (3 h)

Table 7-3 lists catalyst systems described in the literature for the hydrogenation of similar nitro compounds.

None of the known examples met the requirements for high yields at short reaction times with environmentally friendly reagents. Therfore, various supported noble metal catalysts and solvents were tested under the same reaction conditions in a catalyst screening program.

It is known that the hydrogenation activity of supported Pd catalysts is the least affected by different substituents and changes in the reaction conditions. In suspension, aromatic nitro compounds are generally hydrogenated in the temperature range 50–150 °C at pressures of 1–25 bar and with catalyst concentrations of 0.1–1 %. In the screening tests the following reaction conditions were kept constant: pressure, temperature, catalyst concentration, starting material concentration, and stirring speed. The results are summarized in Tables 7-4 and 7-5.

Table 7-4. Catalyst screening in the hydrogenation of substituted 2-nitrobenzonitrile

Experiment	Catalyst	2-Aminobenzonitrile yield [%]
1	Pd/C (1)	87.2
2	Pd/C (2)	85.2
3	Pd/C (3)	90.0 (incl. 10% dimer as intermediate product)
4	Pd/$BaSO_4$	84.1
5	Pt/C	34.1
6	Raney Ni	6.3
7	Rh/C	19.7

Constant reaction conditions: 5 mL stirred autoclave, 0.1 g starting material, 1.0 mL ethanol, 25 °C, 1 bar H_2 pressure, 20 mg catalyst, 120 min reaction time, stirring speed 700 rpm, catalysts 1–3: commercial 5% Pd/activated carbon catalysts.

The best catalyst proved to be the supported Pd catalyst (3) since the dimer can be regarded as an intermediate product. This catalyst was then used for the subsequent solvent tests (Table 7-5).

In all screening tests it is important that stirring be carried out with a high rate of over 600 rpm to ensure that the reactions proceed under kinetic control.

Table 7-5. Solvent screening in the hydrogenation of substituted 2-nitrobenzonitrile

Experiment	Solvent	2-Aminobenzonitrile yield [%]
1	dioxane	60
2	methanol	59
3	acetic acid	33
4	ethanol	90
5	*tert*-butyl methyl ether	74
6	toluene	53
7	ethyl acetate	24
8	dichloromethane	75
9	hexane	33
10	acetic anhydride	21
11	isopropanol	72
12	DMF	77

Reaction conditions: Table 7-4.

Ethanol proved to be the best solvent and was used for subsequent tests. The successful screening program was followed by reactor optimization with a special testing plan (see Section 7.3.3).

7.3.2 Catalyst Test Reactors for Reaction Engineering Investigations [27, 28]

Heterogeneously catalyzed gas-phase reactions play a very important role in industrial chemistry. Therefore, this chapter deals with how kinetic data are obtained for such reactions.

In principle the kinetics of an industrial process can be measured on the laboratory scale or in a pilot plant. Apart from the small amounts of material involved, on the laboratory scale the test conditions can be chosen such that chemical and transport phenomena (microkinetic and macrokinetic effects), which are equally important in an industrial process, can be investigated in isolation [T26].

The two types of laboratory reactor shown in Figure 7-1 have proved to be the most suitable for reaction engineering investigations on heterogeneously catalyzed gas-phase reactions [27].

The concentration-controlled, gradientless differential circulating reactor is best suited for kinetic measurements. Such modern laboratory reactors are now of major importance. They allow kinetic data to be measured and evaluated practically free of distortion by heat- and mass-transport effects [17]. Depending on the material flow, a distiction is made between reactors with outer and inner circulation. Evaluation of the kinetic measurements is straightforward because the simple

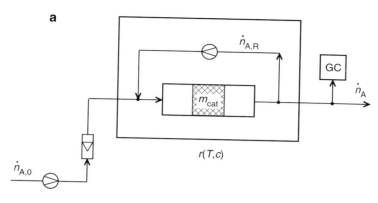

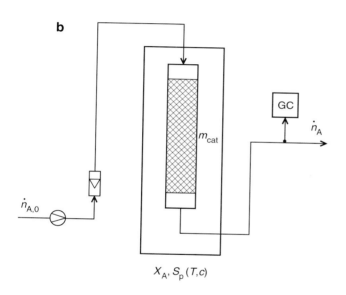

Fig. 7-1. Catalyst test reactors

	(a) Gradientless reactor (differential circulating reactor)	(b) Laboratory or pilot integral reactor
Information:	r and kinetic parameters differential values measured under transport-free conditions	r_{eff} integral measures of activity (X_A) and selectivity (S_P) behavior over the lifetime of the catalyst
Evaluation:	algebraic equation	differential equation
Behavior:	like a continuous stirred tank approximately isothermal	like a flow tube temperature profile

algebraic balance equation for a stirred tank reactor (Eq. 7-7) can be applied (prerequisite: high recycle ratio R). In practice it is found that recycle ratios of $R = 10\text{--}25$ are sufficient to achieve practically ideal stirred tank behavior [8].

$$-r'_A = \frac{\dot{n}_{A,0} - \dot{n}_A}{m_{cat.}} = \frac{\dot{n}_{A,0} X_A}{m_{cat.}} \tag{7-7}$$

Before a series of tests is carried out for a particular reaction, the suitability of the differential circulating reactor for kinetic investigations should be proven. The following should be tested:

- The gradient-free operation of the reactor: no influence of the rate of rotation of the reactor drive or the gas-delivery system on the reaction rate
- Exclusion of pore diffusion: tests with different catalyst particle sizes and pressures
- Ideal stirred trank behavior: residence time measurements, e.g., by step-injection tracer experiments
- Influence of so-called blank test reactions, e.g., reactor-wall catalysis above ca. 450 °C

A very simple variant of the differential circulating reactor is the so-called jet loop reactor shown schematically in Figure 7-2. Combined with online analysis of the product stream, the apparatus can be used to investigate commercially available or specially manufactured heterogeneous catalysts for gas-phase reactions (Fig. 7-4). After passing through a nozzle, the gas (e.g., CO/H_2 in methanol synthesis) flows through the inner tube and carries recycle gas with it in the direction of flow shown in the figure. On the catalyst bed, which consists of about two layers of pellets (ca. 25 g), the synthesis gas is converted into methanol in accordance with the thermodynamic equilibrium. A gas stream is removed from the reactor in an amount corresponding to the feed stream and analyzed by GC. Typical results are exemplified by the gas chromatograms shown in Figure 7-3.

A commercial CuO-based methanol catalyst was preformed with synthesis gas (Fig. 7-3 a). After a short time, CO_2 and H_2O are found in the gas mixture as a result of catalyst reduction. Above ca. 200 °C methanol is formed, and eventually the stationary equilibrium with ca. 20 % methanol at 220 °C is reached (Fig. 7-3 b).

Such a reactor, designed for temperatures up to 500 °C and pressures up to 400 bar, was used for exact kinetic modeling of methanol synthesis [26].

Reaction Conditions:

12 g Cu catalyst, cylindrical pellets,
diameter = height = 5 mm
225–265 °C, 20–80 bar;
Feed stream   7–25 % CO
              1–15 % CO_2
              60–90 % H_2

Throughput 0.2–1.2 m³/h
Recycle ratio $R \geqslant 30$
$r_{CH_3OH} = 0.01 - 0.09$ kmol(kg cat.)$^{-1}$h^{-1}

Langmuir–Hinshelwood kinetics were determined, and the rate-determining step is reduction of the intermediate formaldehyde (Eqs. 7-8 and 7-9).

$$CO + H_2 \rightleftharpoons HCHO \qquad (7\text{-}8)$$

$$HCHO + H_2 \rightleftharpoons CH_3OH \qquad (7\text{-}9)$$

Another versatile catalyst test reactor for the investigation of multiphase reactions is equipped with a fixed catalyst basket or a catalyst basket that rotates in the reaction medium (Fig. 7-5). In such reactors, rotational velocities in excess of 750 rpm ensure very good mass transfer between the catalyst, the gas bubbles, and the liquid, and an internal circulation is generated in the reactor [4].

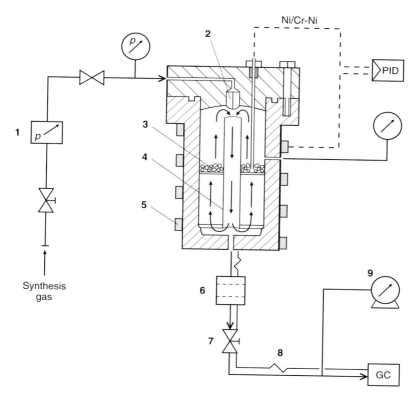

Fig. 7-2. Jet loop reactor for catalyst investigations (high-pressure laboratory, FH Mannheim) 1) Thermal mass flow controller (up to 200 bar); 2) Nozzle, interchangeable; 3) Catalyst pellets on wire mesh; 4) Central tube; 5) Heating band 500 W; 6) Microfilter; 7) Precision feed valve; 8) Supplementary heating; 9) Gas meter

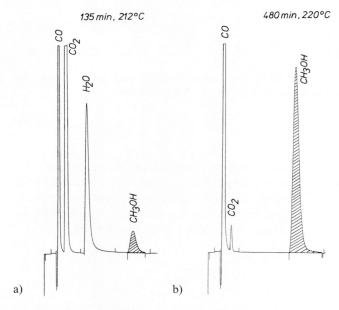

Fig. 7-3. Gas chromatogram of methanol synthesis in the jet loop reactor
Reaction conditions: $CO/H_2 = 1/2$, 40 bar, 25 g cat., $\dot{V}_0 = 500$ mL/min, nozzle diameter 0.1 mm

Fig. 7-4. Jet loop reactor (high-pressure laboratory, FH Mannheim)

Fig. 7-5. Catalyst test reactor with fixed or rotating catalyst basket
(company Autoclave Engineers; high-pressure laboratory, FH Mannheim)

Numerous kinetic studies with such reactors have been reported in the literature, including:

– Hydrodesulfurization of model substances such as dibenzothiophene [20]
– Hydrogenation of olefins
– Dehydrocyclization reactions

The fewest experimental problems are caused by the integral reactor due to its simple construction and straightforward operation. It consists of a flow tube, 20–50 cm in length and ca. 2 cm in diameter, filled with catalyst. The conversion achieved in such reactors is high and can readily be determined by comparing the initial and final concentrations of a reactant. A test series is carried out with variation of the values of the catalyst mass $m_{cat.}$ or the feed flow rate $\dot{n}_{A,0}$, thus covering a wide range of conversions.

The favored method is to evaluate the data with a differential form of the design equation of the tubular reactor (Eq. 7-10).

$$-r'_A = \frac{dX_A}{d\left(m_{cat.}/\dot{n}_{A,0}\right)} \quad (7\text{-}10)$$

The rate r'_A can be obtained directly from the individual measurements by graphical differentiation (Fig. 7-6). The slope of the tangent of the conversion–time factor curve corresponds to the momentary reaction rate under the given test conditions.

The disadvantage of the integral reactor is that it can not be operated isothermally and that the measured overall conversion is generally the result of a complex interplay between transport phenomena and chemical reaction. Hence the integral reactor is mainly used for comparitive catalyst studies and lifetime tests. Its advantages are:

– Rapid, empirical, and practice-relevant process development
– Conclusions about catalyst activity from changes in temperature and concentration profiles
– Catalyst deactivation can be followed
– Relatively simple scale-up

In reaction engineering investigations it is generally not sufficient to draw conclusions about the activity and selectivity of a catalyst on the basis of conversion and yield. Transport limitations and hence the structure of the individual catalyst particles (shell catalyst/bulk catalyst, molded catalyst/extruded catalyst, etc.) must also be taken into account. The determination of the parameters and the selection of models for the quantitative kinetic description of the catalyst should be followed by the simulation of industrial reactors in order to obtain more information on the practical suitability of the chosen catalyst.

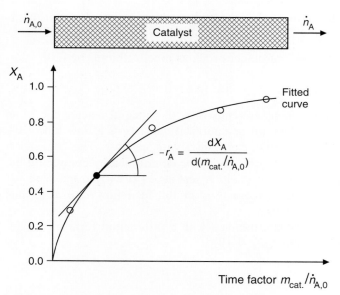

Fig. 7-6. Evaluation of the data from an integral reactor

7.3.3 Statistical Test Planning and Optimization [6, 21]

Statistical test planning is an effective aid to recognizing significant quantities that influence chemical reactions. A systematic process for searching for suitable catalysts and optimizing them is especially helpful in the case of catalytic reactions, with their numerous test parameters.

In this chapter we shall largely dispense with the mathematical basis of statistical test planning and we will illustrate the method with the aid of some simple practical examples.

Factorial Test Plans

A 2^n factorial design is the simplest complete test plan for investigating the influence of n variables on the test result.

Definitions:

Variable: independent quantity of arbitrary magnitude, assumed to have an influence on the test result
Levels: settings of the parameters, e.g., temperature as reaction parameter 30 and 60 °C

Designation of the variables:
— variable at the lower level
+ variable at the higher level
(1) all variables at the lower level
a variable A at higher level, all other variables at lower level
A, B variables (effects)
AB, AC interaction effects

A factorial design with three variables and two levels would lead to a test plan with eight experiments (Table 7-6). In a three-dimensional depiction, these eight tests occupy the corners of a cube (Fig. 7-7).

Table 7-6. 2^3 factorial design (eight experiments, three factors)

Experiment designation	A	B	C
(1)	−	−	−
a	+	−	−
b	−	+	−
ab	+	+	−
c	−	−	+
ac	+	−	+
bc	−	+	+
abc	+	+	+

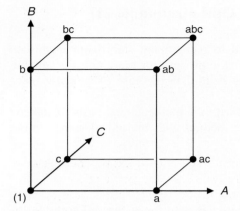

Fig. 7-7. 2^3 factorial design with designation of the experiments

Factorial designs should be preferentially used when:

- The effect of many variables in a limited area has to be tested rapidly
- The interaction effects between several variables are unknown
- Initial tests for the selection of variables are to be carried out
- Several target quantities have to be simultaneously distinguished

Evaluation of 2^n Factorial Designs:

The following questions have to be answered:
1) Which variables have an influence on the target quantity?
2) Which variables interact with one another?

1) Four pairs of results are influenced only by the variable A:

$$A \text{ low} \left| \begin{array}{l} (1)\ldots\ldots\ldots a \\ b \ldots\ldots\ldots ab \\ c \ldots\ldots\ldots ac \\ bc \ldots\ldots\ldots abc \end{array} \right| A \text{ high}$$

Difference between two results determined only by A

The effects of the variables are expressed relative to the mean value of all the measurements and half of the difference between levels, for example:

$A = 1/8 \,[(a-1) + (ab-b) + (ac-c) + (abc-bc)]$

The values calculated in this way allow the effects to be compared with one another. However, whether an effect is measurable depends on the scatter of the tests (significance tests).

2) A distinction is made between twofold and multifold interactions. For example, the interaction AB is defined as the difference in effect A with B high and effect A with B low; hence:

$$AB = 1/8 \left[(ab - b) + (abc - bc) - (a - 1 + ac - c) \right]$$

All effects and interactions can be calculated rapidly by using the Yates scheme [25]. The tests are arranged in the standard order. Then the first and second, third and fourth, etc., values are added together to give the top half of column (1). Now the first value is subtracted from the second, the third from the fourth, and so on, to give the bottom half of column (1).

The calculation is continued until n columns are obtained (i.e., equal to the number of variables). The last column gives the "total" and 2^n times the effects and interactions. The "total" is the 2^n-fold mean test result that is theoretically obtained under average test conditions.

The procedure will now be explained for the example of oxo synthesis. Conjugated dienes are converted into mono- and dialdehydes by phosphine-modified rhodium catalysts [12]. The target quantity, in this case the extent of dialdehyde formation, depends mainly on the three reaction parameters temperature (A), cocatalyst ratio (B), and total pressure (C). A 2^3 factorial design was carried out (Table 7-7). The evaluation of the test results by the Yates scheme is shown in Table 7-8.

Table 7-7. 2^3 factorial design for oxo synthesis

Factors	Levels							
Temp. [°C] A	90				120			
Cocatalyst ratio B	6		30		6		30	
Pressure [bar] C	100	150	100	150	100	150	100	150
Experiment	(1)	c	b	bc	a	ac	ab	abc

Interpretation of the Results:

The average value of the yield is 6.2% and is theoretically attained under average reaction conditions, that is

    Reaction temperature   105 °C

    Cocatalyst ratio   18

    Pressure   125 bar

Table 7-8. Evaluation of the 2^3 factorial design

Experiment	Dialdehyde yield [%]	(1)	(2)	(3)	Effect or interaction	Meaning
(1)	5.6	8.5	21.1	49.5	6.2	Total
a	2.9	12.6	28.4	−1.7	−0.21	A
b	6.0	12.9	−2.3	6.7	0.84	B
ab	6.6	15.5	0.6	3.3	0.41	AB
c	6.3	−2.7	4.1	7.3	0.91	C
ac	6.6	0.6	2.6	2.9	0.36	AC
bc	7.6	0.3	3.3	−1.5	−0.19	BC
abc	7.9	0.3	0	−3.3	−0.41	ABC

Effect $A = -0.21$ means that the yield decreases by 0.21 % when the reaction temperature is raised by 15 °C, and so on.

The predictions are valid only for the measurement range, but it has to be questioned whether the smallest effects are at all meaningful. In the laboratory the effects are assessed in terms of the experimental scatter or the precision of the instrumentation. An effect must differ significantly from the experimental scatter. Test results are assessed by carrying out significance tests.

Terms:

a) Experimental error variance s^2: a measure of the scatter

$$s^2 = \frac{1}{n-1} \sum_{i=1}^{n} (x_i - \bar{x})^2 = \frac{\text{Sum of squares}}{\text{Degrees of freedom}} \tag{7-11}$$

$$\bar{x} = \text{mean value} = \frac{1}{n} \sum_{i=1}^{n} x_i \tag{7-12}$$

b) Sample standard deviation s: has the same dimensions as the measured quantity

$$s = \sqrt{s^2}$$

c) Sample: a randomly selected part of the total number of measurements

d) Normal distribution: the fact that the results of a measurement are scattered around a mean value is well known. If the number of times a particular value occurs within a certain interval is plotted, then the distribution shown in Figure 7-8 is obtained.

Dividing the actual frequencies by the total number of measurements of the sample gives the relative frequency distribution. If the relative probability distri-

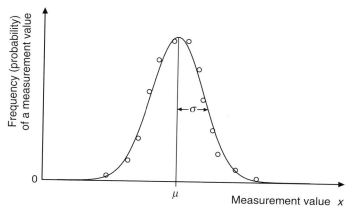

Fig. 7-8. Gaussian or normal distribution

bution is calculated by using the total number of measurements, the probability distribution is obtained (normal or Gaussian distribution; Eq. 7-13).

$$N(\mu, \sigma) = \frac{1}{\sigma\sqrt{2\pi}} e^{-(x-\mu)^2/2\sigma^2} \tag{7-13}$$

The normal distribution has a mean value μ and a variance σ^2. The sample average $\bar{x}$ and the sample variance s^2 are estimates for μ and σ^2 of the total number of measurements, which have no systematic errors.

The frequency, i.e., probability, with which the measurements occur at a given distance from the mean value have been tabulated. Such tables are used to determine whether experimental results support a hypothesis [3, 21].

Let us now check the significance of the results of our factorial design (Table 7-9). The standard deviation is known from earlier investigations to be $\sigma = 0.92$.

The variance of the effects is calculated from the variance of the measurements by means of the error-propagation law (Eq. 7-14).

$$Var_{\text{Eff}} = \frac{1}{2^n} \sigma^2 \tag{7-14}$$

From Equation 7-14 we obtain:

$$\sigma_{\text{Eff}} = \sqrt{1/8} \cdot \sigma = 0.32$$

Step 1: H_0: the effects $A, B, \ldots$ belong to a normal distribution with $\mu = 0$ and $\sigma = 0.32$

Step 2: chosen level of confidence = 95% (5% level of risk)

Step 3: test quantity $z = \dfrac{\text{Eff} - \mu}{\sigma_{\text{Eff}}} = \dfrac{\text{Eff} - 0}{0.32}$

Step 4: significance number $c = 2$ (from Gaussian distribution table; two-sided statistical decision)

Result: only the effects B and C are significant (see Table 7-9).

Table 7-9. Significance test on the results of the factorial design

Effects		$z = \frac{\text{effect}}{0.32}$	$z > c$
A	$= -0.21$	0.66	no
B	$= 0.84$	2.6	yes, significant
AB	$= 0.41$	1.3	no
C	$= 0.91$	2.9	yes, significant
AC	$= 0.36$	1.1	no
BC	$= -0.19$	0.59	no
ABC	$= -0.41$	1.3	no

Plackett–Burman Plan [14, 22]

The Plackett–Burman plan, which is based on statistics and combinatorial analysis, allows $N-1$ effects of variables to be determined simultaneously in N tests. In this highly simplified test plan, only the main effects can be determined numerically; at the same time, error estimation is performed by means of a blank variable. Interactions between the variables can not be determined. A test matrix for seven variables is shown in Table 7-10.

Table 7-10. Plackett–Burman test plan with seven factors

Exp. no.	A	B	C	D	E	F	G
1	+	+	+	−	+	−	−
2	+	+	−	+	−	−	+
3	+	−	+	−	−	+	+
4	−	+	−	−	+	+	+
5	+	−	−	+	+	+	−
6	−	−	+	+	+	−	+
7	−	+	+	+	−	+	−
8	−	−	−	−	−	−	−

To calculate the effects the measurements are added or subtracted with the signs listed in the columns, and the sum is divided by four.

The blank effects should be zero. Usually the blank effects are regarded as the experimental error and are used to test the significance of the main effects. This is done by performing a t-test: the mean sum of squares of the blank effects is calculated as the variance (Eq. 7-15).

$$s^2 = \frac{\Sigma \text{ effects}^2 \text{ of the blank variables}}{\text{sum of blank variables}} \qquad (7\text{-}15)$$

The variance of an effect is determined by calculating the test quantity t = effect/s.

An example is the identification of the significant reaction parameters in the bis-hydroformylation of 1,3-pentadiene (Table 7-11). The results of the hydroformylation experiments are summarized in Table 7-12. The results were evaluated by the method described above (Table 7-13).

Table 7-11. Experimental parameters and reactions conditions

Variables	Levels	
	−	+
(A) Temperature [°C]	100	120
(B) Total pressure [bar]	100	150
(C) CO content synthesis gas [%]	30	70
(D) Catalyst quantity [HRh(CO)(PPh$_3$)$_3$] [mg]	50	200
(E) Solvent	ether	benzene
(F) Solvent quantity [mL]	20	50
(G) Blank variable	−	−

10.2 g 1,3-pentadiene (0.15 mol), 1 g PPh$_3$, 150 mL rocking autoclave

Table 7-12. Plackett–Burman plan for the hydroformylation of 1,3-pentadiene

Experiment	Reaction time [h]	Yield [%]	
		Monoaldehydes	Dialdehydes (target quantity)
1	18	53.3	8.4
2	3	43.2	23.4
3	22	43.8	6.5
4	11	48.3	23.1
5	6	51.6	6.8
6	7	36.6	2.5
7	9	63.3	12.9
8	13	49.6	4.0

The blank variable G also exhibited an effect and therefore could not be used for estimating the standard deviation. The reason for this is probably that the highly simplified test plan did not take any interaction effects into account.

Table 7-13. Evaluation of the experimental matrix (target quantity: dialdehyde yield)

	A	B	C	D	E	F	G
Column total	2.6	48.0	−27.0	3.60	−6.0	11.0	23.4
Effects (Σ/4)	0.65	12.0	−6.8	0.9	−1.5	2.8	5.9
t = Effects/s	0.7	13.3	−7.6	1.0	−1.7	3.1	—

Hence the standard deviation of $s = 0.9$ known from other test series was used for the significance test (t-test):

t-values (from statistics tables)

99%	$t = 63.7$
95%	$t = 12.7$
90%	$t = 6.3$

Only the variables B (total pressure) and C (CO content of synthesis gas), with degrees of confidence of 95 and 90%, respectively, are significant.

Experimental Optimization by the Simplex Method [16, 24, 25]

This simple search method allows multidimensional optimization to be carried out experimentally; the functional dependence of the target function on the individual parameters need not be known. The value of the target function (e.g., the yield of a product) is determined experimentally and is the criterium for deciding whether further search steps should be carried out or the procedure ended after successful optimization.

Procedure of the Simplex Method

At the start of the search $(n + 1)$ points (i.e., in two-dimensional space, three points) are fixed so that they form the corners of a regular simplex. In the example of Figure 7-9, this is an equilateral triangle. The value of the function is then determined for each of these points, and the search is then begun according to the following rules:

1) Determination of the target quantity at the corners of the simplex; $(n + 1)$ experiments
2) Selection of the "worst" corner $\vec{x}^{(...)}$
3) Generation of a new corner $\vec{x}_{n+...}$ by reflection of the triangle about the side opposite to the worst corner
4) Determination of the target quantity y in the new corner
5) Replace the result of the worst corner by y
6) If y is worse than all other results, go to point 7, otherwise point 2

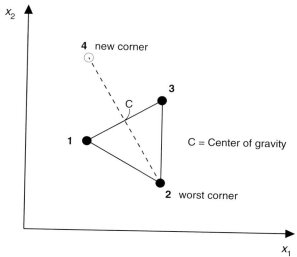

Fig. 7-9. Simplex method with two variables

7) Selection of the second worst corner as $\vec{x}^{(\cdots)}$, then go to point 3
8) The procedure is terminated when the target quantity can no longer be improved

$$\vec{x}_{n+2} = \frac{2}{n}\sum_{i=1}^{n+1}\vec{x}_i - \left(1+\frac{2}{n}\right)\vec{x}^{(1)} \quad \text{for the first reflected point} \quad (7\text{-}16)$$

$$\vec{x}_{n+k+1} = \left(1+\frac{2}{n}\right)\vec{x}_{n+k} - \left(1+\frac{2}{n}\right)\vec{x}^{(k)} + \vec{x}^{(k-1)} \quad \text{for the } k\text{th reflected point} \quad (7\text{-}17)$$

$\vec{x}$ = positional vector of the simplex corners
n = dimension of the factor space (number of variables to be investigated)
$\vec{x}^{(n)}$ = corner with the worst result

An example is the experimental optimization of the bis-hydroformylation of 2,4-hexadiene [12]. In the bis-hydroformylation of the diene with the catalyst [HRh(CO)(PPH$_3$)$_3$] under the usual reaction conditions, the highest yield of dialdehyde was 54%. A systematic simplex search with several significant parameters was carried out with the aim of improving this result. The three variables reaction temperature, total pressure, and H$_2$ content of the synthesis gas were chosen, and the search was begun with an unsymmetrical tetrahedron. The calculated points and the experimental results are listed in Table 7-14.

Reaction conditions:

8.2 g diene (0.1 mol), 80 mL benzene, 25 mg Rh$_2$O$_3$ (0.1 mmol), 1 g PPh$_3$ (co-catalyst); rocking autoclave.

Table 7-14. Bis-hydroformylation of 2,4-hexadiene: experimental optimization by the simplex method

Experiment	T [°C]	$P_{tot.}$ [bar]	H_2 [%]	Dialdehyde yield [%]	Notes
1	100	130	50	51.4*	
2	110	130	50	52.4**	
3	107	130	54	55.5***	
4	105	150	52	57.4	
5	114	143	55	58.0	1st reflected point
6	107	152	58.3	58.7	2nd reflected point
7	110	166.6	57.2	60.2	Optimum
8	115.3	157.7	62.7	59.0	

* Worst corner of 1st simplex. ** Worst corner of 2nd simplex. *** Worst corner of 3rd simplex.

When experiment 8 gave a worse result, the optimization process was terminated. It remains an open question whether with this catalyst system the yield of dialdehyde could be further increased by using higher pressures, other solvents, or more favorable reactor types (e. g., continuous operation).

Transferring these optimum experimental conditions to other dienes is not possible since the experimental parameters presumably depend on other factors such as the structure of the diene and substituents.

Statistical Test Planning with a Computer Program

Statistical test planning can be carried out advantageously by expert systems which design the test plan, evaluate the results, and optimize the process in a single logically constructed sequence. An example is the program APO (Analysis Process Optimization) [30].

The program
– prepares test plans on the basis of a given working hypothesis
– evaluates and assesses the experimental results
– calculates optimal parameter settings
– analyzes weaknesses in the model and systematic errors in the conduction of the process
– analyzes the influence of many parameters on the target quantities

APO presents the results in tabular and graphical form and provides detailed hints for the further development and improvement of the working hypothesis.

To develop a test plan the program only requires information on the independent variables. After input of the variables (experimental parameters) and their levels, the program calculates those points in a multidimensional space that provide the best predictions in a subsequent modeling process. The input levels of a variable determine the range of values, which should have been determined in preliminary investigations and catalyst screening. In general at least 3–5 levels per variable should be chosen. The maximum number of experiments is limited to 80 for nine independent variables.

As an example of the application of the program, we shall use the previously discussed example of the selective hydrogenation of a substituded o-cyano aromatic nitro compound with a supported Pd catalyst (see Section 7.3.1).

For hydrogenation in suspension, the following seven influencing quantities are of importance: temperature, pressure (H_2 partial pressure), type and quantity of catalyst, starting material concentration, stirring speed, and solvent.

On the basis of preliminary investigations, four of the seven variables were kept constant: starting material concentration (5%), stirring speed (710 rpm, kinetic region), the catalyst (5% Pd/activated carbon), and the solvent (ethanol). For the remaining three variables, the following ranges of values were chosen, whereby the reaction engineering conditions were also taken into account:

- Temperature: 0–80 °C, five levels
- Pressure (H_2 partial pressure): 1–40 bar, five levels
- Catalyst quantity (relative to starting material): 1–20%, four levels

The input menu of the program APO was then filled in as presented in Table 7-15.

Table 7-15. Input menu for producing the test plan

Details of the statistical test plan	Selection
(a) Special features of the model	none
(b) Take second-order effects into account	yes
(c) Number and names of independent variables	3
(d) Number of levels of the independent variables	5/5/4
(e) Number of restrictions	0
(f) Generation of restrictions	–
(g) Desired number of experiments	15
(h) Rating of edge zones	3
(i) Number of given experiments	0
(j) Number of randomly generated experiments	0
(k) Calculate test plan	

The program checks that the generated test plan fulfills certain mathematical criteria (e.g., correlations, homogeneity) and provides comments.

Table 7-16 lists the distributions of the experimental combinations in the entire variable space and the corresponding experimental results (percentage amine

Table 7-16. Hydrogenation of substituted o-cyanonitrobenzene: test plan and modeling by APO [13]

Exp.	Reaction conditions				Amine yield [%]		
	Temp. [°C]	H_2 pressure [bar]	Cat.-quantity [%]*	Time [min]	Experiment	Calculated by APO 1st model	2nd model
1	0	1	1	128	5.3	7.4	5.4
2	80	40	20	13	5.9	5.4	4.9
3	50	5	5	62	84.9	80.9	82.4
4	10	25	10	59	95.6	97.3	98.7
5	25	10	20	22	79.1	73.3	79.6
6	80	40	1	231	34.1	35.8	34.4
7	0	25	10	87	92.2	98.1	94.1
8	50	1	5	137	85.6	82.1	85.4
9	25	5	20	36	81.7	78.8	82.5
10	10	10	10	86	93.2	95.4	91.4
11	0	40	5	396	87.6	85.9	87.4
12	80	1	1	205	54.6	57.8	52.4
13	50	5	20	46	74.3	81.9	72.5
14	10	25	5	141	79.8	74.3	72.1
15	25	10	1	387	29.2	28.7	36.0
16	80	1	11,4	44	88.0	126.0	92.1
17	7	40	14,0	35	96.5	118.8	119.0

* Relative to mass of starting material.
Exp. 1–15: test plan according to APO; 16, 17: optimization.

yields). The experiments and the experimental results can also be plotted on the surfaces of the four geometric bodies depicted in Figure 7-10.

On closer inspection it can be seen that each plane of these geometric bodies is defined by at least three points. In the appropriate system correlation, these 15 experimental settings can cover the entire variable space for a parameter optimization procedure. The quasiorthogonal planning method ensures that various optimality criteria (e.g., edge-zone weighting) are taken into account.

The results of the hydrogenation experiments in the APO test plan and the values calculated from model equations are listed in Table 7-16, experiments 1–15. These results were then used for calculation and analysis of a model. APO provides a detailed commentary on the model analysis with the following statements:

– Standard error for the target quantity product: 5.25
– Error variance: 16.63% of the total variance of the target quantity
– Degree of determination: 98.42% (independence of all coefficients of the model equation)
– Normal distribution of the residuals: 70% probability

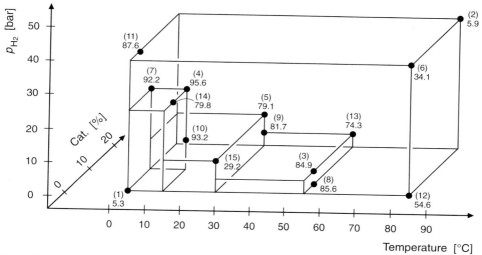

Fig. 7-10. Representation of the experiments and the results in the variable space

– Temperature: considerable error fluctuations in the variable
– Hydrogen partial pressure: neither systematic errors nor weaknesses in the model
– Catalyst: weakness in the model, which, however, is superimposed by considerably inhomogeneous variances

One should not attach too much importance to the not unexpected criticism regarding the variable temperature, since the maintainance of the levels in the highly exothermic reaction in an autoclave can sometimes be problematic. The model weakness catalyst may be due to inhomogeneous distribution in the autoclave.

Next a model analysis was carried with the aim of determining the optimum experimental parameters from the model-space representation. In order to rapidly obtain an overview, the isoline representation was chosen. Here contour lines are used to depict the calculated product yield as a function of the corresponding combination of adjustable parameters (e. g., pressure and catalyst quantity) with one constant quantity (temperature); see Figure 7-11. The program calculates the coordinates of the extreme values of the target parameters, i.e., the optimum.

Numerous isoline diagrams revealed that the optimum experimental conditions are low temperature (<20 °C), 10–14% catalyst, and high pressure (>25 bar). There is also a secondary local maximum at low pressure (1 bar) and higher temperature (80 °C). Simultaneous optimization by model-space analysis gave the eperimental settings: 80 °C, 1 bar H_2 partial pressure, 11.4% catalyst.

However, in the experiment with these settings, only 88% yield was determined, i.e., the optimum was not attained (experiment 16). In order to make use of this data, the result was added to the previous 15 experiments as a given experiment. Repeating the model analysis now led to an improvement in the model.

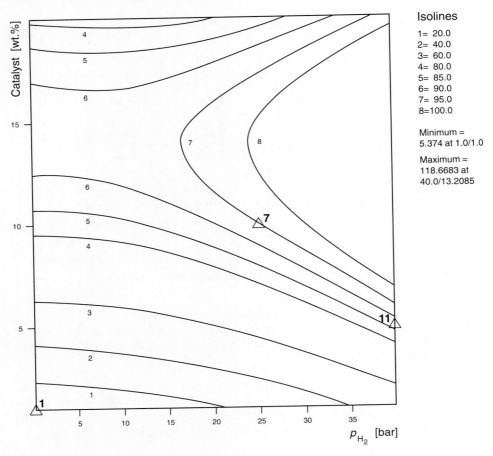

Fig. 7-11. Contour line depiction of the target quantity product at 0 °C (2nd APO model)

As can be seen from Table 7-16, the values of the target parameters calculated from the second model equation describe the experimental results more exactly. The percentage effects of the variables after the extension of the test plan were as follows:

Temperature	$X(1) = 20.3\%$
H_2 pressure	$X(2) = 22.8\%$
Catalyst quantity	$X(3) = 56.8\%$

The variables temperature and pressure have almost the same weighting in the calculation of the target quantity. The repeated simultaneous optimization gave the following optimal settings: 7.0 °C, 40.0 bar H_2 partial pressure, 14.0% catalyst. For these parameters APO calculated a target quantity value of 119 (no upper limit). The experiment (no. 17) gave a reproducible amine yield of 96.5%. The optimization procedure was ended with this very good result. The clear improvement in the mathematical model resulting from the additional empirical value can also be seen in the isoline diagram at 0 °C (Figure 7-11). All investigated results lie within the ranges calculated with the model equations.

It would seem logical to use such programs in the field of heterogeneous catalysis in particular. Numerous influencing quantities can be taken into account without knowledge of the generally complex kinetics. Experimental optimization on the basis of statistical modeling replaces the usual intuitive approach and leads to the goal with a minimum of effort. However, this does not mean that the experience and creativity of the experts can be dispensed with.

7.3.4 Kinetic Modeling and Simulation [5, 18]

This chapter describes how a catalytic reaction can be modeled on the basis of kinetic measurements and how the resulting model equations can be used in reactor simulation and design (see also Scheme 7-6). This detailed analysis of the catalytic behavior requires a major measurement and data-evaluation effort and is rarely carried out in industrial practice.

We shall discuss the process for the example of the hydrogenation of benzaldehyde in various reactors [15]. The heterogeneously catalyzed hydrogenation of benzaldehyde is a model reaction for the hydrogenation of aromatic aldehydes. The main reactions are shown in Equation 7-18.

$$\text{PhCHO} + H_2 \longrightarrow \text{PhCH}_2\text{OH} \xrightarrow{+H_2} \text{PhCH}_3 + H_2O \tag{7-18}$$

Supported Pd/C catalysts, Raney nickel, and nickel boride are good catalysts for the hydrogenation of benzaldehyde. By measuring the take up of hydrogen in a batch reactor, it was found that the reaction is zero order in the reactants benzaldehyde and hydrogen at pressures above 3 bar and aldehyde concentrations in excess of 1 mol/L. With the catalyst 3% Pd/C a reaction rate of 1.6×10^{-2} mol g^{-1} min^{-1} was measured at 22 °C and was independent of the solvent [2].

Other authors carried out measurements with Raney nickel at 70 °C and 6 bar and found that the reaction rate was strongly dependent on the reactant/catalyst ratio, the following range being given:

$$r = 1.7 \times 10^{-4} \text{ to } 1.3 \times 10^{-3} \text{ mol g}^{-1} \text{min}^{-1}$$

No statements were made about the selectivity of product formation. In a more recent study kinetic measurements were made in a suspension process carried out in an autoclave operating in the batch mode, and the results were used for the simulation of a trickle-bed reactor [15].

The kinetic study was carried out under the following reaction conditions:

Raney nickel (Engelhard): 68% Ni, specific surface area 130 m²/g, pellet density 1.72 g/cm³, porosity 0.67, particle size range 35–70 μm
Catalyst concentration: 25–50 kg/m³
Temperature: 343–373 K
Pressure: 1.7–11.2 bar
300 mL stirred autoclave, batch operation

Figure 7-12 shows the typical course of a hydrogenation reaction.

The benzaldehyde concentration is plotted as a function of reaction time. Above 1.5 mol/L there is a linear dependence that apparently reflects zero reaction order with respect to the aldehyde. At lower concentrations the reaction is apparently first order. The kinetic data were only determined in the region of zero reaction order at a stirrer speed of about 2000 rpm. Measurements with stirring speeds in the range 1200–2000 rpm gave constant reaction rates, which means that gas–liquid mass transfer is negligible.

A linear relationship was also found between the reaction rate and the catalyst concentration. Numerous measurements with smaller catalyst particle sizes (10 μm) gave comparable reaction rates. This means that mass transfer within the particle also plays no role in the reaction. An activation energy of the hydrogenation reaction of 55.4 kJ/mol (4.5 bar, 2.5% catalyst) was measured.

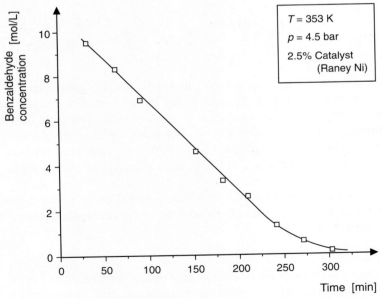

Fig. 7-12. Kinetic study of the hydrogenation of benzaldehyde in a stirred autoclave operating in suspension mode [15] (with permission of Elsevier, Amsterdam)

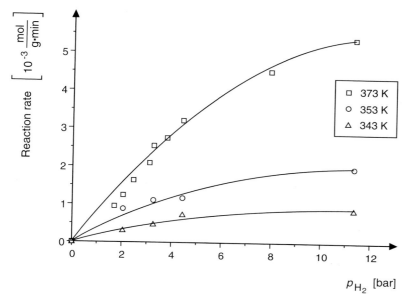

Fig. 7-13. Hydrogenation of benzaldehyde in a stirred autoclave: dependence of reaction rate on H$_2$ pressure [15] (with permission of Elsevier, Amsterdam)

The influence of the hydrogen concentration on the reaction rate was investigated at various hydrogen pressures (Fig. 7-13).

Futhermore, it was found that the benzyl alcohol formed in the reaction has no influence on the reaction rate. From all the experimental results, a Langmuir–Hinshelwood model was deduced, with a rate-determining influence of the surface reaction according to Equation 7-19.

$$r = \frac{k K_B c_B K_H p_H}{1 + K_B c_B (1 + \sqrt{K_H p_H})^2} \tag{7-19}$$

c_B = benzaldehyde concentration
p_H = hydrogen partial pressure
K_B, K_H = adsorption constants of benzaldehyde and hydrogen

In the investigated range, K_B is on the order of 1 L/mol, and this results in a reaction order of zero at high benzaldehyde concentrations (Eq. 7-20).

$$r = k \frac{K_H p_H}{(1 + \sqrt{K_H p_H})^2} = k_0 \tag{7-20}$$

This model was used to analyze the reaction rate data (Fig. 7-13). The following temperature dependences were found for k and K_H:

$k = 2.18 \times 10^8 \exp(-10000/T)$ kmol kg^{-1} s^{-1}
$K_H = 1.85 \times 10^{-10} \exp(5500/T)$ kPa^{-1}

As expected, k increases with increasing temperature while K_H decreases. The conditions of the trickle-bed reactor study are summarized in the following:

– Discontinuously operated trickle-bed reactor with recycle, 1 inch diameter
– Liquid flow rate: 0.004 m/s
– Gas flow rate: 0.004–0.008 m/s
– Temperature: 353–373 °C
– Pressure: 2.2–5.8 bar
– 170 g catalyst and 1 L liquid

The limiting factor in this reactor is the hydrogen, and this must be taken into account in all mass-transfer resistances. Hydrogen transfer from the gas into the liquid, onto the surface of the pellet, and into the interior of the pellet is the same provided the pellet is completely wetted. The total process can then be described by Equation 7-21 [11].

$$k_L a_L (p_H/H - c_{H,L}) = k_S a_S (c_{H,L} - c_{H,S}) = \eta \frac{k K_H H c_{H,S}}{(1 + \sqrt{K_H H c_{H,S}})^2} = r_0 \quad (7\text{-}21)$$

$c_{H,L}$ = H$_2$ concentration in the liquid
$c_{H,S}$ = H$_2$ concentration on the outer catalyst particle surface
H = Henry's constant
$k_L a_L$ = gas–liquid mass-transfer coefficient
$k_S a_S$ = liquid–solid mass-transfer coefficient
η = catalyst effectiveness factor (function of the Thiele modulus)

Details of the calculation can be found in the literature [15, 23].
For the calculation of r_0 the effective diffusion coefficient D_{eff} is required (Eq. 7-22).

$$D_{eff} = \frac{\varepsilon_p D_H}{\tau} \quad (7\text{-}22)$$

ε_p = pellet porosity
D_H = diffusion coefficient of hydrogen
τ = tortuosity factor

Table 7-17 compares the reaction rates measured in the trickle-bed reactor with those predicted by the model calculation. The predicted values are in good agreement with the experimental data. The effectiveness factor of the catalyst is very low, and this means that transport resistance in the pores is highly significant. As expected, gas–liquid transport resistance is also very important, but liquid–solid transport resistance is negligible.

Table 7-17. Hydrogenation of benzaldehyde in a trickle-bed reactor: measured values and model calculations [15]

T [K]	p [kPa]	$-\Delta c/\Delta t$ [kmol m^{-3} s^{-1}] measured	$-\Delta c/\Delta t$ [kmol m^{-3} s^{-1}] calculated	η calculated
353	360	7.8×10^{-2}	8.2×10^{-2}	0.042
	580	1.3×10^{-1}	1.1×10^{-1}	0.041
373	220	8.8×10^{-2}	9.0×10^{-2}	0.043
	360	1.5×10^{-1}	1.4×10^{-1}	0.041
	580	2.5×10^{-1}	2.2×10^{-1}	0.040

Parameter values: $k_L a_L = 0.12$ s^{-1}, $k_S a_S = 0.70$ s^{-1}, $D_H = 8 \times 10^{-9}$ m^2/s, $\tau = 3$, $H = 2.3 \times 10^4$ kPa kmol^{-1} m^3

The most important result is that the reaction is about 50 times faster in suspension than in the trickle-bed reactor. Therefore, at such high reaction rates the suspension reactor is preferred, although gas–liquid mass transfer and separation of the catalyst can cause problems.

Thus, benzaldehyde hydrogenation was tested under practice-relevant conditions in a catalyst test reactor of simple design, and parameter studies were carried out. The construction of the laboratory plant is shown schematically in Figure 7-14. Since we are dealing with an integral reactor, in spite of the relatively small amount of catalyst in the trickle-bed reactor, only comparitive measurements were carried out.

Continuous hydrogenation of benzaldehyde in the solvents hexane and isopropanol:

– Reactor: Catatest plant (VINCI technologies, Fachhochschule Mannheim)
– Substrate concentration: 10% benzaldehyde
– Throughput: 0.125 L/h benzaldehyde solution
– Reaction conditions:
   25 bar, molar ratio H$_2$/aldehyde = 40/1 (isopropanol)
   15 bar, molar ratio H$_2$/aldehyde = 20/1 (hexane)
– Catalyst: 13.6 g of 0.3% Pd/Al$_2$O$_3$ (HO-22, BASF)

The Catatest plant allows the reaction parameters pressure, temperature, and liquid and gas feed to be varied over wide ranges; Figure 7-15 shows just a few results. The reaction products in both test series were benzyl alcohol and toluene. Considerable influence of the solvent and the temperature on the product distribution and the conversion were found.

In the polar solvent isopropanol, benzyl alcohol is predominantly formed at low temperatures, and the amount of toluene formed increases continuously with increasing temperature. In contrast, in the nonpolar solvent hexane, toluene is the predominant final product of the hydrogenation, in spite of the small excess of hydrogen and the low pressure. Between 120 and 130 °C the selectivity with respect

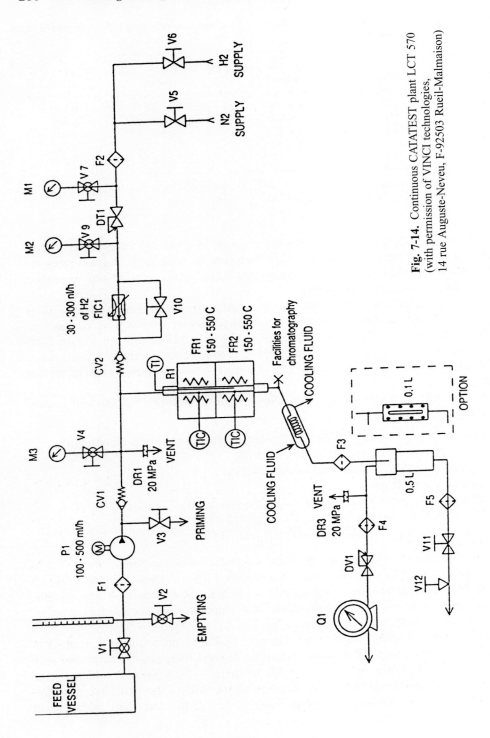

Fig. 7-14. Continuous CATATEST plant LCT 570 (with permission of VINCI technologies, 14 rue Auguste-Neveu, F-92503 Rueil-Malmaison)

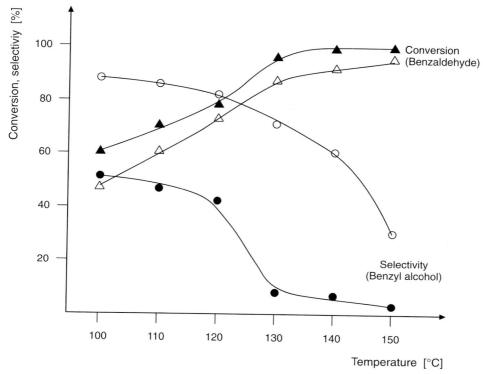

Fig. 7-15. Continuous hydrogenation of benzaldehyde: influence of temperature and solvent on conversion and selectivity
△ ○ Isopropanol, $p = 25$ bar, H_2/aldehyde = 40/1
▲ ● Hexane, $p = 15$ bar, H_2/aldehyde = 20/1

to benzyl alcohol decreases drastically, and at 150 °C in hexane, exclusively toluene is obtained with quantitative conversion (Fig. 7-15).

Another practical example is the modeling of a trickle-bed reactor [11, 23]. The reaction investigated was the high-pressure hydrogenation of a lactone to a diol on a Cu–Zn mixed oxide catalyst [19].

Initially the kinetics were investigated in a stirred autoclave in order to develop a microkinetic model. Also of interest were the fluid-dynamic conditions and the axial dispersion (residence-time behavior). Here we shall only deal with the measured residence-time distributions.

As can be seen from Figure 7-16, the residence-time distributions of the trickle-bed reactor, as determined by pulse injection, showed that considerable backmixing takes place in the reactor. As expected, this decreases with increasing liquid feed. The external holdup was calculated from the residence-time curves.

A simulation was carried out to determine to what extent the conversion in the reactor is influenced by:

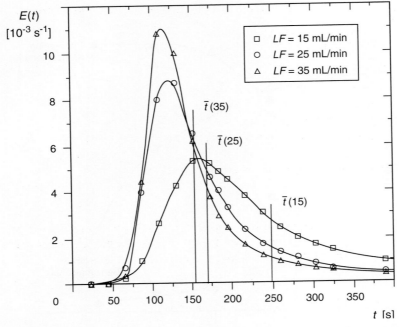

Fig. 7-16. Residence-time spectra in a trickle-bed reactor as a function of liquid flow (LF) [19]. Reactor length 1 m, diameter, 25.4 mm, Cu/Zn mixed oxide catalyst (tablets 6 × 3 mm), liquid phase *tert*-butanol, gas flow 10 L/min, 100 bar H_2 pressure, 25 °C

- The microkinetics
- Liquid feed and holdup
- Mass transfer: film diffusion of hydrogen and of the substrate

The ideal plug-flow model has to be corrected when applied to a trickle-bed reactor, since the conversion in the reactor is not determined by the reaction rate per unit mass but by a corrected value relative to the void volume in the reactor that is occupied by liquid.

The application of the reaction rate from the suspension reactor would only be justified if the catalyst in the fixed bed were as completely wetted by liquid as in the stirred autoclave. A semi-empirical model was used to estimate the conversion (Eq. 7-23)

$$\frac{dc}{dz} = -r(H_{ex}/LF)\rho_K A \qquad (7\text{-}23)$$

LF = liquid feed, mL/min
H_{ex} = external holdup
$\rho_{cat.}$ = pellet density of the catalyst
A = cross-sectional area of the reactor
z = tube length (independent variable)

The ratio H_{ex}/LF is the effective mean residence time of the liquid in the reactor. The reaction rate is determined by three simultaneous processes:

1) Diffusion of the substrate molecules to the catalyst surface

$$-r = k_S a_S (c_{S,L} - c_{S,S}) \qquad (7\text{-}24)$$

k_S = liquid–solid mass-transfer coefficient, m/s
a_S = specific surface area, m²/kg
$c_{S,L}$ = substrate concentration in the liquid
$c_{S,S}$ = substrate concentration on the catalyst surface

2) Diffusion of hydrogen from the gas phase to the catalyst

$$-r = k_{tot} a_S (c_{H,L} - c_{H,S}) \qquad (7\text{-}25)$$

k_{tot} = total transfer coefficient for H_2
$c_{H,L}$ = the equilibrium concentration of hydrogen in the liquid phase corresponding to a given H_2 partial pressure. It is related to the gas-phase concentration by the modified Henry's law (Eq. 7-26).

$$c_{H,G} = H_m c_{H,L} \qquad (7\text{-}26)$$

H_m = modified Henry constant

3) Surface reaction (microkinetics). A five-parameter Langmuir–Hinshelwood model proved to be best suited for describing the kinetics (Eq. 7-27). The criteria were:

- Surface reaction is rate-determining
- Selective adsorption
- Adsorption and desorption equilibrium attained
- Molecular adsorption

$$-r = \frac{k[K_H K_S c_{H,S} c_{S,S} - c_{D,S}/(K_D K_R)]}{(1 + K_S c_{S,S} + c_{D,S}/K_D)(1 + K_H c_{H,S})} \qquad (7\text{-}27)$$

k = rate constant of the reaction
K_H = adsorption constant of hydrogen
K_S = adsorption constant of the substrate
K_D = desorption constant of the diol (product)
K_R = equilibrium constant of the reaction

To calculate the diol concentration, the simplifying assumption of selective hydrogenation was made (Eq. 7-28).

$$c_{D,S} = c_{S,0} X \qquad (7\text{-}28)$$

Since r is contained in two nonlinear expressions, a global equation for the kinetics can not be derived. The equations can, however, be solved simultaneously by dynamic modeling with the program ISIM [18]. The corresponding ISIM program is shown in Figure 7-17.

All calculations were made for a reactor operating at 200 bar and 150 °C under the assumption of isothermal conditions, since the adiabatic temperature difference under the given concentration conditions was only a few degrees and the reactor was operated in a polytropic mode. The target quantity was the conversion,

```
 1 constant ksas = 0.588          : Transport coefficient substrate
 2 constant kgas = 0.04           : Transport coefficient hydrogen
 3 constant k = 0.017713          : Rate constant
 4 constant Kh = 0.411495         : Adsorption constant hydrogen
 5 constant Ks = 55.3381          : Adsorption constant substrate
 6 constant Kd = 0.073762         : Desorption constant product
 7 constant Kr = 3.74371          : Equilibrium constant of the reaction
 8 constant Dk = 2.56             : Catalyst density
 9 constant A = 5.06              : Cross-sectional area of reactor
10 constant LF = 10               : Liquid feed in mL/min
11 constant Hex = 0.2             : External holdup
12 :
13 cint = 1
14 tfin = 100
15 ca = 0.12382                   : Initial concentration substrate
16 chO = 0.960                    : Equilibrium concentration H₂ in the
17                                : Liquid phase at 150° and 200 bar
18 :
19 DO 1 LF = 10,15,5
20 Reset
21 1 sim
22 :
23 initial
24 c = ca
25 cs = 0
26 zfin = tfin
27 :
28 dynamic
29 Hex = 3.9065E−02 + 1.4157E−02*LF−3.25E−04*LF*LF
30 ch = chO + (r/kgas)
31 cs = c + (r/ksas)
32 cd = ca*U/100
33 N = (1 + Ks*cs + cd/Kd)*(1 + Kh*ch)
34 r = −k*(Kh*Ks*ch*cs−cd/(Kd*Kr))/N
35 c' = r*Dk*(A/LF)*Hex
36 U = 100*(1−(c/ca))
37 z = t
38 prepare z, U, r, cs, cd, ch
```

Fig. 7-17. ISIM program for the simulation of a trickle-bed reactor for the high-pressure hydrogenation of a lactone [19]

Influence of Liquid Feed and Holdup

In these simulation runs, the liquid feed LF was varied between 10 and 25 mL/min (Fig. 7-18). The substrate concentration was 5%, and the dimensions of the reactor in the program correspond to those of the test reactor.

In the first simulation (model **A**) the liquid holdup was regarded as a constant (relative to $LF = 10$ mL/min). In a second calculation (model **B**), the external holdup was input as a function of the liquid feed. From the residence time measurements in the trickle-bed reactor, the correlation of Equation 7-29 was obtained.

$$H_{ex} = 3{,}91 \exp(-2) + 1.42 \exp(-2LF) - 3.25 \exp(-4LF^2) \tag{7-29}$$

This equation was used as input for the DYNAMIC part of the program. The conversion profiles were then calculated by simulation (Figure 7-19).

Table 7-18 lists the results of both simulations and the experimentally determined conversions. The results obtained by model **B** with variable holdup agree quite well with the experimental data.

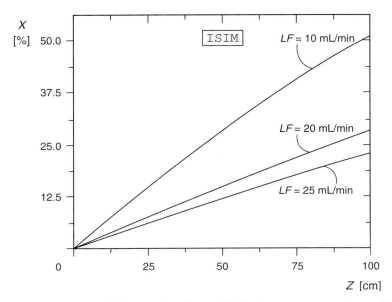

Fig. 7-18. Lactone hydrogenation in a trickle-bed reactor: conversion profiles as a function of liquid flow at constant liquid holdup [19]

292   7 Planning, Development, and Testing of Catalysts

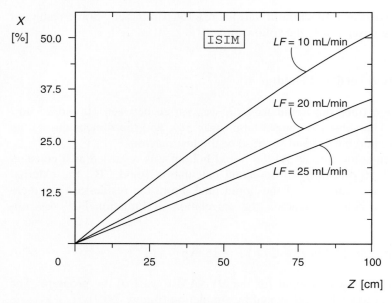

Fig. 7-19. Lactone hydrogenation in a trickle-bed reactor: conversion profiles taking into account the external liquid holdup [19]

Table 7-18. Measured and calculated conversion in the trickle-bed reactor; influence of the external holdup

Liquid feed LF [mL/min]	Conversion [%]		
	Experimental	Simulation **A** (H_{ex} = const.)	Simulation **B** (H_{ex} = f(LF))
10	49.3	50.6	50.6
15	42.3	36.5	42.7
20	33.9	28.4	35.8
25	28.0	23.2	29.1
30	22.5	19.6	22.5

Constant reaction conditions:
p_{H_2} = 200 bar, 150 °C, Cu/Zn mixed oxide catalyst

Influence of the Reactor Length

This calculation was intended to show how the conversion changes in the same reactor with increasing tube length and various liquid feeds (Fig. 7-20). As Figure 7-20 shows, tube lengths in the rage of about 2–3 m give satisfactory yields.

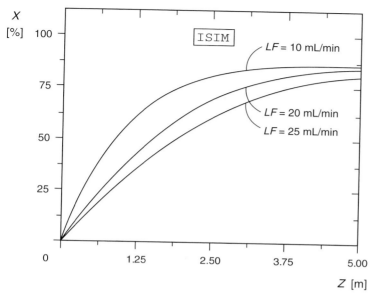

Fig. 7-20. Lactone hydrogenation in a trickle-bed reactor: conversion profiles as a function of liquid flow and reactor length [19]

Influence of Mass Transfer

The calculated mass-transfer coefficients used in the program show that the transport resistance for hydrogen is higher than that for the substrate. Therefore, in the following simulation calculations, a high value and a much lower, but nevertheless realistic value of the global mass-transfer coefficient were introduced into Equation 7-25. The aim was to find out whether the reaction rate is mainly determined by mass transfer or by the surface reaction. As can be seen from Figures 7-21 and 7-22, the process is essentially determined by the intrinsic kinetics.

An assumed value of $k_{tot} a_s = 0.001$ corresponds to about one-tenth of the true value in the lactone hydrogenation, and would hence correspond to an extremely high mass-transfer limitation. Figure 7-22 shows that for low mass-transfer limitation ($k_{tot} a_s = 1.0$) the reaction rate increases in the entry region of the reactor. This is because rapid diffusion of the hydrogen increases its concentration on the catalyst surface up to saturation. The associated high consumption of substrate leads to a steeper gradient of the curve, whereby at a reactor length of ca. 1.3 m, the substrate supply becomes limiting, and the reaction rate even drops below the curve with higher hydrogen-transport limitation. The results suggest that for an initial estimate of the conversion, the relationships for substrate diffusion and hydrogen transport from the gas phase can be neglected.

The simulation results show that in addition to precise microkinetics, knowledge of the fluid dynamics, especially the liquid holdup, is an essential prerequi-

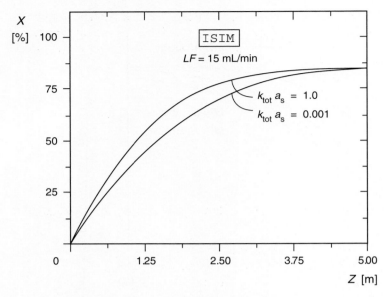

Fig. 7-21. Lactone hydrogenation in a trickle-bed reactor: conversion profiles as a function of the global mass-transfer coefficient of hydrogen [19]

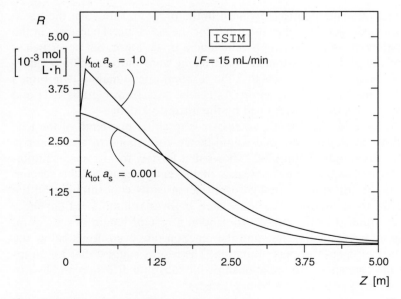

Fig. 7-22. Lactone hydrogenation in a trickle-bed reactor: reaction rate profiles as a function of mass transfer [19]

site for modeling a trickle-bed reactor. The advantage of a simulation program is not so much the calculation of the conversion for a concrete situation; rather, it is the splitting of a complex problem into individual steps, which allows parameter studies to be carried out [10].

Exercises for Chapter 7

Exercise 7.1

Decide whether the following statements are true (t) or false (f). A differential circulating reactor:

☐ Should be operated with a low recycle ratio
☐ Has a residence time spectrum like a plug flow reactor
☐ Has a residence time spectrum like a continuous stirred tank
☐ Exhibits temperature and pressure gradients
☐ Provides a product stream with high conversion

Exercise 7.2

A batch hydrogenation depends on three parameters. Set up a test matrix for a 2^3 factorial design with the following conditions:

Variables	Levels	
Reaction time [min]	30	45
Temperature [°C]	50	80
Cat. concentration [%]	0.25	0.40

Exercise 7.3

The influence of reaction time, temperature, and catalyst concentration on the yield of a chemical reaction was investigated in a 2^3 factorial design. The following table lists the eight experiments that were carried out together with the yields achieved:

Experiment	A [min]	B [°C]	C [%]	Yield [%]
(1)	20	55	0.1	15
a	30	55	0.1	18
b	20	65	0.1	20
ab	30	65	0.1	26
c	20	55	0.2	11
ac	30	55	0.2	29
bc	20	65	0.2	23
abc	30	65	0.2	33

a) Calculate the effects and their interactions.
How can the result be interpreted?

b) The standard deviation of the target quantity is known from earlier tests: $\sigma = 4.2$. Which effects and interactions are significant (confidence level 95%, $c = 2.0$)?

Exercise 7.4

In a hydrogenation process operated in suspension mode, the influence of five variables on the yield of product was of interest. The five variables and their levels were:

Variable		Low level (−)	High level (+)
A	Temperature [°C]	−18	0
B	Starting material conc. [%]	5.0	10.0
C	Cat. conc. [%]	0.2	0.4
D	Purity of solvent	tech.	chem. pure
E	Blank variable	—	—
F	Stirring speed [min^{-1}]	100	500
G	Blank variable	—	—

The experiments were carried out according to a Plackett–Burman plan. The results are summarized in the following matrix:

Experiment	A Temp.	B St. M.	C Cat.	D Solv.	E –	F Stirrer	G –	Product yield
1	+	+	+	–	+	–	–	0.315
2	+	+	–	+	–	–	+	0.336
3	+	–	+	–	–	+	+	0.330
4	–	+	–	–	+	+	+	0.464
5	+	–	–	+	+	+	–	0.322
6	–	–	+	+	+	–	+	0.507
7	–	+	+	+	–	+	–	0.482
8	–	–	–	–	–	–	–	0.463

Determine the significance of the effects by perfoming a *t*-test. The number of degrees of freedom corresponds to the number of blind variables, in this case $n = 2$.

Probability	Two-sided test
80%	$t = 1.89$
95%	4.30
99%	9.9

Exercise 7.5

In an experimental optimization by the simplex method, the yield of a process was determined as a function of two variables x_1 and x_2. The following table shows the results of the initial simplex with three experiments:

Experiment	x_1	x_2	Yield [%]
1	37	21.0	40.0
2	39	21.0	41.0
3 $\vec{x}_{n+1}$	38	22.8	41.4

a) Calculate the coordinates of the next optimization experiment 4.
b) Experiment 4 led to a considerably improved yield of 42.4%. Under which reaction conditions should the fifth experiment be performed?

8 Heterogeneously Catalyzed Processes in Industry

8.1 Overview [T23, T31, T41]

Modern industrial chemistry is based on catalytic processes. Heterogeneous catalysts are used on a large scale in the following areas:

- Production of organic and inorganic chemicals
- Crude oil refining and petrochemistry
- Environmental protection
- Energy conversion processes

We will now give an overview of the most important catalytic processes and the process conditions.

Production of Inorganic Chemicals [19]

The production of hydrogen and synthesis gas mixtures (CO/H_2) from methane and higher hydrocarbons by steam reforming involves numerous reaction steps with different catalysts. The synthesis of ammonia and the oxidation of SO_2 to SO_3 are long-known equilibrium reactions in which the target product is removed from the product stream and the unchanged starting material is recycled.

The oxidation of ammonia to nitrous gases is a fast high-temperature reaction for the production of nitric acid (Ostwald process). The Claus process is an important petrochemical process for obtaining sulfur from H_2S, which results from the desulfurization of petroleum and natural gas (hydrodesulfurization). One-third of the H_2S is combusted to SO_2, which reacts with the remaining H_2S (see Table 8-1).

Production of Organic Chemicals [6, 13]

Heterogeneous catalysts are used on a large scale in the production of organic chemicals. The processes can be classified according to reaction type (Table 8-2).

Catalytic hydrogenations are preferably carried out with metal catalysts based on Ni, Co, Pd, or Pt. In selective hydrogenations, undesired side reactions such as double bond isomerization or *cis/trans* isomerization in fat-hardening processes must be avoided. The hydrogenation of CO to methanol is of major industrial importance.

Table 8-1. Heterogeneous catalysis for the production of industrial gases and inorganic chemicals [T41]

Process or product	Catalyst (main components)	Conditions
Steam reforming of methane $H_2O + CH_4 \rightarrow 3\,H_2 + CO$	Ni/Al$_2$O$_3$	750–950 °C, 30–35 bar
CO conversion $CO + H_2O \rightleftharpoons H_2 + CO_2$	Fe/Cr oxides Cu/Zn oxides	350–450 °C 140–260 °C
Methanization (SNG) $CO + 3\,H_2 \rightarrow CH_4 + H_2O$	Ni/Al$_2$O$_3$	500–700 °C, 20–40 bar
Ammonia synthesis	Fe$_3$O$_4$ (K$_2$O, Al$_2$O$_3$)	450–500 °C, 250–400 bar
Oxidation of SO$_2$ to SO$_3$	V$_2$O$_5$/support	400–500 °C
Oxidation of NH$_3$ to NO (nitric acid)	Pt/Rh nets	ca. 900 °C
Claus process (sulfur) $2\,H_2S + SO_2 \rightarrow 3\,S + 2\,H_2O$	bauxite, Al$_2$O$_3$	300–350 °C

Dehydrogenation is the reverse reaction of hydrogenation. It is preferably carried out with metal oxide catalysts, but metal catalysts are also used at low temperatures since they favor the hydrogenolysis of C–C bonds.

In *oxidation* processes heterogeneous catalysts are mainly used in gas-phase processes. In the oxidation of ethylene to ethylene oxide, supported silver catalysts are used; in the other examples, reducible metal oxide catalysts are used. In *ammoxidation* nitriles and HCN are obtained by using NH$_3$/O$_2$ mixtures. The *oxychlorination* of ethylene with HCl/O$_2$ is used for the production of vinyl chloride.

Acid catalysts are mainly used in alkylation processes, but also for hydration, dehydration, and condensation reactions. In olefin reactions, heterogeneous catalysts are mainly employed for metathesis reactions and the production of polymers.

Refinery Processes

In crude oil processing, catalytic processes are used to produce products such as gasoline, diesel, kerosene, heating oil, aromatic compounds, and liquefied petroleum gas (LPG) in high yield and good quality.

Bifunctional catalysts with acidid and metallic components are used in reforming, hydrocracking, and isomerization; acid catalysts in cracking; and supported metal oxide/sulfide catalysts in hydrorefining for the removal of S, N, and O (hydrofining, hydrotreating). The most important processes in refinery technology are listed in Table 8-3.

Table 8-2. Heterogeneously catalyzed processes for the production of organic chemicals [T41]

Process or product	Catalyst	Conditions
Hydrogenation		
Methanol synthesis $CO + 2H_2 \rightarrow CH_3OH$	$ZnO-Cr_2O_3$ $CuO-ZnO-Cr_2O_3$	250–400 °C, 200–300 bar 230–280 °C, 60 bar
Fat hardening	Ni/Cu	150–200 °C, 5–15 bar
Benzene to cyclohexane	Raney Ni noble metals	liquid phase 200–225 °C, 50 bar gas phase 400 °C, 25–30 bar
Aldehydes and ketones to alcohols	Ni, Cu, Pt	100–150 °C, bis 30 bar
Esters to alcohols	$CuCr_2O_4$	250–300 °C, 250–500 bar
Nitriles to amines	Co or Ni on Al_2O_3	100–200 °C, 200–400 bar
Dehydrogenation		
Ethylbenzene to styrene	Fe_3O_4 (Cr, K oxide)	500–600 °C, 1.4 bar
Butane to butadiene	Cr_2O_3/Al_2O_3	500–600 °C, 1 bar
Oxidation		
Ethylene to ethylene oxide	Ag/support	200–250 °C, 10–22 bar
Methanol to formaldehyde	Ag cryst.	ca. 600 °C
Benzene or butene to maleic anhydride	V_2O_5/support	400–450 °C, 1–2 bar
o-Xylene or naphthalene to phthalic anhydride	V_2O_5/TiO_2 $V_2O_5-K_2S_2O_7/SiO_2$	400–450 °C, 1.2 bar
Propene to acrolein	Bi/Mo oxides	350–450 °C, 1.5 bar
Ammoxidation		
Propene to acrylonitrile	Bi molybdate (U, Sb oxides)	400–450 °C, 10–30 bar
Methane to HCN	Pt/Rh nets	800–1400 °C, 1 bar
Oxychlorination		
Vinyl chloride from ethylene + HCl/O_2	$CuCl_2/Al_2O_3$	200–240 °C, 2–5 bar
Alkylation		
Cumene from benzene and propene	H_3PO_4/SiO_2	300 °C, 40–60 bar
Ethylbenzene from benzene and ethylene	Al_2O_3/SiO_2 or H_3PO_4/SiO_2	300 °C, 40–60 bar
Olefin reactions		
Polymerization of ethene (polyethylene)	Cr_2O_3/MoO_3 Cr_2O_3/SiO_2	50–150 °C, 20–80 bar

Table 8-3. Heterogeneously catalyzed processes in refinery technology [T41]

Process or product	Catalyst	Conditions
Cracking of kerosene and residues of atmospheric crude oil distillation to produce gasoline	Al_2O_3/SiO_2 zeolites	500–550 °C, 1–20 bar
Hydrocracking of vacuum distillates to produce gasoline and other fuels	$MoO_3/CoO/Al_2O_3$ Ni/SiO_2-Al_2O_3 Pd zeolites	320–420 °C, 100–200 bar
Hydrodesulfurization of crude oil fractions	$NiS/WS_2/Al_2O_3$ $CoS/MoS_2/Al_2O_3$	300–450 °C, 100 bar H_2
Catalytic reforming of naphtha (high-octane gasoline, aromatics, LPG)	Pt/Al_2O_3 bimetal/Al_2O_3	470–530 °C, 13–40 bar H_2
Isomerization of light gasoline (alkanes) and of m-xylene to o/p-xylene	Pt/Al_2O_3 $Pt/Al_2O_3/SiO_2$	400–500 °C, 20–40 bar
Demethylation of toluene to benzene	MoO_3/Al_2O_3	500–600 °C, 20–40 bar
Disproportionation of toluene to benzene and xylenes	$Pt/Al_2O_3/SiO_2$	420–550 °C, 5–30 bar
Oligomerization of olefins to produce gasoline	H_3PO_4/kieselguhr H_3PO_4/activated carbon	200–240 °C, 20–60 bar

Catalysts in Environmental Protection [7, 9]

As early as the 1940s, supported Pt/Al_2O_3 catalysts were used in the USA for the catalytic purification of off-gases by oxidation. In 1975 the purification of automobile exhaust emissions became required by law, and similar laws were later introduced in western Europe and Japan. The catalytic converters are monolithic supports coated with platinum and other noble metals (Fig. 8-1). Of major importance is the catalytic purification of the flue gases from power stations, in which the nitrous gases are converted to nitrogen and water by treatment with ammonia. Heterogeneous catalysts also have numerous applications in the catalytic afterburning of impurities or odoriferous components in industrial off-gases. Table 8-4 lists some examples.

Catalytic afterburning can solve various emission problems without generating secondary pollutants. There are numerous examples of off-gas purification in the chemical industry, the textile and furniture industry, and in printing works. Catalytic afterburning units (Fig. 8-2) are also successfully used for removing odors, e.g., in the foodstuffs industry.

Fig. 8-1. Supported metal catalyst with large reaction surface (Doduco)

Table 8-4. Heterogeneous catalysts in environmental protection [T32]

Process	Catalyst	Conditions
Automobile exhaust control (C_nH_m, CO, NO_x)	Pt, Pd, Rh, washcoat Al_2O_3, ceramic monolithes, rare earth oxide promoters	400–500 °C, 1000 °C short-term
Flue gas purification (SCR): removal of NO_x with NH_3	Ti, W, V mixed oxides as honeycomb bulk catalysts Ti, W, V oxides on inert honeycomb supports	hot denitrification (400 °C) cold denitrification (300 °C
Combined denitrification and desulfurization (DESONOX process)	SCR catalyst + V_2O_5 honeycomb catalyst, catalyst bed	up to 450 °C
Catalytic afterburning (off-gas purification)	Pt/Pd; $LaCeCoO_3$ (perovskite); oxides of V, W, Cu, Mn, Fe; supported catalyst (honeycomb monolith or catalyst bed) or bulk catalyst	150–400 °C 200–700 °C

Fig. 8-2. Catalytic pyrolysis and post-combustion unit (COMETT plant from the company Nabertherm, catalysis laboratory, FH Mannheim)

8.2 Examples of Industrial Processes

8.2.1 Ammonia Synthesis [8, 14]

The synthesis of ammonia from nitrogen and hydrogen is one of the most important processes in the chemical industry; over 100×10^6 t/a of ammonia is produced worldwide. The Haber–Bosch process, introduced in 1913, was the first high-pressure industrial process [12]. Ammonia synthesis is carried out at ca. 300 bar and 500 °C on iron catalysts with small amounts of the promoters Al_2O_3, K_2O, and CaO.

Extensive investigations of the mechanism of ammonia synthesis have shown that the rate-determining step is the dissociation of coordinatively bound nitrogen molecules on the catalyst surface. Hydrogen is much more readily dissociated on the catalyst surface. The adsorbed species then undergo a series of insertion steps, in which ammonia is formed stepwise and is finally desorbed (Scheme 8-1).

At moderately high pressures the reaction rate is independent of the hydrogen pressure and first order with respect to nitrogen. The stationary occupation by N* atoms is low, and this indicates that the dissociative adsorption of N_2 is rate-determining. At higher hydrogen pressures, there is a fractional reaction order in H_2 corresponding to displacement of the rate-determining step towards

$$N^* + H^* \longrightarrow NH^*$$

1) Dissociative chemisorption of starting materials

$N_{2,G} \longrightarrow N_2^* \longrightarrow 2N^*$

$H_{2,G} \longrightarrow 2H^*$

2) Reaction of adsorbed atoms

$N^* + H^* \longrightarrow NH^* \xrightarrow{H^*} NH_2^* \xrightarrow{H^*} NH_3^*$

3) Desorption of product

$NH_3^* \longrightarrow NH_{3,G}$

Scheme 8-1. Simplified mechanism of ammonia synthesis

The elucidation of the reaction mechanism has occupied catalysis researchers up to the present day [8]. Over 20 000 catalysts have been tested, but none has been found that operates at room temperature. Catalytic activity is exhibited by metals that chemisorb N_2 dissociatively with relatively strong binding, especially the metals of Groups 6–8 with d gaps, on which large amounts of H_2 are also rapidly chemisorbed. The activity increases with increasing heat of adsorption of N_2 in the order:

$Cr < Mn < Fe, Mo < Ru$ and $W < Re < Os$

Other metals are more active in cleaving the $N \equiv N$ bond (e.g., Li) but the resulting metal nitrides are too stable to take part in a catalytic cycle.

For economic reasons, industrial catalysts consist of smelted iron oxides (60–70% Fe) mixed with oxides of Al, Ca, Mg, and K, ground to 6–20 mm. During the activation of the catalyst by reduction, iron crystallites are formed with an interconnected pore system and an inner surface area of 10–20 m^2/g. The surface is partially covered by promoter oxides.

The industrial production of ammonia from natural gas involves eight different catalytic steps (Scheme 8-2). The overall reaction equation is given in Equation (8-1).

$$3 CH_4 + 2 N_2 + 3 O_2 \longrightarrow 4 NH_3 + 3 CO_2 \tag{8-1}$$

The thermodynamic energy requirement is 2×10^7 kJ/t NH_3, which represents the theoretical minimum for all conceivable processes. Modern processes for the production of ammonia from natural gas have energy consumptions of around 3×10^7 kJ/t NH_3, i.e., only 1.5 times the theoretical minimum energy consumption. Today much of the energy requirement can be covered by means of heat recovery. Modern ammonia plants produce up to 2000 t/d.

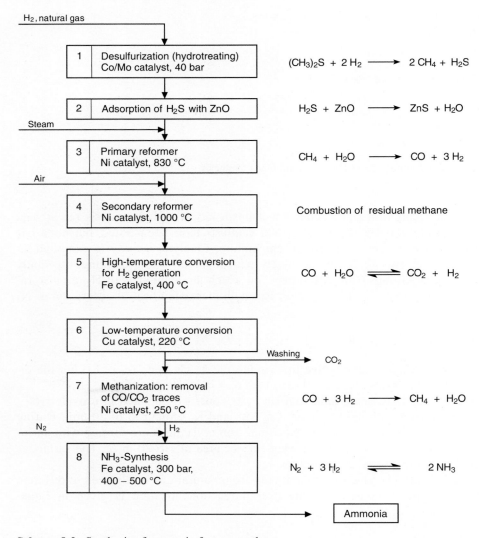

Scheme 8-2. Synthesis of ammonia from natural gas

8.2.2 Hydrogenation [2, 4]

With increasing crude oil prices, there is a growing trend towards renewable raw materials. Fats (triglycerides) are being used in increasing quantities as raw materials in the chemical industry. The glycerides are oxidized, hydrogenated, and aminated to remove undesired functional groups, to shorten the chain length, or to introduce other functional groups. Many of these steps are carried out catalytically, and for economic reasons should take place at low temperatures and pressures in order to attain high selectivities [3, 10].

An important process in the foods industry is the hardening of vegetable oils, for example, the production of margarine by hydrogenation of double bonds. In this process an oil is converted to a solid that should have high stability towards oxidation, which leads to rancidity. The main aim in an industrial process is to remove linolenic acid (three C–C double bonds) as completely as possible while minimizing conversion of the desired oleic acid (one C–C double bond) to the saturated stearic acid.

The hydrogenation is carried out at ca. 3 bar hydrogen pressure with a suspension of supported Ni/SiO$_2$ catalyst in the liquid phase at 200–210 °C, usually in a batch process. The reaction is terminated by stopping the supply of hydrogen and lowering the temperature to ca. 100 °C. The strength of adsorption on the catalyst decreases with increasing degree of saturation of the fats. Therefore, the more highly unsaturated side chains are preferentially hydrogenated. The rate of hydrogenation is generally zero order with respect to the concentration of the oil and first order with respect to the hydrogen pressure. This indicates that the nickel surface is largely covered with unsaturated molecules, whereas hydrogen is only weakly adsorbed. A highly simplified reaction mechanism is shown in Scheme 8-3; the last step is rate-determining.

The highly unsaturated triglycerides can be hydrogenated to oleic acid with high selectivity by using copper catalysts, but traces of copper remain in the margarine, which excludes the industrial use of copper catalysts. A typical hydrogenation plant hydrogenates oil in a stirred tank in batches of up to 15 t and produces about 90 t/d of hardened fat.

Noble metal catalysts such as Pd and Pd are being increasingly used in oil and fat treatment processes since they have a less pronounced tendency to remove

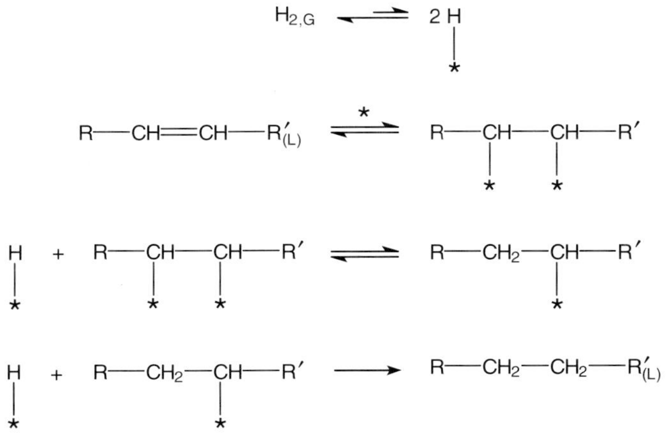

Scheme 8-3. Reaction steps in the hydrogenation of fats [T22]

functional groups. The choice of catalyst and its optimization are of growing importance in this area. For example, a Pd shell catalyst is used advantageously for the hydrogenation of soya oil to edible oil. In this process the content of linolenic acid must be reduced to less than 2% without hydrogenating the other unsaturated fatty acids. In spite of the small metal surface area, the shell catalyst is the most suitable, since the metal is most readily accessible for the bulky triglyceride molecules. The Pd catalysts have higher activity and can be more easily recovered. The high cost of the noble metal is made up for by the low metal concentration. While nickel catalysts are used in concentrations of 0.04% relative to the oil to be hydrogenated, concentrations of platinum of less than 0.005% are sufficient [10].

Another area of major industrial importance is the production of oleochemical raw materials such as fatty acids, fatty acid methyl esters, fatty alcohols, and glycerol. The company Henkel is the world's largest processor of renewable fats and oils, with a capacity of 10^6 t. Tailor-made catalysts are used in most oleochemical reactions.

A successful catalyst development in the last few years has made possible the direct hydrogenation of fats to fatty alcohols in a one-stage process. The laborious transesterification of the the fats can be dispensed with. Beside the high-quality coconut and palm oils, lower quality, acid-containing fats and oils can now be hydrogenated by using new acid-stable copper chromite spinel catalysts.

Production of Fine Chemicals [10, 11]

Numerous, very different groups of products such as pharmaceuticals, dyes, food additives, cosmetics, vitamins, and photochemicals are classified as fine chemicals. Many products and intermediates are produced by catalytic hydrogenation under mild conditions. Here we can only discuss a few basic principles.

The starting materials often have complex structures and high molecular masses and are temperature-sensitive. For this reason, most of these chemicals are hydrogenated in solution. Possible processes are suspension processes with powder catalysts and reactions on fixed-bed catalysts; diffusion control often plays a decisive role.

The hydrogenation of double bonds is the most important reaction in the production of fine chemicals [1]. Since the starting materials often have several reactive groups, hydrogenation can give rise to various products, only one of which is usually desired. Careful adjustment of all experimental parameters, including type and amount of catalyst, solvent, temperature, pressure, and degree of mixing (e.g., stirring speed), is necessary to maximize the yield of the desired product (Fig. 8-3).

Some unsaturated groups differ in their reactivity so much that selective hydrogenation is readily possible. For example, the benzene ring is normally not easily hydrogenated, and this makes many selective conversions possible, for example,

Fig. 8-3. Operation of a continuous high-pressure hydrogenation plant (CATATEST plant, VINCI technologies, France; high-pressure laboratory, FH Mannheim)

styrene to ethylbenzene, nitrobenzene to aniline, benzaldehyde to benzyl alcohol, and benzonitrile to benzylamine. However, the benzene ring can be hydrogenated by using special catalysts, as shown by the example of the hydrogenation of an ester with different catalysts (Scheme 8-4).

The most important factor is the choice of the appropriate catalyst. Although the skeletal catalyst Raney nickel is often used in practice, milder reaction conditions can be achieved with noble metal catalysts. The relatively cheap palladium catalysts are most widely used. They can often hydrogenate aromatic ring systems at room temperature and higher pressure.

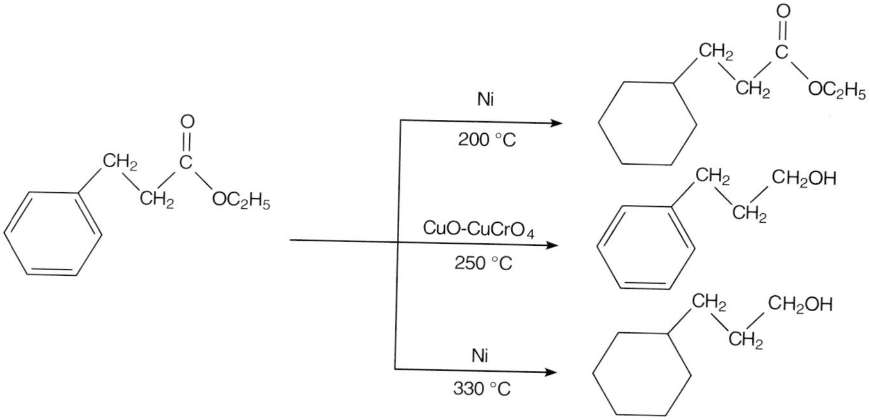

Scheme 8-4. Selective hydrogenation of an ester

The solvent is chosen on a more empirical basis, and neutral solvents such as methanol, ethanol, hexane, and cyclohexane are commonly used. It should, however, be noted that the solubility of hydrogen in hydrocarbons is lower than in alcohols and this can lead to reactions that are more diffusion controlled. Acidic solvents such as acetic acid are useful in cases where the product is a nitrogen base that is strongly adsorbed on the catalyst. Basic solvents are preferred in the case of acidic products.

The choice of catalyst becomes problematic if hydrogenolysis reactions can occur with intermediate products or other functional groups in the molecule. Hydroxyl groups, alkoxyl groups, and halogen substituents are often cleaved from aromatic rings by hydrogenolysis. The best catalyst can only be found by experiment (Fig. 8-4).

Fig. 8-4. A hydrogenation catalyst is introduced into a pilot plant in order to test it under process-relevant conditions (BASF, Ludwigshafen, Germany)

Catalytic reactions often give surprising results. An interesting result is the reductive amination of aldehydes and ketones. The keto compounds react with primary or secondary amines or ammonia in the presence of H_2 and a suitable catalyst (e.g., Pd) to give a new amine. Thus, as expected, 2-methylcyclohexanone is converted to 2-methylcyclohexylamine (Eq. 8-2).

(8-2)

Under the same reaction conditions, the unsubstituted cyclohexanone forms dicyclohexylamine (Eq. 8-3). The reasons for this behavior have not yet beeen completely elucidated [T20].

$$\text{cyclohexanone} \xrightarrow{\text{NH}_3/\text{H}_2, \text{Cat.}} \text{dicyclohexylamine} \tag{8-3}$$

8.2.3 Methanol Synthesis [T22, T41]

The synthesis of methanol from CO and H_2 (Eq. 8-4) has been known since the early 1920s. Mittasch found oxygen-containing compounds during investigations of ammonia synthesis at BASF. In 1923 the first large-scale methanol plant operating with synthesis gas was erected in Germany.

$$CO + 2H_2 \rightleftharpoons CH_3OH \quad \Delta H_R = -92 \text{ kJ/mol} \tag{8-4}$$

The high-pressure process, which used to be exclusively operated, is carried out with ZnO/Cr_2O_3 catalyst at 250–350 bar and 350–400 °C. The development of more active, copper-based catalysts allowed the process to be carried out in the pressure range 50–100 bar and at lower temperatures. This improved the economics of the process. The low-pressure processes were developed by ICI and Lurgi and introduced in the mid-1960s.

Let us examine the mechanism of methanol synthesis [16]. In 1962 the activating effect of CO_2 in the synthesis gas was discovered. When cracked gas (CO + $3H_2$) from methane-rich natural gas is used, CO_2 is added to the synthesis gas and it consumes more H_2 than the CO (Eq. 8-5).

$$CO_2 + 3H_2 \rightleftharpoons CH_3OH + H_2O \quad \Delta H_R = -50 \text{ kJ/mol} \tag{8-5}$$

Another side reaction is the water-gas shift equilibrium (Eq. 8-6).

$$CO_2 + H_2 \rightleftharpoons CO + H_2O \quad \Delta H_R = 41{,}3 \text{ kJ/mol} \tag{8-6}$$

Thus the question of what is the actual carbon source in methanol synthesis can not be unambiguously answered. Two mechanisms have been suggested to explain the formation of methanol on the heterogeneous catalyst. In the first mechanism (Eq. 8-7), adsorbed CO reacts on active copper centers of the surface with dissociatively adsorbed hydrogen in a series of hydrogenation steps to give methanol.

$$\underset{M}{\overset{|}{CO}} \xrightarrow{M-H} \underset{M}{\overset{|}{\underset{C}{H\diagdown\diagup O}}} \xrightarrow{M-H} \underset{M}{\overset{|}{\underset{C}{H\diagdown\diagup OH}}} \xrightarrow{M-H} \underset{M}{\overset{|}{CH_2OH}} \xrightarrow{M-H} M + CH_3OH \qquad (8\text{-}7)$$

In the second mechanism (Eq. 8-8), the first step is the insertion of CO into a surface OH group with formation of a surface formate. This is followed by further hydrogenation steps and dehydration to give a surface methoxyl group, from which methanol is formed.

In this mechanism the intermediates are bound to the surface through oxygen. Support for this assumption is provided by the fact that CO is known to react with strongly basic hydroxides such as NaOH to form formate ions. The methanol catalyst contains the strongly basic component ZnO. Furthermore, it is known that copper is a highly active catalyst for the hydrogenation of formate to methanol.

However, more recent investigations with $CO/CO_2/H_2$ mixtures have shown that the active catalyst is finely divided copper on the surface of the catalyst and that ZnO plays no particular role in the industrial catalyst. Carbon dioxide plays a key role here (Scheme 8-5) [T22]. All steps take place on copper surfaces. The hydrogenolysis of formate on copper surfaces has been proven. It is now assumed that the function of CO is to remove atomically bound oxygen from the surface with formation of CO_2, which then act as the primary reactant. This example shows once again that it is not possible to formulate a mechanism simply by combining apparently plausible reaction steps.

$$\underset{M}{\overset{|}{OH}} + CO \longrightarrow \underset{M}{\overset{|}{\underset{\overset{|}{O}}{\overset{O}{\underset{\|}{C}}\diagdown H}}} \xrightarrow{2\,M-H} \underset{M}{\overset{|}{\underset{O}{\diagdown CH_2OH}}} \xrightarrow[-H_2O]{2\,M-H} \underset{M}{\overset{|}{OCH_3}}$$

$$\xrightarrow{M-H} CH_3OH + M \xrightarrow{H_2O} M-OH \qquad (8\text{-}8)$$

Methanol synthesis is carried out in a recycle process similar to ammonia synthesis. The strong temperature dependence of the methanol equilibrium and the increasing extent of side reactions at higher temperature require rapid heat removal or cooling by introduction of fresh gas. Methanol plants are usually operated at a partial conversion of 15–18% of the CO starting material with recycle of the unchanged synthesis gas. Large quantities of gas must be circulated (ca. 40 000 $m^3 h^{-1} m^{-3}$ catalyst). Figure 8-5 shows a modern methanol plant.

In order to control heat removal and therefore the catalyst temperature, multiple-tube reactors (Lurgi process) or quench reactors with several catalyst layers

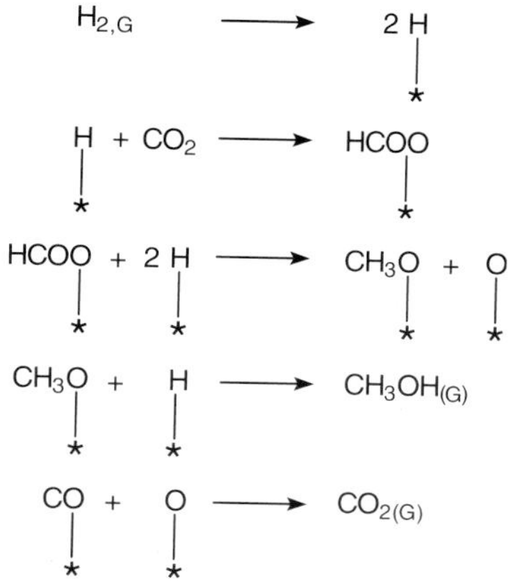

$$H_{2,G} \longrightarrow 2\,\underset{*}{H}$$

$$\underset{*}{H} + CO_2 \longrightarrow \underset{*}{HCOO}$$

$$\underset{*}{HCOO} + 2\,\underset{*}{H} \longrightarrow \underset{*}{CH_3O} + \underset{*}{O}$$

$$\underset{*}{CH_3O} + \underset{*}{H} \longrightarrow CH_3OH_{(G)}$$

$$\underset{*}{CO} + \underset{*}{O} \longrightarrow CO_{2(G)}$$

Scheme 8-5. Mechanism of methanol synthesis

Fig. 8-5. Methanol plant (BASF, Ludwigshafen, Germany)

and introduction of cold gas (ICI process) are mainly used. Catalyst performance in modern larger reactors is 1.3–1.5 kg of methanol per liter per hour, and large-scale plants have capacities of up to 10^6 t/a, which reflects the position of methanol as a key product of C_1 chemistry.

8.2.4 Selective Oxidation of Propene [5, 13]

The heterogeneously catalyzed gas-phase oxidations of unsaturated hydrocarbons are large-scale industrial processes. The best known processes are:

- Oxidation of ethylene to ethylene oxide
- Oxidation of propene to acrolein and ammoxidation to acrylonitrile
- Oxidation of *n*-butane, butenes, or benzene to maleic anhydride
- Oxidation of *o*-xylene to phthalic anhydride

Economic operation of these processes requires a selectivity of at least 60%. In the last few decades the industrial oxidation catalysts have been so much improved that selectivities of over 90% are achieved in some cases. Thus the space–time yields of the processes could be improved and better use made of the raw materials.

Selective oxidation still offers interesting development possibilities for the chemical engineer [5]. Here we shall consider the oxidation and ammoxidation of propene, which both proceed by a similar mechanism, in more detail.

In the selective oxidation of propene, metal oxides are mainly used as catalysts, and many different products are obtained (Scheme 8-6), depending on the catalyst used [17].

The catalytic oxidation of propene leads preferentially to formation of acrolein (Eq. 8-9).

$$H_2C=CH-CH_3 + O_2 \longrightarrow H_2C=CH-CHO + H_2O \qquad (8\text{-}9)$$

$$\Delta H_R = -368 \text{ kJ/mol}$$

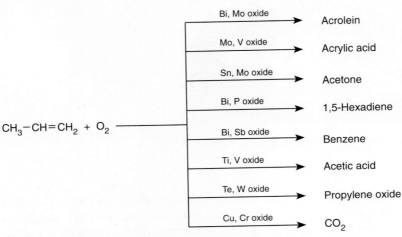

Scheme 8-6. Oxidation of propene on various metal oxide catalysts

Carbon dioxide, acetaldehyde, and acrylic acid are formed as side products. A technical breakthrough was achieved by Standard Oil of Ohio (SOHIO) with the discovery of the bimetallic bismuth molybdate and bismuth phosphomolybdate catalysts. Propene is oxidized with air on a Bi_2O_3/MoO_3 catalyst at 300–400 °C and 1–2 bar in a fixed-bed tubular reactor, which allows effective removal of heat from the exothermic reaction [13].

The mechanism of the allyl oxidation of propene is explained in terms of a reaction cycle [17]. As shown in Scheme 8-7, propene and air do not react directly with one another. Instead, the propene initially forms a π complex **A** with an Mo center of the bismuth molybdate catalyst. Hydrogen abstraction by an oxo oxygen atom on bismuth leads to formation of a hydroxyl group and a π-allyl complex at Mo **B**, whereby one electron flows into the lattice. Transfer of oxygen to the allyl group forms an Mo–alkylate bond, and a further hydrogen abstraction (**C**) on the same Mo center leads to formation of acrolein, which is desorbed from the catalyst surface (**D**). In these steps, three electrons flow into the lattice, and what remains is an oxygen-deficient bismuth molybdate with OH groups (**E**). This reacts with atmospheric oxygen with cleavage of water (**F**) and re-formation of the original catalyst (**G**). In the reoxidation of the catalyst, an O_2 molecule is reduced to O^{2-} ions by four electrons, available in the lattice. The oxide ions then diffuse to the lattice vacancies.

How can the side products of the oxidation reaction be explained? It can be assumed that the allylmolybdenum complex (**B**) is cleaved into C_1 and C_2 frag-

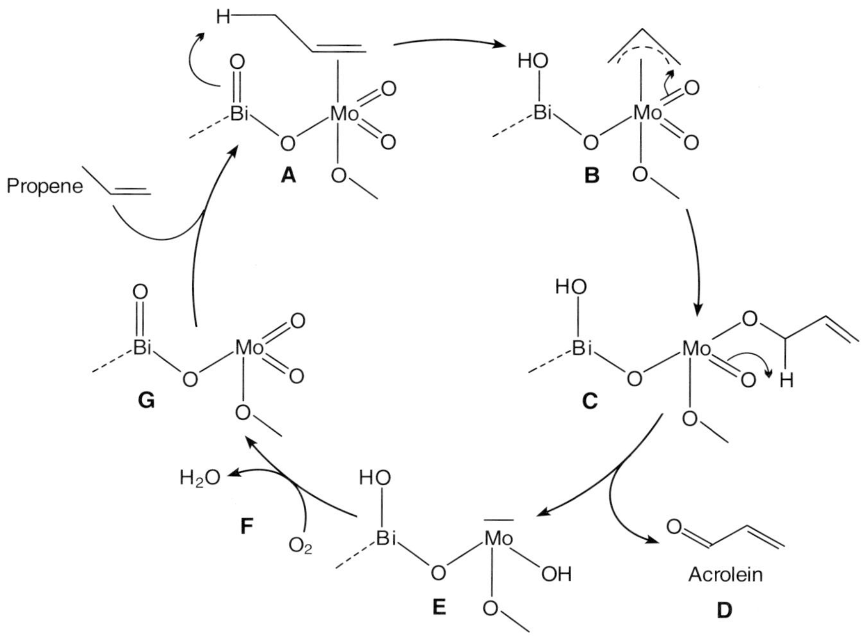

Scheme 8-7. Oxidation of propene to acrolein on Bi/Mo catalysts [17]

ments, which result in acetaldehyde and CO_2, the latter presumably via formaldehyde as intermediate. Carbon dioxide can, however, also be formed by total oxidation of propene.

The true SOHIO process is in fact the ammoxidation of propene with NH_3 and atmospheric oxygen in a highly exothermic reaction to give acrylonitrile (Eq. 8-10).

$$CH_2=CH-CH_3 + 3/2\,O_2 + NH_3 \longrightarrow H_2C=CH-CN + 3\,H_2O \qquad (8\text{-}10)$$

$$\Delta H_R = -502 \text{ kJ/mol}$$

In this process the activated methyl group in the allylic position is converted to a nitrile. In the SOHIO process, stoichiometric quantities of propene and ammonia are treated with an excess of oxygen in a fluidized-bed reactor at ca. 450 °C and 1–2 bar. The catalysts used today contain several multivalent main group metals (Bi^{3+}, Sb^{3+}, Te^{4+}), molybdenum, and a redox component ($Fe^{2+/3+}$, $Ce^{3+/4+}$, $U^{5+/6+}$) in a solid oxide matrix. However, most research investigations deal with the standard bismuth molybdate catalyst, which can be simply formulated as $Bi_2O_3 \cdot n\,MoO_3$ [20].

Besides oxidizing propene and being regenerable by atmospheric oxygen, the catalyst must also activate ammonia. In spite of numerous experimental findings, the mechanism of ammoxidation is still largely speculative. According to Scheme 8-8 the BiO group abstracts hydrogen from the alkane, and this leads to formation of an π-allyl complex at the Mo center. The actual allyl oxidation and the NH_3 activation then take place on the molybdenum side. It is assumed that by reaction with ammonia the oxomolybdenum groups are partly converted to imino-

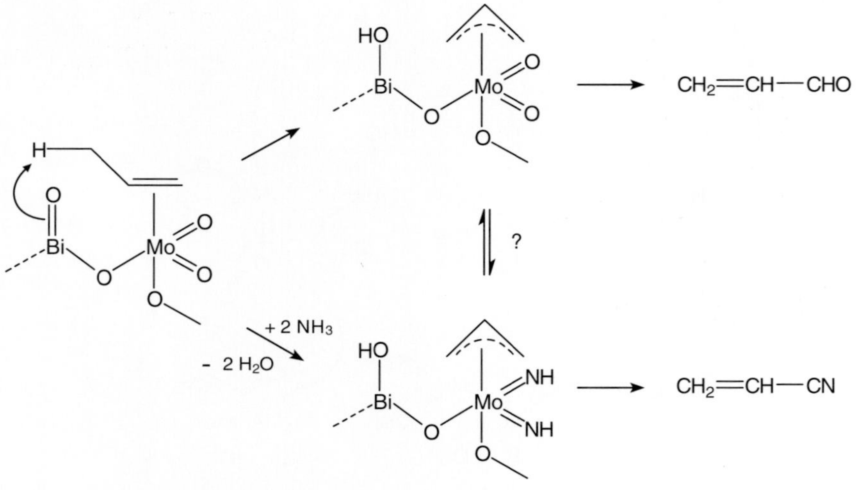

Scheme 8-8. Postulated reaction mechanism for the ammoxidation of propene [13]

molybdenum groups, which are responsible for the C–N bond-forming reaction to give acrylonitrile.

The oxidation is a six-electron process, and it is regarded as certain that the α-H abstraction with formation of the π-allyl complex is the rate-determining step. Both the dioxomolybdenum cations and the diimino species are believed to be involved in the formation of acrylonitrile. On the basis of selectivity measurements, the stoichiometry shown in Equation 8-11 was assumed [L32].

$$2\ CH_3-CH=CH_2 + \underset{\substack{\\ \text{HN}\diagdown\ \diagup\text{NH} \\ \text{Mo} \\ + 2\ \overset{\text{O}\diagdown\ \diagup\text{O}}{\text{Mo}}}}{} \longrightarrow 2\ CH_2=CH-CN + 4\ H_2O + 3\ Mo \qquad (8\text{-}11)$$

In the industrial process the acrylonitrile selectivity is greater than 70%, and the side products are acetonitrile (3–4%), HCN (ca. 15%), CO_2, acrolein, and acetaldehyde. After washing with water, the acrylonitrile is purified by multistage distillation to give a purity of >99%, as is required for the production of fibers. The acetonitrile byproduct is isolable but is usually incinerated. Figure 8-6 shows a scheme of the SOHIO process.

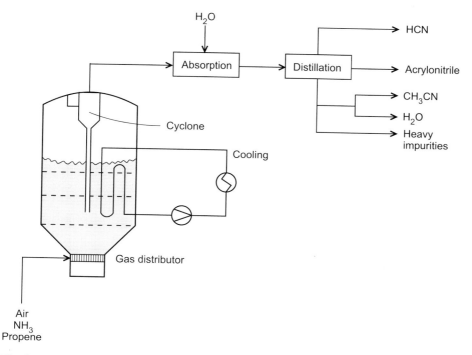

Fig. 8-6. SOHIO process for the ammoxidation of propene

The SOHIO ammoxidation process was developed since 1957. Production capacity for acrylonitrile, the most important product derived from propene, is greater than 4×10^6 t/a, of which over 70% is produced by the SOHIO process. Plants are constructed with capacities of up to 180 000 t/a. There are numerous variants of ammoxidation, the following products also being produced by this process:

- Methacrylonitrile from isobutene
- Hydrogen cyanide from methane
- Phthalodinitrile from o-xylene
- Nicotine nitrile from β-picoline

8.2.5 Selective Catalytic Reduction of Nitrogen Oxides [9]

Selective catalytic reduction (SCR; DENOX process) is the reduction of NO and NO_2 (NO_x) by ammonia in the presence of oxygen to give molecular nitrogen. Since the 1970s SCR processes have been used to an increasing extent for the catalytic after-treatment of flue gases from power stations and furnaces. In Germany the limiting NO_x value for new coal-fired plants with a power output of 300 MW is 200 mg/m^3. Such low NO_x levels can only be achieved by applying secondary measures. The 3–12% oxygen in the flue gas also takes part in the reaction, as shown for NO in Equation 8-12.

$$NH_3 + NO + 1/4\ O_2 \longrightarrow N_2 + 3/2\ H_2O \tag{8-12}$$

It can be assumed that in the presence of an excess of oxygen, NO reacts with an equimolar quantity of NH_3 to give N_2 and H_2O.

The catalysts must be designed so that side reactions such as the oxidation of ammonia by oxygen (Eq. 8-13) and the formation of N_2O (Eq. 8-14) are suppressed. The oxidation of SO_2 to SO_3 must also be avoided.

$$NH_3 + 3/4\ O_2 \longrightarrow 1/2\ N_2 + 3/2\ H_2O \tag{8-13}$$

$$NH_3 + O_2 \longrightarrow 1/2\ N_2O + 3/2\ H_2O \tag{8-14}$$

Transition metal oxides on ceramic supports have proved be particularly suitable as catalysts; for example: support: TiO_2 (ca. 90%), active components: V_2O_5 (1.5–5%), WO_2 (5–10%), MoO_3, GeO_2. Sheet or honeycomb catalysts are used industrially, and the usual operating temperature is 350–400 °C.

A mechanistic proposal (Scheme 8-9) explains the formation of N_2 besides N_2O and H_2O [21]. On the hydroxyl-group-containing surface of the oxide, ammonia is adsorbed on Brønsted acid centers with formation of an ammonium structure (step 1). In step 2, NO undergoes addition to the ammonium complex

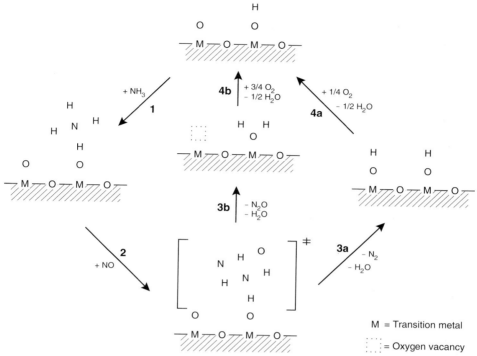

Scheme 8-9. Mechanism of the selective catalytic reduction of NO by NH_3 [21]

according to the Eley–Rideal mechanism. The resulting complex has two possibile decomposition routes. In the major route (step 3a), an $N \equiv N$ bond is formed and N_2 and H_2O are cleaved off. In the following reaction (step 4a), oxygen is filled up and water is released by the catalyst surface. In the minor route (step 3b), lattice oxygen is abstracted from the catalyst, and N_2O and H_2O are formed. In step 4b the oxygen vacancy is filled and water is cleaved off to regenerate the original catalyst.

In industry, two variants of the process compete with one another [7]. In the first variant, the SCR reactor is located in the high-dust high-temperature region directly after the boiler on the raw-gas side. The flue gas enters with a temperature of 300–450 °C and a dust content of 10–30 mg/m³ (high-dust configuration). Since this variant involves strong abrasion and more rapid poisoning of the catalyst, bulk catalysts on the basis of V, W, or Ti oxides are used. In the second variant, the SCR reactor is located after the flue gas purification and desulfurization stages (low-dust configuration). Since abrasion and poisoning are much lower in this case, honeycomb and sheet catalysts can also be used. A disadvantage is that the flue gas leaving the desulfurization stage at 50–70 °C must be heated to the reaction temperature of 300–350 °C.

The SCR processes have become established in western Europe, and the required TiO_2/V_2O_5 based honeycomb catalysts are produced by various European catalyst producers under a Japanese license.

8.2.6 Olefin Polymerization [13]

The polymerization of olefins has been carried out industrially for decades and can be performed by various mechanisms. The high-pressure radical polymerization of ethylene leads to low-density polyethylene (LDPE, $\rho = 0.92–0.93$ g/cm^3). In the mid-1950s Ziegler achieved the low-pressure polymerization of ethylene and propylene (up to 10 bar, 50–150 °C) by using organometallic catalysts based on $TiCl_4/Al(C_2H_5)_3$. The Ziegler catalysts give less branched, linear high-molecular polyethylene (high-density polyethylene, HDPE; $\rho = 0.94–0.97$ g/cm^3). Using this catalyst system, Natta succeeded in manufacturing crystalline, isotactic polypropylene, and around the same time, the company Phillips in the USA developed silica-supported chromium catalysts.

In the following we shall take a closer look at supported catalysts for the polymerization of olefins [T22]. Oxides of Cr and Ti on various support materials have high activities for the polymerization of ethylene to linear chains (HDPE). The processes operate at relatively low ethylene pressure (20–30 bar) in the temperature range 130–150 °C (solution polymerization) or 80–100 °C (suspension and gas-phase polymerization).

The Phillips catalysts are manufactured by impregnating amorphous silica gel with chromates up to a metal loading of ca. 1 %. The material is then dried and calcinated at 500–1000 °C. The surface silanol groups react with the chromate groups to give a disperse monolayer of chromate and dichromate esters (Eq. 8-15).

$$\text{Si-OH} \quad \text{Si'-OH} \xrightarrow{Cr(VI)} \text{Si-O-Cr(=O)_2-O-Si'} + \text{Si-O-Cr(=O)(O)-O-Cr(=O)(O)-Si'} \tag{8-15}$$

However, it is assumed that the active centers are coordinatively unsaturated Cr^{II} or Cr^{III} centers that are generated by reaction with ethylene (Eq. 8-16). It is also possible to convert the chromate deposited on the silica surface to an active form by high-temperature reduction with CO. In an alternative method of catalyst production, low-valent organochromium compounds such as chromocene and tris(η^3-allyl)chromium are used as catalyst precursors.

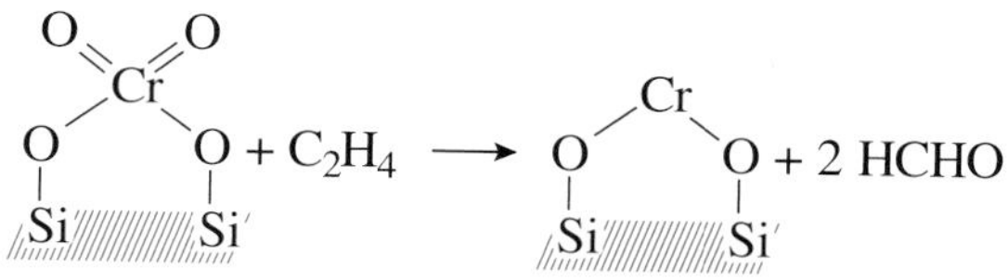

$$\underset{\text{Si}\quad\text{Si}'}{\overset{\text{O}\diagdown\underset{\diagup}{\text{Cr}}\diagup\text{O}}{\text{O}\quad\text{O}}} + C_2H_4 \longrightarrow \underset{\text{Si}\quad\text{Si}'}{\overset{\text{Cr}}{\text{O}\quad\text{O}}} + 2\ HCHO \qquad (8\text{-}16)$$

Similar to the polymerization of ethylene on Ziegler catalysts, the first reaction step is the coordination of an ethylene molecule at a Cr^{II} center. The initiator of the polymerization reaction is thought to be a Cr–H group, into which ethylene inserts to form an ethyl ligand (Eq. 8-17). It was shown that only isolated Cr centers on the surface are catalytically active.

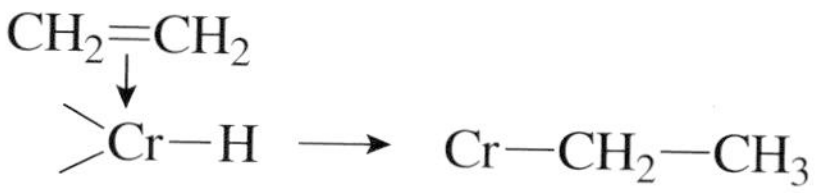

$$\overset{\overset{CH_2=CH_2}{\downarrow}}{>Cr-H} \longrightarrow Cr-CH_2-CH_3 \qquad (8\text{-}17)$$

Coordinatively unsaturated transition metal centers are the prerequisite for olefin polymerization in both Phillips and Ziegler–Natta catalysts, and this makes it possible to simultaneously bind the monomer and the growing chain. This does not occur by a redox reaction, since the transition metal does not change its oxidation state during the polymerization process. The chain-growth process can be described as shown in Scheme 8-10.

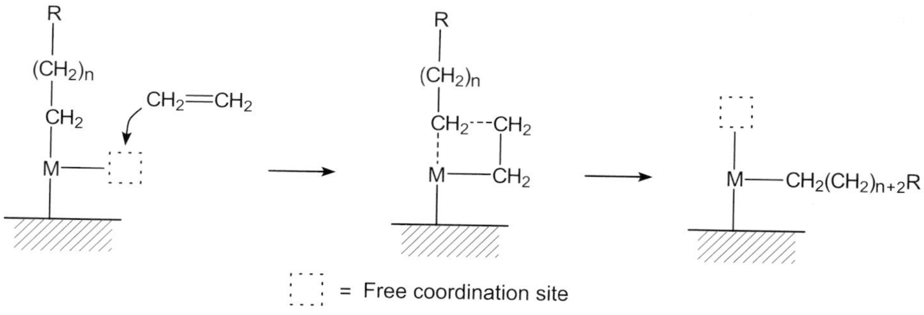

⌞⁝⌟ = Free coordination site

Scheme 8-10. Polymerization of ethylene at a metal center (M)

The chain-growth mechanism involves the insertion of two CH_2 units between the metal center M and the original alkyl chain, so that chain branching can not occur in the growing polymer. Thus the vacant site on the transition metal atom simply changes its position. Chain branching can be introduced in a controlled fashion by copolymerizing ethylene with short-chain terminal alkenes, which leads to modified polymer properties.

Exercises for Chapter 8

Exercise 8.1

The following industrial processes are to be carried out. Which of the catalysts A–L is potentialy suitable for which reaction?

1) Alkylation of benzene to ethylbenzene
2) Cracking of higher hydrocarbons
3) Dehydration of amides to amines
4) Dehydration of ethylbenzene to styrene
5) Esterification
6) Hydrogenation of CO to methanol
7) Hydrogenation of vegetable oils
8) Isomerization of pentane to isopentane
9) Oxidation of ammonia to nitrogen oxides
10) Oxidation of SO_2 to SO_3
11) Reforming processes for the production of aromatics
12) Oxidation of methanol to formaldehyde

Pt/support (A), zinc chromite (B), V_2O_5 (C), Pt (D), Al_2O_3 (E), Ag (F), aluminosilicates (G), zeolites (H), Ni (I), iron oxides/promoter (J), ion-exchange resins (K), CuO (L).

Exercise 8.2

Cinnamaldehyde is hydrogenated with a supported Pd catalyst. Which parts of the molecule are attacked under the usual conditions (A, B, C, or several)?

Ph–CH=CH–CHO
A   B   C

Exercise 8.3

Name the important differences between gas-phase and liquid-phase hydrogenation.

Exercise 8.4

How is propene converted to 2-ethylhexanol?

Exercise 8.5

In a publication, the hydrogenation of 1-propen-1-ol on a supported rhodium catalyst with a side reaction is formulated as follows:

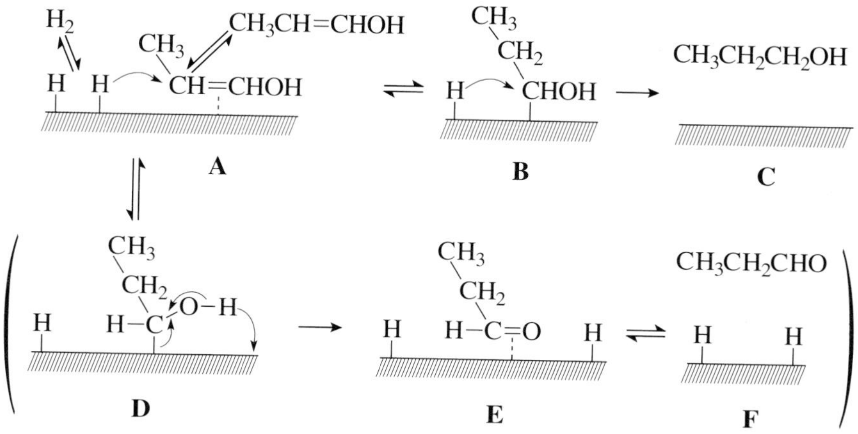

[Reaction scheme with intermediates A–F showing surface-bound species on a supported rhodium catalyst: A = CH$_3$CH=CHOH adsorbed; B = CH$_3$CH$_2$CHOH adsorbed; C = CH$_3$CH$_2$CH$_2$OH; D = CH$_3$CH$_2$-CH(O-H) surface species; E = CH$_3$CH$_2$-CHO adsorbed; F = CH$_3$CH$_2$CHO]

a) Explain the elementary steps of both reactions with the intermediates **A–F**.
b) How would the side reaction be designated?

Exercise 8.6

In the Fischer–Tropsch synthesis, CO/H$_2$ mixtures are converted to hydrocarbons on Co or Fe catalysts. A common concept of the course of the process is depicted in a highly simplified form in the following scheme:

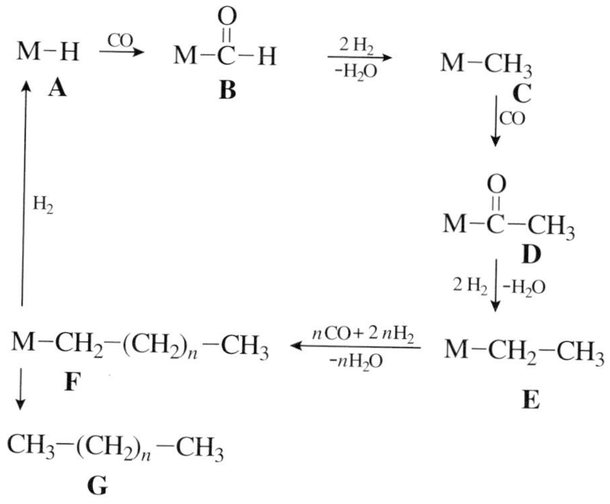

$$M-H \xrightarrow{CO} M-\overset{O}{\underset{\|}{C}}-H \xrightarrow[-H_2O]{2\,H_2} M-CH_3$$
$$\quad\;\; A \qquad\qquad B \qquad\qquad\qquad C$$

$$\downarrow CO$$

$$M-\overset{O}{\underset{\|}{C}}-CH_3$$
$$\qquad D$$

$$\Big\downarrow 2\,H_2 \;\; |-H_2O$$

$$M-CH_2-(CH_2)_n-CH_3 \xleftarrow[-nH_2O]{n\,CO+\,2\,n H_2} M-CH_2-CH_3$$
$$\quad\;\; F \qquad\qquad\qquad\qquad\qquad\qquad E$$

$$\downarrow$$

$$CH_3-(CH_2)_n-CH_3$$
$$\qquad G$$

(with H$_2$ arrow from A upward from the F side)

Explain the individual reactions to give the products **A–G**.

Exercise 8.7

The following mechanism is given for the oxidation of propene to acrolein on bismuth molybdate catalysts:

$$\text{H-CH}_2\text{-CH=CH}_2 \quad \text{over} \quad \text{-Bi-O-Mo(=O)(O)-O} \quad \xrightarrow{\text{①}} \quad \text{H}_2\text{C(-CH-)CH}_2,\ \text{OH} \quad \text{over} \quad \text{-Bi-O-Mo(=O)(O)-O}$$

$$\big\Updownarrow \text{②}$$

$$\text{-Bi-O-Mo=O with OH and H}_2\text{C=CH-CH(H)-O} \quad \xrightarrow{\text{③}} \quad \text{OH, OH on -Bi-O-Mo- + H}_2\text{C=CH-CHO}$$

$$\text{④}\ -\text{H}_2\text{O} \mid +\text{O}_2$$

Explain the course of the reaction with the steps 1–4.

Exercise 8.8

The oxidation of benzaldehyde on an SnO_2/V_2O_5 catalyst is described as follows ($\square$ represents an anion vacancy in the catalyst surface):

$$\text{C}_6\text{H}_5\text{-C(=O)-H} \text{ over } O^{2-}\ \square\ O^{2-},\ M^{n+} \quad \xrightarrow{\text{①}} \quad \text{C}_6\text{H}_5\text{-C} \text{ (Surface) over } O^{2-}\ O\ (O)^-\ \text{H},\ M^{(n-1)+}$$

$$O_2 \uparrow \text{④} \qquad\qquad \big\downarrow \text{②}$$

$$\text{C}_6\text{H}_5\text{-C(=O)(OH)} \text{ over } \square\ \square\ O^{2-},\ M^{(n-2)+} \quad \xleftarrow{\text{③}} \quad \text{C}_6\text{H}_5\text{-C} \text{ over } O^{2-}\ O^{2-}\ OH^-$$

Explain the course of the reaction with the steps 1–4.

Exercise 8.9

A BASF process poceeds according to the following equation:
$$4\,NO + 4\,NH_3 + O_2 \longrightarrow 4\,N_2 + 6\,H_2O$$

a) What is the significance of the process and what is it called.
b) Catalysts and temperature range?

9 Catalysis Reactors

The selection and design of a catalysis reactor depends on the type of process and fundamental process variables such as residence time, temperature, pressure, mass transfer between different phases, the properties of the reactants, and the available catalysts [16].

The prerequisites for successful reactor design are the coupling of the actual microkinetics of the reaction with the mass and energy transfer and the determination of fluid-dynamic influences such as backmixing, residence time distribution, etc. The factors that influence the modeling of a reactor are summarized in Figure 9-1 [11].

The choice and calculation of the reactor for a specific chemical reaction involves solving the following problems, on the basis of theoretical knowledge or by more empirical considerations:

- Choice of the reactor type according to the flow behavior of the fluid
- Heat removal
- Heat and mass transfer
- Fluid dynamics

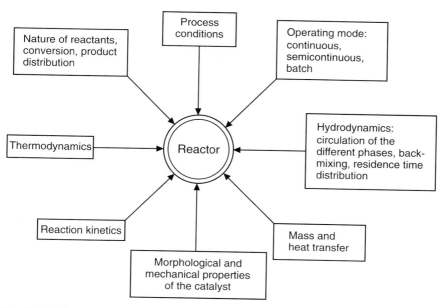

Fig. 9-1. Influences on the design of catalysis reactors

Catalysis reactors can be classified according to their phase conditions. The most important industrial reactor types are two-phase reactors for the system gas/solid and three-phase reactors for the system gas/solid/liquid [13, 15].

Understanding the fundamental reactor types requires knowledge of the design equations of reaction engineering, which will be treated in short form here. Details can be found in text books dealing with chemical reaction engineering [6, T26].

Catalytic gas-phase reactions are generally carried out in continuous fixed-bed reactors, which in the ideal case operate without backmixing. The model reactor is the ideal plug flow reactor, the design equation of which is derived from the mass-balance equation. As we have already learnt, in heterogeneous catalysis the effective reaction rate is usually expressed relative to the catalyst mass m_{cat}, which gives Equation (9-1). The left side of this equation is known as the time factor; the quotient is proportional to the residence time on the catalyst.

$$\frac{m_{cat}}{\dot{n}_{A,0}} = \int_0^{X_A} \frac{dX_A}{r_{eff}} \qquad (9\text{-}1)$$

m_{cat} = catalyst mass, kg
$\dot{n}_{A,0}$ = feed rate of starting material A
X_A = conversion of A

In an ideal tube with plug flow profile, the reaction rate is not constant; it varies in the direction of flow. Therefore, a pronounced temperature profile develops along the length of the reactor. Because the mathematical expression for r_{eff} is often complex, the integral in Equation (9-1) must generally be solved numerically. The feed rate $\dot{n}_{A,0}$ can be determined from the known production capacity of the reactor. Thus, Equation 9-1 allows the catalyst mass and therefore the reactor volume to be calculated from the target quantity conversion and the kinetics. This shows the fundamental importance of kinetics in reaction engineering.

The counterpart of the ideal plug flow reactor is the ideal continuous stirred-tank reactor with complete backmixing of the rection mass. Because of the ideal mixing, the reaction rate is constant, and a simple design equation is obtained for the catalysis reactor (Eq. 9-2).

$$\frac{m_{cat}}{\dot{n}_{A,0}} = \frac{X_A}{r_{eff}} \qquad (9\text{-}2)$$

The graphical depiction of the two design equations in Figure 9-2 clearly shows the advantages of the tubular reactor compared to the stirred tank. By plotting $1/r_{eff}$ against X_A, the time factor can obtained as the area under the curve for the tubular reactor or the corresponding straight line of the continuous stirred tank. While the catalyst mass or reactor volume is proportional to the area under

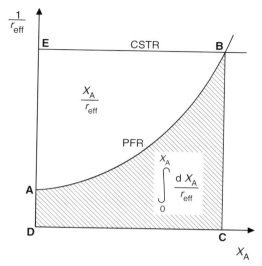

Fig. 9-2. Comparison of a continous stirred tank reactor (CSTR) with a plug flow reactor (PFR)

the curve ABCD for the plug flow reactor, the much larger rectangular area BCDE applies for the continuous stirred tank. In the majority of cases of simple reactions, the stirred tank requires a larger reactor volume than the tubular reactor, and the ratio becomes increasingly unfavorable with increasing conversion. Thus, the degree of backmixing is a decisive quantity in the design of catalysis reactors. However, the continuous stirred tank has the advantage that it can be operated isothermally, and in contrast to the tubular reactor there is no temperature profile in the homogeneous reaction space.

9.1 Two-Phase Reactors [13, 15]

Gas-phase reactions in the presence of solid catalysts have numerous technical advantages. They can generally be carried out continuously at low to medium pressure. In comparison to liquid-phase processes, they usually require higher reaction temperatures and therefore thermally stable starting materials, products, and catalysts. For this reason, the selectivity of gas-phase processes is often lower than that of liquid-phase processes.

Of major importance in this type of reaction is a large surface area of the solid. Depending on the nature of the solid (particle size, porosity, etc.), the required residence time, the mass-flow mode, and the heat transfer, a wide range of different reactors are used (Fig. 9-3).

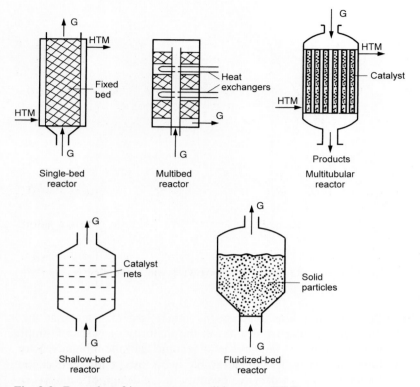

Fig. 9-3. Examples of important gas–solid reactors [T26]

The most important factors to be considered in the design of such reactors are:

1) Residence-time distribution: influence on conversion and selectivity.
2) Temperature control: maintenance of temperature limits, axially and radially; minimum temperature difference between reaction medium and catalyst surface, as well as within the catalyst particle.
3) Catalyst lifetime and catalyst regeneration.
4) Pressure drop as a function of catalyst shape and gas velocity.

The most widely used types of reactor for heterogeneously catalyzed reactions in the chemical and petrochemical industries are fixed-bed and fluidized-bed reactors [T26]. The most important reactors for heterogeneously catalyzed reactions are the fixed-bed reactors. They can be classified according to the manner in which the temperature is controlled into reactors with adiabatic reaction control, reactors with autothermal reaction control, and those with reaction control by removal or supply of heat in the reactor. Some of the well-known reactor designs are discussed below.

Single-Bed Reactor

The single-bed reactor is the simplest catalysis reactor. It is completely filled with catalyst and is mainly used for thermally neutral and autothermal gas reactions. Owing to its design, the pressure drop is high, and the residence-time distribution has a major influence on the selectivity and conversion of the reaction. Of particular importance is the maintenance of temperature limits, both axially and radially, as heat removal is naturally poor. An advantage is the ease of catalyst regeneration.

Process Examples:

- Isomerization of light gasoline: 400–500 °C, 20–40 bar H_2 pressure, Pt/Al_2O_3 catalyst.
- Catalytic reforming of solvent naphtha: cascade of 3–5 single-bed reactors, 450–550 °C, 20–25 bar H_2 pressure, $Cr_2O_3/Al_2O_3/K_2O$ catalyst.
- Hydrocracking of heavy hydrocarbons: 400–500 °C, 20–60 bar H_2 pressure, oxidic or sulfidic hydrogenation catalysts (Mo/W, Co/W) on acidic supports.

Multibed Reactor

This type of reactor contains several separate, adiabatically operated catalyst beds, allowing defined temperature control. Several methods of cooling are possible: internal or external heat exchangers or direct cooling by introduction of cold gas (quench reactor). The multibed reactor is particularly suitable for high production capacities.

Process Examples:

- Ammonia synthesis: several adiabatically operated catalyst layers with interstage cooling, 400–500 °C, 200–300 bar, iron oxide catalyst.
- Methanol synthesis by the high-pressure process: CO/H_2, 350–400 °C, 200–300 bar, Zn/Cr oxide catalyst, quench reactor.
- Contact process: oxidation of SO_2 to SO_3, 450–500 °C, V_2O_5 catalyst, external heat exchanger.

Multitubular Reactors

In these reactors the catalyst is located in a bundle of thin tubes (diameter: 1.5–6 cm), around which the heat-transfer medium (boiling liquid, high-pressure water, salt melt) flows, giving intensive heat exchange. Multitubular reactors with up to 20 000 or more parallel tubes are used preferentially for strongly endo- or exothermic reactions. The high flow rate in the tubes leads to a relatively uniform residence time, so that the reactors can be modeled as almost ideal tubes. Due to the nonadiabatic process control, a characteristic axial temperature gradient becomes established. The radial temperature profile in the catalyst bed must also be taken into account. Owing to the design of the reactor, changing the catalyst is a laborious process that can be carried out at most twice per year.

Process Examples:

- Low-pressure methanol synthesis: 260–280 °C, 45–55 bar, Cu/ZnO catalysts.
- Oxidation of ethylene to ethylene oxide: 200–250 °C, supported silver catalyst.
- Hydrogenation of benzene to cyclohexane: 250 °C, 35 bar H_2, Ni catalysts.
- Dehydrogenation of ethylbenzene to styrene: 500–600 °C, endothermic, Fe_3O_4 catalysts.

Shallow-Bed Reactors

In these reactors the catalyst is present in the reactor in the form of a thin packed bed or metal net. They are used for reactions with very short residence times and mostly for autothermally operated, heterogeneously catalyzed gas reactions at high temperatures.

Process Examples:

- Dehydrogenation of methanol to formaldehyde: methanol, air and steam are passed over a 5–10 cm high bed of silver crystals.
- Combustion of ammonia to form nitrous gases (Ostwald nitric acid process): cold air and ammonia are introduced, with an excess of air such that the heat of reaction is consumed in heating the initial mixture; 900 °C, Pt/Rh nets.
- Ammoxidation of methane (Andrussow process for the production of HCN): methane, ammonia, and air are passed over a Pt or Pt/Rh net at 800–1000 °C.

Fluidized-Bed Reactors

In a fluidized-bed reactor, finely divided catalyst particles of diameter 0.01–1 mm are held in suspension by flowing gas. This widely used technique allows large volumes of solid to be handled in a continuous process. The factors for the formation of the fluidized state are the gas velocity and the diameter of the particles. Large-scale industrial reactors are operated with fine-grained catalysts and high gas velocities to give a large solid–gas exchange area and high throughput.

The thorough mixing of the solid leads to effective gas–solid heat exchange with an excellent heat-transfer characteristic and hence a uniform temperature distribution in the reaction space. Heat-transfer coefficients are typically 100–400 $kJ\,m^{-2}\,h^{-1}\,K^{-1}$ and for small particles can be as high as 800 $kJ\,m^{-2}\,h^{-1}\,K^{-1}$.

For fine particles and at high reaction rates, circulating fluidized-bed reactors with separation and recycling of the solid are particularly suitable.

To give a conversion comparable to that of a fixed-bed reactor, a fluidized-bed reactor must be considerably larger. Disadvantages are the broad residence-time distribution of the gas, which favors side reactions; attrition of the catalyst particles; and the difficult scale-up and modeling of this type of reactor.

Process Examples:

– Ammoxidation of propene to acrylonitrile (SOHIO process): a mixture of propene, ammonia, and air reacts on a Bi/Mo oxide catalyst; 400–500 °C, 0.3–2 bar, high gas throughput, small catalyst particles (mean particle diameter ca. 50 μm); the high heat of reaction is removed by cooling coils incorporated in the fluidized bed.
– Oxidation of naphthalene or *o*-xylene to phthalic anhydride: The liquid starting material is injected into the fluidized bed, which has a temperature of 350–380 °C; large excess of air, V_2O_5/silica gel catalyst, low gas throughput, catalyst particle diameter of up to 300 μm.
– Catalytic cracking of kerosene or vacuum distillate to produce gasoline: capacities up to 3×10^6 t/a, 450–550 °C, aluminosilicate catalysts.

9.2 Three-Phase Reactors [7, 12]

The reaction of gases, liquids, and dissolved reactants on solid catalysts requires intensive mixing to ensure fast mass transfer from the gas phase to the liquid phase and from the liquid phase to the catalyst surface. Three-phase reactions between gaseous and liquid reactants and solid catalysts are often encountered in industrial chemistry. A well-known example is the hydrogenation of a liquid on a noble metal catalyst. Conducting the process in the liquid phase has advantages and disadvantages, which we will briefly discuss.

The generally low reaction temperatures allow the production of heat-sensitive compounds and the use of thermally less stable but particularly active or selective catalysts such as:

– Solid–liquid phase (SLP) catalysts
– Ion-exchange catalysts
– Immobilized transition metal complex catalysts

Liquid-phase processes generally give higher space–time yields than gas-phase processes. The higher heat capacity and thermal conductivity of liquids leads to better heat transfer in the catalyst layer and in the heat exchangers. Heat can be removed very effectively by evaporative cooling. With liquids, the reactivity can

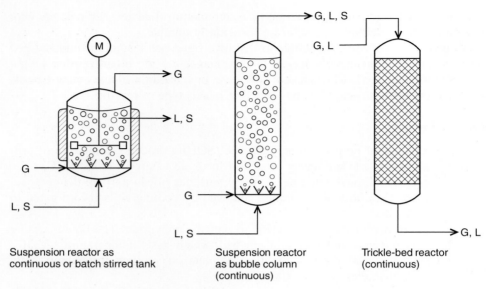

Fig. 9-4. Three-phase reactors

be influenced by, for example, suppressing secondary reactions in the liquid phase and by modification of the active centers of the catalyst.

Disadvantages of liquid-phase processes are:
- Separation and purification of the product streams is laborious
- Separation of suspended catalyst from the reaction products is often difficult
- Mass transfer is hindered by the liquid phase, and the necessary intensive mixing of the material streams requires mechanically stable catalysts and supports and often high pressures

Depending on the arrangement of the catalyst, three-phase reactors can be classsified as:

- Fixed-bed reactors with a stationary catalyst packing
- Suspension reactors, in which the catalyst is finely dispersed in the liquid (Fig. 9-4)

9.2.1 Fixed-Bed Reactors

Fixed-bed reactors contain a bed of catalyst pellets (diameter 3–50 mm). The catalyst lifetime in these reactors is greater than three months. The best known design is the trickle-bed reactor [8, 10].

In a trickle-bed reactor the liquid flows downwards through a packed catalyst bed, while the gas can flow cocurrently or countercurrently to the liquid. The gas phase, which is present in excess, is the continuous phase. In the cocurrent trickle-bed reactor (Fig. 9-5), the gas/liquid mixture leaving the bottom of the reactor is separated, and the gas is recycled. The advantages of this type of reactor are the good residence-time behavior of the liquid and gas streams and the possibility of operating with high liquid flows. In the simplest case the flow of the liquid phase can be desribed as plug flow (ideal tube). Backmixing is not a problem provided the catalyst bed is sufficiently long (at least 1 m).

Average values for the liquid flow are $10-30 \text{ m}^3 \text{m}^{-2} \text{h}^{-1}$, and for the gas flow $300-1000 \text{ m}^3 \text{m}^{-2} \text{h}^{-1}$. Solid–liquid separation is not necessary. Disadvantages are the poor heat removal and the occurrence of hot spots with potential instabilities. However, since the reactors are generally operated adiabatically, the relatively poor heat removal is not necessarily a problem.

Stream formation in large-diameter reactors and wall channeling in small-diameter reactors can lower reactor performance. Often the catalyst is not fully exploited owing to incomplete wetting by the liquid and low mass-transfer rates together with low residence times within the catalyst pellets.

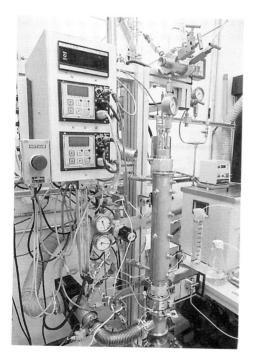

Fig. 9-5. Pilot plant with 0.2 L trickle-bed reactor (Hoffmann-La Roche, Kaiseraugst, Switzerland)

Trickle-bed reactors are widely used in petrochemical hydrogenation processes and in the production of basic products. They are being used increasingly for the manufacture of fine chemicals.

Process Examples [14, T26]:

- Petrochemistry: desulfurization, hydrocracking, refining of cude oil products (e. g., hydrogenative refining of tar fractions from low-temperature carbonization), 300–350 °C, 220 bar, $NiS/WS_2/Al_2O_3$ catalysts).
- Synthesis of butynediol from acetylene and formaldehyde: reactor height 18 m, diameter 1.5 m, 100 °C, 3 bar, copper acetylide catalyst, introduction of cold acetylene at various points in the reactor.
- Selective hydrogenation (cold hydrogenation) of acetylene and allene contained in C_4 fractions: up to 50 °C, 5–20 bar, supported Pd catalysts.
- Hydrogenation of aldehydes and ketones to alcohols: 100–150 °C, up to 30 bar, Ni, Pd, Pt catalysts.
- Hydrogenation of butynediol, adiponitrile, and fatty acid esters.
- Reduction of adiponitrile to hexamethylenediamine: 100–200 °C, 200–400 bar, Co or Ni on Al_2O_3.
- Fine chemicals: hydrogenation of quinones, sugars, lactones, substituted aromatic compounds.

Small trickle-bed reactors, operated in batch mode by recycling the liquid phase, are also used, for example, for the hydrogenation of trifluoroacetic acid [11].

9.2.2 Suspension Reactors [2, 4]

In suspension reactors, gas and catalyst particles are distributed in a relatively large volume of liquid. Catalyst concentrations are typically less than 3% with particle sizes of less than 0.2 mm. In general, the reactants (L and G) are introduced into the lower part of the reactor together with the catalyst (S), which is suspended in the liquid. In the upper part of the reactor, the unconsumed gas is separated or removed together with the liquid product (L) and the suspended catalyst (S). In this system, the liquid is the continuous phase, in which the gas is dispersed as bubbles. Suspension reactors behave largely as gas–liquid systems, and little energy is required for suspension.

Suspension reactors can be regarded as isothermal with a behavior that approximates that of an ideal stirred tank. The reactors are followed by a phase-separation unit in which the liquid is separated from the catalyst and the gas. The gas and the catalyst can be partially recycled.

The advantage of suspension reactors is the effective exploitation of the catalyst, which is completely wetted by the liquid. Because of the small particle size, diffusion processes within the catalyst play no role, and owing to the good temperature control, local overheating can not occur. This type of reactor is particularly suitable for rapidly deactivated catalysts since rapid catalyst replacement is possible.

Disadvantages are potential problems in separating the catalyst and the risk of fractionation and sedimentation of the catalyst in the reactor. Since the residence-time behavior is similar to that of a continuous ideal stirred tank, lower conversions are attained compared to a fixed-bed reactor. A comparison of the two most important three-phase reactors — the trickle-bed reactor and the suspension reactor — is given in Table 9-1.

The majority of suspension reactors are stirred tanks and bubble columns (see Fig. 9-4). Other industrially important variants of the suspension reactor are the loop and Buss (jet) loop reactors, which achieve better exploitation of the catalyst by recirculating it in a loop (Fig. 9-6).

Table 9-1. Comparison of trickle-bed and suspension reactors

Characteristic	Trickle-bed reactor	Suspension reactor
Process mode	continuous	mostly batch
Degree of automation	high	low
Conditions (temperature, pressure)	moderate	mild
Temperature	depends on position	uniform
Pressure drop	high	low
Reactor performance	high	moderate
Plant size	easily extended by tube bundles	limited
Selectivity	low	high
Liquid content	low	high
Residence time behavior		
– liquid	ideal plug flow reactor	ideal stirred tank – plug flow reactor with axial dispersion
– gas	ideal plug flow reactor	plug flow reactor with axial dispersion
Catalyst effectiveness factor	very low	ca. 1
Catalyst performance	low	high
Heat usage	unfavorable	favorable
Applicability	limited (selectivity)	universal
Particular suitability	high liquid feeds	in case of rapid catalyst deactivation

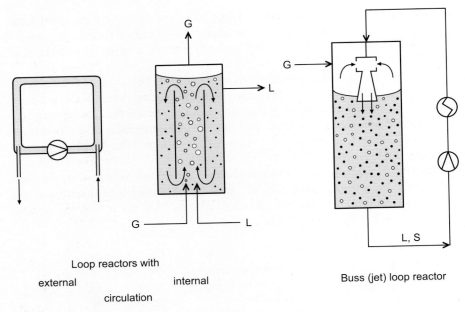

Fig. 9-6. Variants of the suspension reactor

Three-phase bubble columns are operated with the liquid flowing cocurrently with the rising gas. They are used when the mass-transfer resistance lies on the liquid side and the reaction is relatively slow. The gas is introduced at the bottom of the reactor through perforated plates or sintered disks, and the reactors often incorporate sieve trays. Without internals the liquid is almost ideally mixed at high gas velocities, and this results in good heat transfer. The residence-time distribution of the gas and the liquid corresponds approximately to that of a cascade of stirred-tank reactors. The advantages of bubble columns are the simple, inexpensive design and their versatility. Reaction volumes of up to several cubic meters are possible. Bubble columns with internal circulation are also used (loop and air-lift reactors).

In *loop reactors*, the liquid is completely mixed in a relatively small reactor, which gives good heat removal.

In the *Buss loop reactor*, the liquid with the suspended catalyst form a jet that entrains the gas, finely dividing it. The high flow rates lead to intensive turbulence and a high interfacial area between the small gas bubbles and the suspension. An external heat exchanger in the loop allows isothermal operation and very effective removal of the heat of reaction from the system, even with highly exothermic reactions. However, Buss loop reactors can only be operated in a discontinuous mode and require special, highly abrasion resistant catalysts.

Process Examples:

- Liquid-phase hydrogenation of chlorinated aromatic nitro compounds; for example, conversion of *p*-chloronitrobenzene to *p*-chloroaniline in a stirred tank with a powder catalyst (Ni/SiO$_2$ or Pd on activated carbon).
- Continous hydrogenation of fats in a chamber reactor (several stirred chambers one above the other), narrow residence-time spectrum is an advantage, 150–200 °C, 5–15 bar.
- Hydrogenation of benzene to cyclohexane in a bubble column: 200–225 °C, ca. 50 bar, Raney nickel (10–100 µm), removal of heat of reaction from the suspension in an external circuit. Cyclohexane is removed a a gas.
- Hydrogenation of fatty esters to fatty alcohols in a bubble column.
- Hydrogenation of fats and fatty acids in a tank reactor with a turbine stirrer (110–120 rpm); H$_2$ is introduced through a distributor at the bottom, 150–200 °C, up to 30 bar, Ni/Cu catalysts.
- Hydrogenation of oil in a Buss loop reactor.
- Hydrogenation of 2-ethylanthraquinone to 2-ethylanthraquinol: bubble column with parallel chambers, suspended catalyst.

The choice of the "right" reactor for a given catalytic reaction can often not be answered unambiguously, as shown, for example, by the fact that different technologies compete in the high-pressure hydrogenation of adiponitrile in the presence of ammonia (Table 9-2) [11].

In new plants for the hydrogenation of fine chemicals there is currently a trend to use Buss loop reactors rather than conventional stirred tanks. Today, trickle-bed reactors are generally preferred to suspension reactors for hydrogenation processes. The examples of the hydrogenation of glucose to sorbitol and of esters to alcohols demonstrate the dilemma of reactor choice. Formerly, suspension reactors with Raney nickel or copper chromite catalysts were used, but today trickle-bed reactors with novel noble metal catalysts are preferred. The following advantages are claimed for the trickle-bed reactors:

Table 9-2. Various technologies for the hydrogenation of adiponitrile [11]

Company	Reactor	Temperature control
BASF	trickle bed	cooling and partial recycling of liquid phase
Phillips	suspension loop reactor	
DuPont	sump reactor (liquid and gas are passed cocurrent from below into catalyst fixed bed)	several catalyst beds with intermediate cooling
ICI	fixed bed	cooling of recycled off-gas
Vickers-Zimmer	multitubular reactor with downward cocurrent operation	evaporative cooling with inert solvents

- No loss of metal; higher product quality (no contamination)
- Fewer side reactions in the liquid phase due to the lower liquid holdup

The major disadvantage is the risk of poor temperature control owing to the occurrence of hot spots in the catalyst bed. Examples of this are:

- Benzene is formed as a side product in the hydrogenation of cyclohexene
- In the hydrogenation of benzoic acid, decarboxylation of the cyclohexane carboxylic acid product can occur. Therefore, a cascade of stirred-tank reactors is preferred here

The catalyst form is is also decisively influenced by the chosen reactor type. It has been found experimentally that at particle sizes below 0.1 mm pore diffusion is rarely limiting, whereas at particle sizes above 5 mm, pore diffusion is always dominant. This is the reason why shell catalysts are advantageously used in trickle-bed reactors. The diffusion limitation need not always result from the transport of the gas in the pores; it can also be due to the substrate. Such effects are found with long-chain organic molecules, for example:

- Hydrogenation of linoleic esters (C_{18}): a shell catalyst with Pd on activated carbon is recommended
- Hydrogenation of C_{12}–C_{22} nitriles: a large-pore catalyst based on Ni/MgO/SiO_2 is recommended [11]

These few examples show how the complexity of three-phase processes greatly complicate reactor modeling and scale-up. The engineer responsible for reactor design must be familiar with reaction engineering and should also work in close cooperation with the synthetic chemist and the catalyst expert.

9.3 Reactors for Homogeneously Catalyzed Reactions [3, 5]

Homogeneously catalyzed reactions with dissolved transition metal complexes are generally carried out in the usual two-phase reactors for gas–liquid systems. The standard reactor is the batch or continuous stirred tank. Since diffusion problems are rarely encountered in homogeneous catalysis, the reaction engineering is much simpler than for heterogeneously catalyzed reactions.

Efficient mixing of the two phases is important as this determines the exchange surface area between the gas and the liquid. Modern stirred tanks (Fig. 9-7) are often equipped with gasifying stirrers, in which the gas is drawn in at the top of the drive shaft and then finely dispersed by the stirrer blades.

Since we have already become familiar with the most important reactors for gas–liquid reactions, we will restrict ourselves here to a few examples of processes in special reactors.

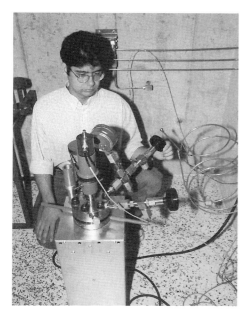

Fig. 9-7. 0.5 L stirred autoclave reactor in a high-pressure cell (FH Mannheim)

Bubble-column Reactors:

- Homogeneously catalyzed air oxidation of hydrocarbons (e.g., of toluene to benzoic acid): 130–150 °C, 1–10 bar, Mn or Co salts as catalyst.
- Oxidation of *p*-xylene to terephthalic acid with Co/Mn salts and bromide at 100–180 °C and 1–10 bar.
- Oxidation of ethylene to acetaldehyde (Wacker process): 100–120 °C, 1–10 bar, $PdCl_2/CuCl_2$ catalyst.
- Oxo synthesis: reaction of ethylene with synthesis gas to form propanal, 100–150 °C, 200–300 bar, propanol solvent, $[HCo(CO)_4]$ catalyst.

Loop Reactors:

- Oxo synthesis.
- Carbonylation of methanol with CO to produce acetic acid, 150 °C, 200–300 bar, CoI_2 catalyst (Rh catalysts preferred nowadays).

Stirred Tanks:

- Hydroformylation of olefins with Co or Rh catalysts.
- Low-pressure hydroformylation of propene to butanals with water-soluble Rh phosphine complexes (Rhône-Poulenc/Ruhrchemie process): 50–150 °C, 10–100 bar, 10–100 ppm Rh.
- Polymerization of ethylene with $TiCl_4/Al(C_2H_5)_3$ at 70–160 °C and 2–25 bar.

Exercises for Chapter 9

Exercise 9.1

The design equation for a catalysis reactor is:

$$\frac{m_{cat}}{\dot{n}_{A,0}} = \int_0^{X_A} \frac{dX_A}{r_{eff}}$$

a) To which quantity is the left side of the equation proportional?
b) Prepare a graphical depiction of the integral.
c) To which ideal reactor does the catalysis reactor correspond?
d) What is meant by the term effective reaction rate?

Exercise 9.2

The driving force of a reaction is much smaller in a reactor with backmixing than in a reactor without backmixing. Why?

Exercise 9.3

The kinetics of a second-order heterogeneously catalyzed gas-phase reaction of the type A→R is investigated in a differential circulating reactor. Under isothermal conditions with a reactor feed stream of $\dot{V}_0 = 1$ L/h and $c_{A,0} = 2$ mol/L and a catalyst quantity of 3 g, an outlet concentration of $c_A = 0.5$ mol/L was obtained.

a) Calculate the rate constant for the reaction.
b) What quantity of catalyst would be required in an integral reactor (ideal plug flow reactor), in which a conversion of 80% is to be achieved for a feed stream of 1000 L/h with a concentration of $c_{A,0} = 1$ mol/L?
c) The same reaction is carried out in a reactor with complete backmixing. What quantity of catalyst is required (conditions as in b).

Discuss the result.

Exercise 9.4

The catalytic dealkylation of toluene is carried out over a bifunctional catalyst at 660 °C and 30 bar:

$$C_6H_5-CH_3 + H_2 \longrightarrow C_6H_6 + CH_4$$
$$\quad\quad T \quad\quad\quad H \quad\quad B$$

The reaction follows a rate law of the Langmuir–Hinshelwood type:

$$r = \frac{kK_T p_T p_H}{1 + K_T p_T + K_B p_B}$$

At 660 °C:

$k = 0.202 \text{ mol kg}^{-1} \text{ bar}^{-1} \text{ h}^{-1}$
$K_T = 0.9 \text{ bar}^{-1}$
$K_b = 1.0 \text{ bar}^{-1}$

The molar ratio of toluene ($M = 92$) to hydrogen in the initial mixture is 1/10. Calculate the catalyst mass for a reactor handling 2000 t/a toluene with 60% conversion (1 year = 8000 operating hours).

Exercise 9.5

Name industrial processes that are carried out in the following reactors (one per reactor):

- Single-bed reactor
- Tubular reactor
- Multibed reactor
- Shallow-bed reactor
- Fluidized-bed reactor

Exercise 9.6

In the oxidation of methane to formaldehyde, CO_2 is the main side product. At the reaction temperatures required to oxidize methane, formaldehyde is unstable and is easily oxidized to CO_2. Since both reactions are exothermic, the catalyst temperature rises, and this favors further oxidation and catalyst sintering.

Which recommendations can be made for the choice of catalyst and reactor?
☐ High porosity
☐ Low porosity
☐ High thermal conductivity
☐ Low thermal conductivity
☐ Tubular reactor
☐ Fluidized-bed reactor
☐ Shallow-bed reactor
☐ Single-bed reactor

Exercise 9.7

Compare trickle-bed and suspension reactors according to the following criteria:
- Temperature distribution
- Selectivity
- Residence-time behavior of the liquid
- Catalyst particle diameter
- Catalyst effectiveness factor
- Catalyst performance

Exercise 9.8

Phthalonitrile is produced industrially from o-xylene and NH_3/O_2.

a) What type of reaction is involved?
b) What type of reactor can be recommended for the process?

Exercise 9.9

In the hydrogenation of α-methylstyrene, varying degrees of catalyst effectiveness factors η were found:

A) Supported Pd/Al_2O_3 catalyst with 0.03 mm particle diameter in a suspension reactor: $\eta = 1$.
B) The same supported catalyst with 8.25 mm particle diameter in a trickle-bed reactor: $\eta = 0.007$.

Explain this dramatic difference.

10 Economic Importance of Catalysts

The modern industrialized world would be inconceivable without catalysts [1]. Around 85% of all chemical products pass through at least one catalytic stage during their production. It has been estimated that about 17% of the US national product was generated with the direct or indirect aid of catalysts. The total commercial value of all catalysts worldwide is over $\$5 \times 10^9$. In crude oil refining processes the catalyst costs amount to only about 0.1% of the product value, and for petrochemicals this value is about 0.22%.

Since the special properties of the catalysts decisively influence the economics of a process, their true economic importance is considerably higher than their "market value". The value of the products that are produced with catalysts are an order of magnitude higher [2,6].

Although market estimates vary widely—for example, there are no figures available for the considerable internal consumption of the chemical industry—the key importance of industrial catalysts can be recognized from the above data. In this chapter we shall treat the catalysts according to their area of use.

The traditional area in which catalysts have been used for over 100 years is the chemical industry [8]. For example, the contact process for the production of sulfuric acid was introduced as early as 1880. In the 1920s and 1930s catalysts for crude oil processing came on the market, initially in the USA and later in Europe, mainly after World War II. Environmental catalysts became of importance from 1970 onwards. They can be divided into automobile and industrial catalysts (Fig. 10-1), the latter being those that purify off-gases from power stations and industrial plants. The environmental catalysts are not part of any wealth-creating process; instead, they contribute to protection of the environment and thus to a generally higher standard of living. Therefore, their importance can scarcely be expressed in monetary terms.

Thus the catalyst market can be divided into four main areas (Fig. 10-2) [3]:

– Industrial and automobile environmental catalysts
– Chemistry catalysts
– Petroleum catalysts

The area of automobile catalysts arose in the 1970s in the USA and Japan due to the introduction of government regulations requiring control for automobile exhaust emissions. Ten years later the European countries followed suit and the market grew rapidly. The automobile catalyst market is by far the fastest growing and today has a value of about $\$2 \times 10^9$, 50% of which is accounted for by noble me-

10 Economic Importance of Catalysts

Fig. 10-1. Catalytic afterburning of the off-gases from a cyclohexanone plant (BASF, Antwerp)

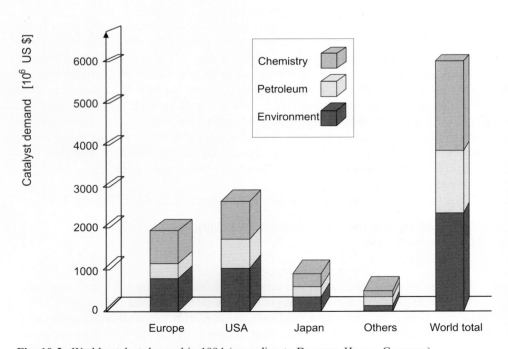

Fig. 10-2. World catalyst demand in 1994 (according to Degussa, Hanau, Germany)

tals. Owing to increasingly strict emission-control laws, further growth can be expected in this area.

The industrial environmental catalysts are also a growth market; their volume has doubled in the last five years. Automobile catalysts are generally honeycomb supports doped with Pt or Pd. For the denitrogenation of power station flue gases by the SCR process, mainly honeycomb catalysts made of V_2O_5, WO_3, MoO_3, and TiO_2 are used. Market data for industrial catalysts only reflect the cost of the catalysts, which account for only about 10% of the cost of a complete off-gas purification plant.

The chemistry and petrochemistry catalysts belong to the traditional well-developed areas. A growth rate of ca. 3%/a is expected. Nevertheless, these catalysts still offer a major development potential [7]. Figure 10-3 shows the market distribution for chemistry catalysts according to type of product [9]. Almost half of the catalysts are used for polymerization processes. About one-quarter of all chemical products are produced in heterogeneously catalyzed oxidation and ammoxidation processes, which corresponds to about 18% of the market value of catalysts. The most important processes are the selective allyl oxidation, epoxidation processes, and the oxidation of alkanes and aromatics. Organic synthesis includes many different reactions, for which an overall growth rate of ca. 6% per year is expected.

The production of synthesis gas and synthesis gas reactions such as methanol synthesis and carbonylation reactions also play an important role in the catalyst business. For hydrogenations, mainly the elements of the platinum group and nickel, but also Co and Cu, are used on a wide range of support materials, such as

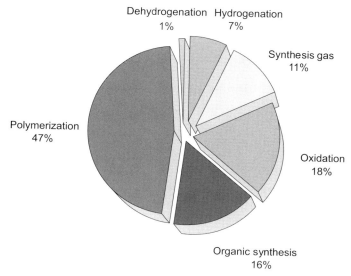

Fig. 10-3. Areas of application of chemistry catalysts (percentage economic value) [9]

activated carbon and Al_2O_3. The Western European market volume for hydrogenation catalysts in 1989 was around $DM\,370 \times 10^6$, which is proportionally higher than on a worldwide basis. The smallest segment in Figure 10-3 is occupied by the dehydrogenation processes, the most important of which are the dehydrogenation of ethylbenzene to styrene and of butane to butadiene.

An important task of catalysts is to lower the raw material and energy requirements of chemical reactions. An important target in catalyst development is to obtain high yields in conjuction with high selectivities, since decreased formation of side products saves raw materials and reduces pollution, it also leads to energy savings in separation processes such as distillation and extraction.

The economic benefits of catalyst development are particularly large in the case of mass products [8], as can be shown by a rough calculation [5]. In the production of styrene from ethylbenzene, selectivities of around 90% are achieved today. If a company producing 500 000 t/a of styrene achieves a 2% higher selectivity by improving the catalyst, then this would lead to a saving of 10 000 t/a of ethylbenzene with a value of $DM\,2 \times 10^6$. With a worldwide catalyst demand of 3000–4000 t/a, this result could not be achieved if the same company manufactured and sold the catalyst, even for a large market share. Thus the actual profit generated by catalysts is much higher than would be expected from their commercial value alone.

In the case of petrochemistry catalysts, the cracking catalysts are the most important in terms of volume and economic value. They are mainly amorphous and crystalline aluminosilicates. The second most important group are the Co/Mo catalysts for the hydrorefining of crude oil fractions of various types and boiling ranges (S and N removal). Other heterogeneous catalysts of lesser importance are:

- Reforming catalysts (Pt, Re)
- Hydrocracking catalysts (Ni–Mo on amorphous aluminosilicates and Pd or Pt on crystalline aluminosilcates)
- Hydrogenation catalysts (Co, Ni, Pt, Pd)

Petroleum processing and petrochemistry plays a more important role in the USA than in Europe (Fig. 10-2).

As we have seen, noble metal catalysts play a special role in all four catalyst segments. It is estimated that ca. 40% of the value of all catalyst sales is accounted for by noble metal catalysts [5].

In the case of molybdenum, annual demand is around 300 000 t/a, of which about 90% is used in metallurgy and only about 8% for chemistry. The corresponding figures for cobalt and tungsten are ca. 9% and ca. 4%, respectively. Thus chemical processes account for only a small fraction of the total consumption of the metals.

Table 10-1 summarizes the use of noble metals for catalysts in the western world [4]. The dominant role of automobile exhaust catalysts can clearly be seen.

Table 10-1. Important noble metals as catalysts [4]

Metal	World production (t)	Fraction for catalysts [%]	
Pt	ca. 100	auto exhaust	35
		chemistry	6
		petroleum industry	2
Pd	ca. 100	auto exhaust	6
Rh	ca. 10	auto exhaust	73
		chemistry	7

The largest catalyst producer and consumer in western Europe is Germany with about one-third of total production. The leading producer of non-noble-metal catalysts is the company Sud-Chemie (Germany), followed by BASF (Germany), ICI (UK), Akzo (Netherlands), Haldor Topsoe (Denmark), and Procatalyse (France). The largest producer of noble metal catalysts is Degussa/Germany), followed by Johnson Matthey (UK), Engelhard (USA), Heraeus (Germany), and Doduco (Germany) [7].

In this chapter we have seen that catalysts play an essential role in industry, not only in economic terms but also in reducing pollution of the environment.

11 Future Development of Catalysis

11.1 Homogeneous Catalysis

Nowadays the broad sprectrum of catalytic processes would be inconceivable without homogeneous transition metal catalysis, the importance of which can be expected to grow in future [6].

The driving force for the introduction of new processes are economic considerations, which are largely influenced by the production costs of the product and product quality. The optimal exploitation of raw materials, energy saving, and the environmental friendliness of processes will still take presidence in future. Selectivity is becoming more and more the decisive factor in industrial processes, mainly as a result of increasing purity demands, for example, in polymer chemistry and in the pharmaceutical sector. Higher selectivity means that better use is made of raw materials and therefore lower formation of side products, which must be removed in expensive separation processes or pollute the environment.

It is apparent that in future, new transition metal catalysts with new ligands, newly discovered reactions, and improvements to existing processes will be introduced into industry [5,7]. Energy and raw materials politics will presumably increasingly determine the future direction of development of industrial organic chemistry. Only a few aspects can be discussed here. Homogeneous catalysis has by no means reached the limits of its potential, but is of course not easy to depart from the well-trodden paths of known technologies.

In the case of basic chemicals the chances for new catalytic processes are small, but they are better for higher value chemicals such as fine and specialty chemicals. Pharmaceuticals and agrochemicals are two areas where homogeneous catalysts have advantages. Complex molecules can often be synthesized in single-step one-pot reactions with the aid of transition metals. This sector has many potential points of overlap with biotechnology, especially enzyme catalysis [3].

Especially noteworthy is the field of asymmetric catalysis. Asymmetric catalytic reactions with transition metal complexes are used advantageously for hydrogenation, cyclization, codimerization, alkylation epoxidation, hydroformylation, hydroesterification, hydrosilylation, hydrocyanation, and isomerization. In many cases, even higher regio- and stereoselectivities are required. Fundamental investigations of the mechanism of chirality transfer are also of interest. New chiral ligands that are suitable for catalytic processes are needed.

A major disadvantage of homogeneous catalysis up to now has been that in general olefins can effectively activated but not alkanes. If it becomes possible to carry out the CH activation of alkanes in homogeneously catalyzed reactions, this would open up cheap new routes to many industrial chemicals. Research in this direction has been carried out for many years, a major target being the exploitation of methane or lower alkanes in catalytic processes. Interesting stoichiometric and also catalytic CH activation reactions have been discovered. The key reaction is the cleavage of the C–H bond with insertion of a metal center. Numerous interesting reactions are then possible, mainly giving oxygen-containing compounds such as alcohols, aldehydes, and carboxylic acids (Fig. 11-1) [7,11]. Another desirable reaction is the direct oxidation of methane to methanol.

Another area that is still of interest is the long-known synthesis gas chemistry, for example, conversion to C_2–C_4 olefins or C_1 and C_2 oxygen compounds such as methyl formate and acetic acid.

Methanol is also an important starting material for further syntheses. Interesting new routes could be based on reactions such as carbonylation, reductive carbonylation, and oxidative carbonylation. Another example is the homologization of methanol to ethanol via acetaldehyde.

A further area of major future potential is CO_2 chemistry. The chemical exploitation of the huge quantities of CO_2 that are released into the atmosphere is of great interest. Although many CO_2 complexes of transition metals and model reactions are already known, so far none has been introduced into industry. The main reactions investigated up to now are the reaction of CO_2 with alcohols and olefins to give esters and lactones and the reduction of CO_2 to CO.

Of particular interest in the long term are catalytic processes on the basis of water and air, operated with solar energy. These include the reduction of atmospheric nitrogen to ammonia or hydrazine, the activation of oxygen for use in fuel

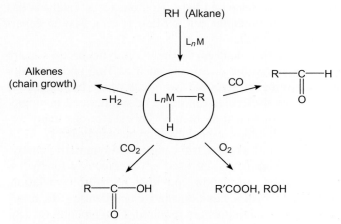

Fig. 11-1. CH activation of alkanes [7]
L_nM = catalyst (M = metal, L = ligand)

cells, and the photochemical cleavage of water to give oxygen and hydrogen. Interesting approaches involve carbonyl catalysts and clusters.

Since only 12 metals have been used as homogeneous catalysts in industrial reactions up to now, a broad-based study of the less well investigated metals (e.g., the lanthanides) is called for [7].

Changing the phase in which a homogeneous catalyst is used also has major development potential. An example is multiphase catalysis, in which the catalyst is dissolved in a solvent in which the substrate or the product is insoluble. The catalyst and product solution can then be separated by a simple phase-separation process. In particular, water-soluble catalysts for use in two-phase processes have very good future prospects. The heterogenization of homogeneous catalysts is another area where improvements are necessary and possible.

These few examples show that although homogeneous transition metal catalysis has achieved remarkable success in the last few years, there is still a very large potential for further development, both in fundamental research and in industrial application.

11.2 Heterogeneous Catalysis

Heterogeneous catalysts are among those products that will continue to exhibit development potential for the next few decades. One reason for this is that scientific knowledge about the individual steps and mechanisms of heterogeneously catalyzed reactions is still incomplete. Another is the increasing necessity to produce chemicals in an economic and environmentally friendly manner. Modern methods for the investigation of surfaces are particularly helpful in the search for new catalysts and the improvement of existing catalysts, and they make a more systematic catalyst research possible [2].

Two main areas of future catalyst development can be expected:

- Improvement of existing processes: increasing the yield and selectivity, energy savings in the production processes
- Development of new processes: use of other raw materials with the aid of new catalysts

Although a decline in research activity in the field of heterogeneous catalysis was predicted in recent years, this in fact did not happen. Instead, stricter environmental requirements and the general trend towards milder reaction conditions have meant that heterogeneous catalysis has increased in importance [3]. This is demonstrated by many examples, including:

- Better removal of harmful impurities from raw materials and intermediates
- Development of processes low in off-gases and wastewater, including those for fine chemicals

- Reducing the number of process steps by activation of simpler raw materials (e. g., alkanes)
- Replacement of expensive and less widely available catalyst components (e. g., Pt by other metals or metal oxides)

Here we will discuss some trends and perspectives of heterogeneous catalysis by means of a few examples [4].

In the beginning, the development of a new product or process is slow until a certain state of knowledge is reached, after which rapid growth is observed. The process is adopted in many sites, tested and developed further. Finally, the development process or the growth in knowledge proceeds slowly, and a state of saturation or maturity is reached (Fig. 11-2).

This S-shaped development cycle applies both to the production of chemicals (especially basic chemicals) and catalyst development. If further progress is to be made, then new routes must be explored before existing technologies reach their limits [10]. Process development often proceeds in technological leaps and bounds, as has been seen in many areas of chemistry/catalysis in the past. Often the decisive improvement to the process is only possible with the aid of catalysts:

- Methanol synthesis: replacement of high-pressure processes by medium- and low-pressure processes
- Oxo synthesis: replacement of Co by Rh catalysts
- Direct oxidation of ethylene to ethylene oxide with supported silver catalysts

Which possibilities can be expected in the future?

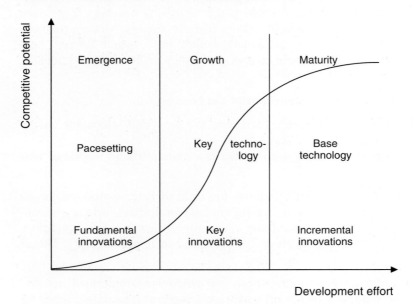

Fig. 11-2. Development cycle of a product or process

Use of Other, Cheaper Raw Materials

Raw material costs decisively influence the total manufacturing costs of many chemical products; a contribution of 70 % by the raw materials is not unusual. Thus the use of less highly refined raw materials is desirable. Therefore, efforts are being made worldwide to produce valuable chemicals from lower alkanes such as methane, ethane, propane, and butane instead of the olefins that are currently used.

Natural gas, the main component of which is methane, is of particular interest as a raw material. Currently, methane is converted to synthesis gas by steam reforming. This route involves high energy costs since a highly endothermic reaction is involved. A promising reaction is the oxidative coupling of methane to give ethane (Eq. 11-1) or ethylene, preferably with alkali or alkaline earth metal catalysts. The reaction temperatures of over 600 °C are still a disadvantage.

$$2\,CH_4 + 0{,}5\,O_2 \xrightarrow{Cat.} C_2H_6 + H_2O \qquad (11\text{-}1)$$

Another possibility is the partial oxidation of methane to oxygen-containing compounds (methanol, higher alcohols, aldehydes) or synthesis gas and dehydrogenative coupling to give aromatic compounds.

The activation of higher alkanes is also being intensively investigated. An example is the oxidative dehydrogenation of ethane, propane, and isobutane to the corresponding alkenes and oligomeric products. A pilot plant for the ammoxidation of propane has been operated by BP. Economically favorable processes would be the direct oxidation of propane to acrylic acid or of isobutane to methacrylic acid [10].

Other good prospects are offered by C_1 chemistry, especially that of methanol, the price of which has continuously fallen and is now only about twice that of ethylene. A major challenge for catalyst research is the oxidative coupling of methanol to produce ethylene glycol.

Catalytic oxidations of hydrocarbons have relatively low selectivities. Reactions with interesting perspectives are the direct oxidation of propylene to propylene oxide, of benzene to phenol, and of propane to isopropanol and acetone.

Catalysts for Energy Generation

Conventional combustion processes generally proceed at high temperatures and lead to formation of undesired nitrous oxides. Combustion catalysts are intended to achieve fast total combustion of the fuel at lower temperatures. Catalytic combustion of methane in a gas turbine has already been developed by a company in Japan, where a research society for catalytic combustion has also been established. However, the complex metal oxide catalysts do not yet have sufficient temperature stability and resistance to catalyst poisons.

Fuel cells with improved catalysts would allow the most efficient use of fossil fuels for the direct generation of electricity. There is major interest in the electrochemical reaction of synthesis gas or, better still, methane. It would be desirable to simultaeously generate energy and to produce valuable oxidation products in a fuel cell. Another interesting fuel is methanol, but technical realization has so far been unsuccessful because of the associated high activation energies.

Emission control is of the greatest importance in energy generation, and new high-performance catalysts play a key role here. For example, new catalysts that can decompose nitrous oxides into N_2 and O_2 would be of interest because the use of ammonia in the SCR process could then be dispensed with. A start has already been made in this direction: in Japan it was found that [Cu]-ZSM-5 zeolites are highly active and stable catalysts for the dissociation of NO [8].

Better Stategies for Catalyst Development

Up to now there has been no general theory for the description and prediction of heterogeneous catalytic processes. The reason is the complexities of real systems, which consist of numerous components, including structure and dispersion stabilizers, dopants, additives for increased selectivity, and many others. Therefore, there is great need for research on the behavior of catalyst surfaces that consist of several components [2]. The fundamental knowledge required, for example, to improve the selectivity of catalysts is also lacking.

The structure–activity relationships of catalysts, i.e., the connections between the production parameters, the structure, and the catalytic properties of solids are generally elucidated by empirical means. Catalyst development usually starts from a working hypothesis that is based on a semi-empirical model of the course of the reaction. However, it would be a mistake to assume that targeted design of catalysts can be achieved by means of modern surface analysis techniques and computer calculations alone.

A successful catalyst development strategy must take the chemical and physical conditions into account, from the outer shape of the catalyst to the pore structure to the active center, and from the chemical composition to the various crystalline phases to the influence of promoters. The reactor type must also be included in an overall view of the process [9].

Mechanistic concepts of the microscopic interaction between catalyst and substrate are becoming increasingly refined, and the possibilities have by no means been exhausted, so that advances are still to be expected. Here we can only discuss a few recent trends.

Shaped catalyst bodies with optimized geometries (e.g., wagonwheels, honeycombs) offer lower resistance to gas flow and lower the pressure loss in reactors. The mechanical and thermal stability of catalysts and supports is being improved. New support materials such as magnesite, silicon carbide, and zircon ($ZrSiO_4$) ceramics with modified pore structures offer new possibilities. Meso- and macropores can be incorporated into solids to accelerate transport processes, and the question of porosity will increasingly be the subject of interest.

Higher starting material purities are being achieved by the use of guard catalysts that remove catalyst poisons such as sulfur and halogen compounds, metals, and organic impurities. Here the zeolites have advantages over conventional adsorbents such as activated carbons [8].

The search for new selectivity promoters will be improved, and more and more unusual elements such as Sc, Y, Ga, Hf, and Ta will be used. The zeolites have major potential, and it is expected that especially the pentasils and the metal-doped zeolites will achieve wider application in organic syntheses. The industrial application of aluminosilicates and sheet silicates is also imminent.

Another promising class of compounds are the heteropolyacids. Depending on the reaction conditions, they can act according to three basic mechanisms: as normal surface catalysts, with involvement of the entire volume, or as pseudo-liquid phase catalysts. They have so far mainly been used in Japan for hydrogenation/dehydrogenation, selective oxidation, and acid/base reactions.

Other catalysts for acid/base reactions will also increase in importance, for example, acid modification of support materials (B/Al_2O_3, Zr/TiO_2, W/ZrO_2) and superacids, combinations of metal sulfates on metal oxides, such as $FeSO_4$ on Fe_2O_3 and $ZrSO_4$ on ZrO_2 or TiO_2 [8].

Colloids and amorphous metals and alloys are further interesting nontraditional catalysts, but there are difficulties in manufacturing them reproducibly. Other development possibilities are represented by transition metal compounds such as Mo and W carbides and nitrides, which are already being tested as potential replacements for the noble metal platinum [3]. Special areas that will develop rapidly are biocatalysis, enzyme catalysis, photocatalysis, and electrocatalysis, to name but a few.

New types of reactor, such as the membrane reactor, will in future be applied to additional areas of application. Already today this type of reactor is being used not only for homogeneous catalysis, but also for selective hydrogenation. Selective oxidation reactions in a membrane reactor appear promising.

Another challenge for the future is the exploitation of new raw materials and energy sources. New poison-resistant catalysts will be required in a few decades in order to economically process heavy crude oils, tar sands, and oil shales. Coal gasification and liquefaction will regain importance. Even if hydrogen technology and high-temperature reactions with solar energy are introduced, other chemical problems will only be solvable with the aid of catalysts [9].

In spite of the above future perspectives, it must be remembered that technological developments are not predictable. Important new developments are generally not the result of a logically designed catalyst development program; often they are surprise discoveries. Nevertheless, with the modern high-performance analytical methods and combinations thereof, new catalysts can be rapidly tested and further developed.

In future, catalysis will remain one of the most important areas of research in academia and technology.

Solutions to the Exercises

Chapter 1

Exercise 1.1

a) Homogeneous catalysis: all reactants and the catalyst (NO) are gaseous.
b) Heterogeneous catalysis: three phases.
c) Homogeneous catalysis in aqueous solution.

Exercise 1.2

	Heterogeneous catalysts	Homogeneous catalysts
Active centers	only surface atoms	all metal atoms
Concentration	high	low
Diffusion problems	yes, mass transfer controlled	mostly none; kinetically controlled reactions
Modifiability	low	high
Catalyst separation	simple	laborious

Exercise 1.3

− Severe reaction conditions and high temperatures are possible
− Wide applicability
− High thermal stability of the catalyst
− Catalyst recycling unnecessary or simple

Exercise 1.4

a) Activity: quantitative measure for the comparison of catalysts
   Selectivity: the fraction of the starting material that is converted to the desired product

b) – Reaction rate as a function of concentration under constant conditions
 – Measurement of the activation energy
 – Achievable yield per unit time and reaction space (space–time yield)

Exercise 1.5

	Homogeneous catalysis	Heterogeneous catalysis
Activation of H_2	oxidative addition	dissociative chemisorption
Activation of olefin	π complex formation	π complex formation (surface complex)

Chapter 2

Exercise 2.1

a) $[\overset{-1}{V}(\overset{0}{CO})_6]^-$

b) $\overset{-3}{Mn}(\overset{+1}{NO})_3\overset{0}{CO}$ In metal carbonyl nitrosyls, the NO ligand is present as the nitrosyl cation NO^+

c) $[\overset{+2}{Pt}(\overset{-1}{SnCl_3})_5]^{3-}$ $SnCl_3^-$ is an anionic π complex ligand with a single negative charge

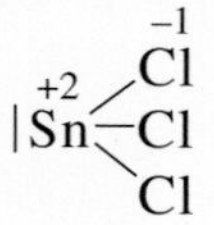

$$|\overset{+2}{Sn}\overset{-1}{\underset{Cl}{\overset{Cl}{-Cl}}}$$

d) $[\overset{+3}{Rh}\overset{-1}{Cl}(\overset{0}{H_2O})_5]^{2+}$

e) $[(\pi\text{-}\overset{-1}{C_5H_5})_2\overset{+3}{Co}]^+$ $C_5H_5^-$ acts as a ligand with a single negative charge

Solutions to the Exercises   361

f) $\overset{+1}{H_2}\overset{-2}{Fe}(\overset{0}{CO})_4$ iron carbonyl dihydride is a strong acid and therefore a hydric compound:

$$H_2Fe(CO)_4 \rightleftharpoons 2\,H^+ + Fe(CO)_4^{2-} \qquad \begin{array}{l} K_1 = 3.6 \times 10^{-5} \\ K_2 = 1 \times 10^{-14} \end{array}$$

The metal center has a formal negative oxidation state

g) $[\overset{-0.5}{Ni_4}(\overset{0}{CO})_9]^{2-}$

h) $\overset{0}{Fe}(\overset{0}{CO})_3(\overset{0}{SbCl_3})_2$ |SbCl$_3$ is a neutral π-acidic ligand

i) $\overset{+0.5}{O_2}[\overset{+5}{Pt}\overset{-1}{F_6}]$ The [PtF$_6$]$^-$ ion stabilizes the dioxygenyl cation

j) $\overset{-1}{H}\overset{+1}{Rh}(\overset{0}{CO})(\overset{0}{PPh_3})_3$ Hydrido compound with neutral π-acidic ligands

Exercise 2.2

a) Pt^{2+} ⟶ Pt^{4+}, oxidative addition of HCl.

b) α-Elimination, permethyltungsten gives a carbene structure.

c) Co^{3+} ⟶ Co$^+$, neutral trimethylphosphite ligands, reductive elimination of H$_2$.

d) Redox reaction:

$$[(\pi\text{-}\overset{-1}{C_5H_5})\overset{0}{W}(\overset{0}{CO})_3]\,\overset{+1}{Na} + \overset{+1}{CH_3}\overset{-1}{I} \longrightarrow (\pi\text{-}\overset{-1}{C_5H_5})\overset{+2}{W}(\overset{0}{CO})_3\overset{-1}{CH_3} + NaI$$

e) Ir$^+$ ⟶ Ir^{3+}, variant of oxidative addition (addition of oxonium salts).

f) Mn retains the oxidation state +1, formation of an olefin π complex with displacement of a CO ligand (coordination number remains unchanged).

g) Mo^0 ⟶ Mo^{2+}, the π-allyl group acts as a ligand with a single negative charge. A π-olefin complex is converted into a π-allyl complex; special case of an oxidative addition reaction.

h) The hydridorhenium compound acts as a metal base and forms a Lewis acid/Lewis base complex with BF$_3$.

Exercise 2.3

a) $\overset{+2}{\text{Co}} \longrightarrow \overset{0}{\text{Co}}$, $\overset{0}{\text{H}} \longrightarrow \overset{+1}{\text{H}}$, redox reaction.

b) $\overset{-2}{\text{Fe}}(\overset{0}{\text{CO}})_2(\overset{+1}{\text{NO}})_2 \longrightarrow [\overset{-1}{\text{Fe}}\overset{-1}{\text{I}}(\overset{+1}{\text{NO}})_2]_2$, $\overset{-2}{\text{Fe}} \longrightarrow \overset{-1}{\text{Fe}}$, $\overset{0}{\text{I}} \longrightarrow \overset{-1}{\text{I}}$, redox reaction.

c) $\overset{-1}{\text{CH}_3}\overset{+2}{\text{Pt}}\overset{-1}{\text{I}}(\overset{0}{\text{PPh}_3})_2$, $\overset{0}{\text{Pt}} \longrightarrow \overset{+2}{\text{Pt}}$ oxidative addition of an alkyl halide to a d^{10} platinum complex with simultaneous loss of a PPh_3 ligand.

d) $[\text{Mn}(\text{CO})_6]^{+\overset{+1}{}}[\text{AlCl}_4]^{-\overset{+3}{}}$ ligand substitution; oxidation states remain unchanged. Removal of chloride ligands by the chloride acceptor $AlCl_3$. CO enters the resulting empty coordination site under pressure to give salts of the hexacarbonyl cation.

e) Platinum retains the oxidation state $+2$, $\overset{-2}{\text{N}}$ (in N_2H_4) $\longrightarrow \overset{0}{\text{N}} + \overset{-3}{\text{N}}$, disproportionation.

Exercise 2.4

a) $\overset{0}{\text{W}}(\overset{0}{\text{CO}})_6 + \overset{+3}{\text{Si}_2}\overset{-1}{\text{Br}_6} \longrightarrow \overset{0}{\text{W}}(\overset{0}{\text{CO}})_5\overset{0}{\text{SiBr}_2} + \overset{+4}{\text{Si}}\overset{-1}{\text{Br}_4} + \text{CO}$

In complexes the neutral $SiBr_2$ group has a silene structure:

$|\overset{+2}{\text{Si}} \begin{matrix} \overset{-1}{\text{Br}} \\ \overset{-1}{\text{Br}} \end{matrix}$, $\overset{+3}{\text{Si}} \longrightarrow \overset{+2}{\text{Si}} + \overset{+4}{\text{Si}}$, disproportionation

b) $\overset{0}{\text{Pt}}(\overset{0}{\text{PPh}_3})_4 + \overset{+3}{\text{Si}_2}\overset{-1}{\text{Cl}_6} \longrightarrow \overset{+2}{\text{Pt}}(\overset{0}{\text{PPh}_3})_2(\overset{-1}{\text{SiCl}_3})_2 + 2 \text{ PPh}_3$

The $SiCl_3$ group acts as an anionic complex ligand

$|\overset{+2}{\text{Si}} \begin{matrix} \overset{-1}{\text{Cl}} \\ \text{Cl} \\ \text{Cl} \end{matrix}$, $\overset{0}{\text{Pt}} \longrightarrow \overset{+2}{\text{Pt}}$, $\overset{+3}{\text{Si}} \longrightarrow \overset{+2}{\text{Si}}$, redox reaction.

c) $\overset{0}{\text{Fe}} \longrightarrow \overset{0}{\text{Fe}}$, nucleophilic substition of a CO ligand.

Exercise 2.5

a) Rh retains the oxidation state $+3$; insertion of ethylene into a transition metal–hydride bond with formation of a σ-alkyl complex.

b) Rh retains the oxidation state $+1$; formation of a π-olefin complex and simultaneous displacement of a PPh_3 ligand.

Solutions to the Exercises   363

Exercise 2.6

a) $IrCl(CO)(PR_3)_2 + SnCl_4 \longrightarrow$

$$\begin{array}{c} SnCl_3 \\ R_3P\diagdown \mid \diagup CO \\ Ir \\ Cl \diagup \mid \diagdown PR_3 \\ Cl \end{array}$$

Oxidative addition of a metal halide to an Ir complex; $d^8 \longrightarrow d^6$, $Ir^{+1} \longrightarrow Ir^{+3}$; $SnCl_4$ is formally cleaved into the two anionic groups Cl^- and $SnCl_3^-$.

b) $(\pi\text{-}C_5H_5)_2(CO)_3WH + CH_2N_2 \xrightarrow{-N_2} (\pi\text{-}C_5H_5)_2(CO)_3W\text{-}CH_3$

Insertion of carbene CH_2 (from diazomethane) into a metal–hydride bond gives a methyl complex.

c) $\left[(\pi\text{-}C_5H_5)(CO)_2Fe\text{---}\|\begin{array}{c}CH\text{-}CH_3\\CH_2\end{array} \right]^+ + BH_4^- \xrightarrow{-BH_3} (\pi\text{-}C_5H_5)(CO)_2Fe\text{-}CH\begin{array}{c}CH_3\\CH_3\end{array}$

Nucleophilic attack of hydride on a cationic olefin complex gives a σ-alkyl complex. Fe retains the oxidation state +2.

d) $[RuCl_2(PPh_3)_3] + H_2 + Et_3N \longrightarrow [RuClH(PPh_3)_3] + Et_3NHCl$

Ru retains the oxidation state +2; heterolytic addition of H_2; the tertiary amine is a strong base that traps the protons and thus supports the reaction.

Exercise 2.7

– Oxidative addition: addition of small covalent molecules to a transition metal in a low oxidation state with an increase of the oxidation state of the central atom by two units (Eq. 2-31)
– Requirement: coordinatively unsaturated compounds
– Reverse: reductive elimination

Exercise 2.8

Insertion of a molecule in a transition metal–X bond (X = H, C, N, O, Cl, metal) without changing the formal oxidation state of the metal (Eq. 2-55).

Example: CO insertion into a cobalt carbonyl complex to give an acyl complex.

$(CO)_4Co(C_2H_5) \longrightarrow (CO)_3Co-\underset{\underset{O}{\|}}{C}-C_2H_5$

Exercise 2.9

Formation of an ethylene π complex is followed by an insertion reaction to give the alkyl complex $[CH_3CH_2Pt(SnCl_3)(CO)(PPh_3)]$.

Exercise 2.10

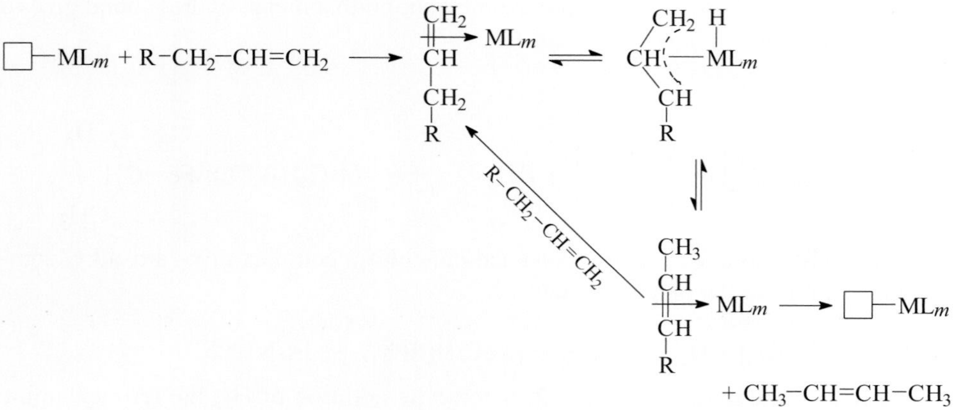

[Scheme showing: □—ML*m* + R—CH₂—CH=CH₂ → π-complex with CH₂=CH—CH₂—R coordinated to ML*m* ⇌ π-allyl hydride intermediate ⇌ crotyl (CH₃—CH=CH—CH₃ coordinated) complex → □—ML*m* + CH₃—CH=CH—CH₃]

Complexation of the alkene is followed by an H abstraction to give a labile π-allyl hydride and then a rearrangement with allyl insertion into the M–H bond.

Exercise 2.11

a) Electrons flow from the metal into the antibonding σ orbital of hydrogen, weakening the bond. Two *cis* M–H bonds can be formed by oxidative addition if two vacant coordination sites are present on the metal center.

b) Heterolysis of H_2 by removal of H^+ in the presence of strong bases.

Exercise 2.12

β-Hydride elimination with formation of a π-allyl complex:

$$L_nPd\underset{H}{-}\overset{CH_2}{\underset{\underset{R}{CH}}{CH}}$$

Exercise 2.13

β-Hydride elimination takes place:

$$Ni-CH_2CH_2CH_2CH_3 \longrightarrow NiH + CH_2=CHCH_2CH_3$$

Exercise 2.14

The addition of PPh_3 lowers the equilibrium concentration of $[RhCl(PPh_3)_2(solvent)]$.

Exercise 2.15

The acac complex can react by an associative mechanism, but this route is blocked for the 18-electron complex $[CpRh(C_2H_4)]$.

$$[(acac)Rh(C_2H_4)_2] \underset{C_2H_4}{\rightleftarrows} [(acac)Rh(C_2H_4)_3] \xrightarrow{} [(acac)Rh(C_2H_4)_2]$$
$$\text{16e} \qquad\qquad \text{18e} \qquad\qquad C_2H_4 \quad \text{16e}$$

Exercise 2.16

a) A $\overset{+1}{Rh}$ 16 e
   B $\overset{+1}{Rh}$ 16 e
   C $\overset{+3}{Rh}$ 18 e

b) Only complex C is coordinatively saturated.

Exercise 2.17

a) Oxidative addition of H_2 to a square-planar Rh^I complex, $d^8 \longrightarrow d^6$, formation of an octahedral Rh^{III} dihydrido complex.

b) Dissociation of a phosphine ligand gives an empty coordination site on the transition metal Rh^{3+}.

c) Formation of a π complex with ethylene, Rh^{3+}.

d) *cis*-Insertion reaction of the ethylene ligand into the rhodium–hydride bond with formation of a σ-alkyl complex, Rh^{3+}.

e) Irreversible reductive elimination of ethane, $Rh^{3+} \longrightarrow Rh^{1+}$.

f) A phosphine ligand coordinates to the coordinatively unsaturated Rh^I complex.

Exercise 2.18

In the complexes $[PtX_4]^{2-}$, Pt^{2+} acts as a soft acid according to the HSAB concept. The ligands X^- become increasingly soft in the given sequence, and the combination soft/soft gives more stable compounds.

Exercise 2.19

The cocatalyst $SnCl_2$ supports the formation of hydrides, and the $SnCl_3^-$ ligands inhibit the reduction of Pd^{II} to the metal. The symbiosis of the two very soft ligands H^- and $SnCl_3^-$ leads to highly active, stable catalysts:

$$PdCl_2(PPh_3)_2 \underset{}{\overset{SnCl_2}{\rightleftharpoons}} Pd(SnCl_3)_2(PPh_3)_2 \underset{}{\overset{H_2}{\rightleftharpoons}} H\,Pd(SnCl_3)(PPh_3)_2$$
s m   s s s
$$+ H^+ + SnCl_3^-$$

(s = soft, m = medium hard)

Exercise 2.20

Bases containing the elements P, As, Sb, Se, and Te are soft compounds according to the HSAB concept and therefore form stable complexes with the soft transition metals and thus deactivate the catalysts. The hard oxygen and nitrogen bases hardly react with the transition metals due to the hard/soft dissymmetry.

Exercise 2.21

a)
H^-	Ir^+	N_2H_4	SO_2	Ti^{4+}	CO_2	CO	$CH_2=CH_2$
s B	s A	h B	m A	h A	h A	s B	s B

b) $C_5H_5^-$ CN^- PPh_3 C_6H_6

c) Sn^{4+}: higher oxidation state
   $P(OC_2H_5)_3$: oxygen in the molecule increases the hardness
   $[Co(NH_3)_5]^{3+}$: the hard NH_3 ligands make the complex harder than the cyano complex. Cobalt has the oxidation state +III in both complexes

Exercise 2.22

a) $[Ni(CO)_4]$: terminal CO groups in a neutral complex

   $[Mn(CO)_6]^+$: in metal carbonyl cations, the positive charge on the metal center increases the CO frequency

   $[V(CO)_6]^-$: in metal carbonyl anions, the CO ligand has to accept more negative charge from the metal, and the CO bands are therefore at lower wavenumbers than in neutral complexes

b) Replacing the phenyl groups by electronegative chloro groups, which are capable of backbonding, increases the CO stretching frequency

Exercise 2.23

The CO stretching frequency is an indication of the metal basicity of carbonyl complexes. It increases with decreasing electron density on the metal, i.e., when the metal acts as base.

Exercise 2.24

A) Metal–metal bond with terminal CO ligands $(CO)_4Co–Co(CO)_4$

B) Bridging CO ligands:

# Chapter 3

Exercise 3.1

H addition to the conjugated diene gives a π-allyl complex, via which the subsequent isomerization/hydrogenation of the diene takes place:

Exercise 3.2

Oxo synthesis converts 1-hexene to heptanal and 2-methylhexanal, which are separated by distillation.

Exercise 3.3

1) Wacker–Hoechst process: oxidation of ethylene to acetaldehyde with Pd/Cu catalysts followed by oxidation to acetic acid.
2) Methanol carbonylation with rhodium iodide catalysts gives acetic acid directly.

Exercise 3.4

1) Oxidative addition of HCl to an Rh^I complex gives the active hydrido rhodium(III) catalyst **A**.
2) Coordination of butadiene to the catalyst followed by insertion of the diene into the Rh–H bond to give the *syn*-π-crotyl complex **B**.
3) Coordination of ethylene and subsequent insertion into the terminal Rh–C bond to the pecursor **C** of the desired diene.
4) Coordination of butadiene to complex **D**, elimination of *trans*-1,4-hexadiene, and regeneration of the π-allyl complex **B**.

Exercise 3.5

The metal ion is alkylated, and ethylene is activated by coordination to the transition metal. Since the metal is present in a relatively high oxidation state, nucleophilic attack of the alkyl group on the neighboring alkene is favored, and a *cis* insertion reaction occurs. This process continues until chain termination occurs.

Exercise 3.6

Many active centers with different structures are present on the surfaces of solids; therefore, the selectivity is low. In homogeneous catalysis there is only one active species, and the ligands can readily be modified.

Chapter 4

Exercise 4.1

Equation (4-7) is transformed into a linear equation.

$$\theta_A = \frac{K_A p_A}{1 + K_A p_A} \quad \text{and} \quad \theta_A = \frac{V}{V_\infty}$$

V_∞ = volume at complete coverage

$$\frac{V}{V_\infty} = \frac{K_A p_A}{1 + K_A p_A}$$

rearrangement gives

$$\frac{1}{V_\infty K_A} + \frac{p_A}{V_\infty} = \frac{p_A}{V}$$

Thus, plotting p_A/V against p_A gives a straight line with slope $1/V_\infty$ and an intersection with the axis of $1/K_A V_\infty$. The following values can be calculated from the experimental measurements:

p_A (mbar)	133	267	400	533	667	800	933
p_A/V	12.9	13.8	14.6	15.6	16.7	17.6	19.4

The calculated values are displayed in the following figure. At high loadings, a straight line is no longer obtained.

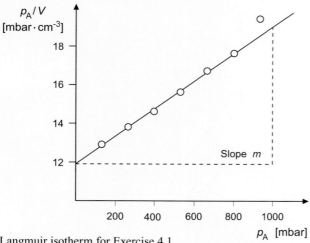

Langmuir isotherm for Exercise 4.1

$$\text{Slope } m = \frac{18.95 - 11.9}{1000} = 0.0071$$

$$V_\infty = 1/m = 141.8 \text{ cm}^3$$

At $p = 0$ we obtain for the ordinate intersection

$$\frac{p_A}{V} = 11.9$$

$$\frac{1}{V_\infty K_A} = 11.9$$

$$K_A = \frac{1}{141.8 \times 11.9} = 5.9 \times 10^{-4} \text{ mbar}^{-1}$$

Exercise 4.2

According to Eq. 4-8

$$-\frac{dp_A}{dt} = k\theta_A = \frac{k K_A p_A}{1 + K_A p_A} \qquad p_A = \text{partial pressure of phosphine}$$

For $K_A p_A \ll 1$

$$-\frac{dp_A}{dt} = k K_A p_A$$

and the reaction is thus first order.

For $K_A p_A \gg 1$ in contrast

$$-\frac{dp_A}{dt} = k$$

i.e., the reaction is zero order.

Exercise 4.3

a) Eley–Rideal mechanism: only one partner (hydrogen) is adsorbed and reacts with the other starting material (CO_2) from the gas phase.

b) $$r_\text{eff} = \frac{\text{kinetic term} \times \text{driving force}}{(\text{chemisorption term})^n} \qquad (4\text{-}16)$$

Exercise 4.4

The kinetic expression

$$r = \frac{k\, K_{IB}\, c_{IB}^2}{1 + K_{IB}\, c_{IB}}$$

describes the findings in the simplest form.

Adsorbed isobutene forms a solvated carbenium ion, which reacts with a molecule from the gas phase. This Eley–Rideal mechanism is often observed for solid acids.

Exercise 4.5

The two components compete for the catalytically active sites. The nitrogen compounds are more strongly adsorbed than the sulfur compounds and displace them from the active surface sites. This explains the low reaction rate of the sulfur compounds in the presence of nitrogen compounds.

The higher reactivity of the nitrogen compounds in the mixture indicates that they occupy the majority of the catalytic centers and hinder the reaction of the sulfur compounds by blocking the adsorption sites,

The results are indicative of Langmuir adsorption on an ideal surface. If the organo-nitrogen compounds are the only strongly adsorbed species, then as a first approximation the rate of hydrodesulfurization r_{HDS} can be described by the equation

$$r_{HDS} = \frac{k\, K_S\, p_S\, K_{H_2}\, p_{H_2}}{1 + K_S\, p_S + K_N\, p_N} \qquad \text{where } K_N\, p_N \gg (1 + K_S\, p_S)$$

Exercise 4.6

1) Dissociative adsorption of oxygen.
2) Eley–Rideal step in which gaseous SO_2 reacts with adsorbed oxygen to give SO_3.

Exercise 4.7

Chemisorption, surface reaction, desorption.

Exercise 4.8

$$r'_{CH_4} = N_{CH_4} \, D \, \frac{1}{M_{Ru}} \cdot \frac{\% \, Ru}{100} \quad \text{(cf. Sect. 1.2)}$$

N is the number of molecules that react at an active surface atom under the given conditions, that is, reacting molecules per atom of catalyst or moles per mole of catalyst. Therefore, for 1 g of catalyst (metal + support), we obtain:

$$r'_{CH_4} = \frac{0.160 \, \text{mol}}{\text{mol Ru s}} \cdot 0.42 \cdot \frac{1 \, \text{mol Ru}}{101.1 \, \text{g}} \cdot 0.005$$

$$r'_{CH_4} = 3.32 \times 10^{-6} \, \text{mol s}^{-1} \, \text{g cat.}^{-1}$$

Exercise 4.9

	Chemisorption	Physisorption
Cause	covalent or electrostatic forces, electron transfer	van der Waals forces, no electron transfer
Adsorption heat	high $\sim \Delta H_R$ 80–600 kJ/mol usually exothermic	low $\sim$ heat of melting $\sim$ 10 kJ/mol always exothermic
Temperature range	generally high	low
No. of adsorbed layers	monolayer	multiple layers

Exercise 4.10

a) Dissociative chemisorption of a saturated hydrocarbon, abstraction of H, alkyl complex formation.
b) Associative chemisorption through the electron lone pair, double alkyl complex.
c) Dissociative chemisorption, H abstraction, π-allyl complex formation.

Exercise 4.11

Fe: At the process temperature, CO is completely dissociated, and the concentration of CH_2 groups is therefore high. The formation of higher hydrocarbons is favored.

Ni: CO dissociation is more difficult, so that the concentration of CH_2 is lower. Hydrogenation with formation of methane dominates.

W, Mo: Poor Fischer–Tropsch catalysts; the metal carbides are probably too stable to readily undergo hydrogenation.

Pt: platinum group metals are not Fischer–Tropsch catalysts.

Exercise 4.12

Titanium nitride is too stable.

Exercise 4.13

CO is strongly chemisorbed and blocks the sites for hydrogen.

Exercise 4.14

The catalyst is porous; pore diffusion resistance lowers the activation energy by a factor of 2.

Exercise 4.15

There are two different adsorption sites, one of which forms a single metal–CO bond, while at the other, the CO molecule dissociates into the atoms C and O, which are individually adsorbed.

Exercise 4.16

a) Lewis acid complex with the base NH_3.
b) Carbonyl complex, linear.
c) Carbonyl complex, bridging, two centers.
b) Dissociative chemisorption of H_2.
e) Dissociative chemisorption of ethane, formation of a σ-alkyl and a hydride bond.

f) Dissociative chemisorption of H_2 on ZnO, heterolytic.
g) π-Olefin complex.
h) Ethylene as double π-alkyl complex, associatively bound.

Exercise 4.17

H–S(H)→* Associative chemisorption; the molecule remains intact. Formation of a complex with the electron lone pair on sulfur.

Exercise 4.18

a) Cube surface   A
   Octahedron surface C
   Prism surface   B

b) Each surface has a particular catalytic activity. The most densely occupied surfaces (especially fcc) are often the most active.
Steps and kinks on the surface have a major influence on the catalytic activity.

Exercise 4.19

a) (1 1 0)   b) (2 3 0)   c) (0 1 0)

Exercise 4.20

a) Miller indices: position of the crystal/atomic planes in the coordinate system.

b) Diagonal (1 1 1) surface is normally of higher activity because it is more densely occupied.

Exercise 4.21

The reaction is structure-insensitive.

Exercise 4.22

To maximize the surface area.

Exercise 4.23

Pd	Al$_2$O$_3$	ZnO	Alumosilicates	MgO	CoO	Zeolites
–	I, at high temperature n-type semiconductor, A	S	A, I	I	S	A

Exercise 4.24

a) p-type semiconductor
b) Cationic chemisorption: $CO \longrightarrow CO^+ + e^-$
c) p-type doping; increased conductivity

Exercise 4.25

VO$_2$	Cu$_2$O	WO$_3$	MnO$_2$	Nb$_2$O$_5$	CoO
p	p	n	n	n	p

Exercise 4.26

Donor reaction: the donor is SO$_2$.

Exercise 4.27

Ni on CoO (c)

Ni (a)

Ni on Al$_2$O$_3$ (b)

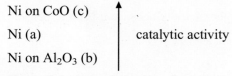

catalytic activity

Flow of electrons from the substrate through the metal to the support is most favorable for the donor reaction of the hydrogen ("rectifier effect").

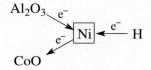

Exercise 4.28

In the donor reaction H $\longrightarrow$ H$^+$ + e$^-$ electrons are donated to the catalyst.

MnO is a semiconductor which readily takes up electrons. The *p*-type conductivity is enhanced by Li$^+$.

Exercise 4.29

The catalyst should function as an electron acceptor or have acidic properties. It should also have a low porosity so that total oxidation of the maleic anhydride is prevented by mass transfer resistance. Vanadium phosphorus oxide complexes in which several V oxidation states are combined are used; mean specific surface area 20 m^2/g.

Exercise 4.30

a) Al$_2$O$_3$ has OH groups on the surface that act as Brønsted acid centers and has Lewis acid centers as electron acceptors (see Fig. 4-32).

b) SiO$_2$/Al$_2$O$_3$ > SiO$_2$ $\gg$ γ-Al$_2$O$_3$ > MgAl$_2$O$_4$ > MgO.

Exercise 4.31

An additional proton is necessary for charge compensation when Si^{4+} is replaced by Al^{3+} in the lattice.

Exercise 4.32

Titration with bases; poisoning with N bases such as NH$_3$, pyridine, quinoline; IR spectroscopic investigations on catalysts with adsorbed pyridine.

Exercise 4.33

$$\text{Si}-\text{OH}-\text{Al} + \text{CR}_2=\text{CR}_2 \longrightarrow \text{R}_2\text{CH}-\overset{+}{\text{C}}\text{R}_2 + \text{Si}-\text{O}^--\text{Al}$$

A C=C bond is protonated by a Brønsted acid center on the surface to give a carbocation that initiates the polymerization.

Exercise 4.34

Electronic interactions; reduced support species on the metal surface; phase formation at interfacial surfaces.

Exercise 4.35

Specific surface area, pore volume, pore structure.

Exercise 4.36

SMSI = strong metal–support interaction.

Exercise 4.37

Activity, selectivity, and stability.

Exercise 4.38

Acidic cracking centers are neutralized by bases; potassium lowers the coking tendency of Al_2O_3 supports.

Exercise 4.39

Decreasing productivity; lower selectivity.

Exercise 4.40

Measurement of the decreasing catalyst activity as a function of time.

Exercise 4.41

Activity losses due to phase changes to oxides with smaller surface area or due to thermal sintering.

Exercise 4.42

Under certain circumstances, formation of highly toxic [Ni(CO)$_4$] could occur.

Exercise 4.43

F < Cl < Br < I

Softer halides form stronger bonds with the soft transition metals.

Exercise 4.44

a) Rapid coking
b) Continuously

Exercise 4.45

A) IR reflection spectroscopy
B) Temperature-controlled desorption
C) BET method
D) SIMS
E) ESCA, ESR
F) X-ray structure analysis
G) Scanning electron microscopy

Exercise 4.46

Number of active surface atoms per gram of catalyst. Degree of dispersion.

Exercise 4.47

At 200 °C surface-bound water is released, and Al^{3+}-bound OH groups remain on the surface. These form hydrogen bonds with pyridine (1540 cm^{-1}). Partial dehydration of the OH groups gives O^{2-} and free Al^{3+}, which forms a Lewis acid–base complex with pyridine (1465 cm^{-1}).

Exercise 4.48

A)
$$\underset{-\mathrm{Ni}-\mathrm{Ni}-}{\overset{\overset{\displaystyle O}{\|}}{C}}$$
Bridged structure; the CO group has double bond character.

B)
$$\underset{\mathrm{Ni}}{\overset{\overset{\displaystyle O}{|||}}{\underset{|}{C}}}$$
Linear carbonyl complex; stronger CO bond.

Exercise 4.49

Dissociative adsorption: $CO_2 \longrightarrow CO + \frac{1}{2} O_2$.

Exercise 4.50

The C=C double bond of ethylene is weakened by π complexation to Pd, and the IR frequency is therefore lower.

Exercise 4.51

LEED = low-energy electron diffraction.

A surface method that uses low-energy electrons to provide diffraction patterns for the upper atomic layers. Can detect surface structures and the occupation of the surface by adsorbed molecules (adsorption complexes).

Chapter 5

Exercise 5.1

Active surface area; pore structure; mechanical strength.

Exercise 5.2

– Pore structure and surface of the catalyst can be controlled.
– Catalysts can be tailor-made with respect to mass-transfer effects.
– More economic, since the content of expensive active components is often low.
– The distribution and crystallite size of the active components can generally be varied over a wide range.
– Multiple impregnation is possible.

Exercise 5.3

Activated carbon, silicagel, Al_2O_3.

Exercise 5.4

– Nonporous, low surface area supports.
– Porous supports with wide pores: silicates, α-Al_2O_3, SiC, ZrO_2, graphite.

Exercise 5.5

Diffusion-controlled reactions; fast reactions.

Exercise 5.6

a) Because they have smooth, nonporous surfaces.
b) Washcoat.

Exercise 5.7

a) Short transport or diffusion paths; pore structure independent of the support; improved heat transfer in the catalyst layer; low coking.
b) Generally α-Al_2O_3 spheres; low specific surface area (<1 m^2/g).

Exercise 5.8

In mononuclear complexes, only one metal center is present, and C–C coupling is not possible.

Exercise 5.9

The steric and electronic properties of the coordinated metal atoms or ions can be better controlled than in conventional heterogeneous catalysts.

Exercise 5.10

$$\longrightarrow \text{┤}-PR_2Fe(CO)_4 + CO$$

Bonding of the metal to the phosphine ligand in a ligand-exchange reacion.

Exercise 5.11

Low mechanical stability; poor heat-transfer properties; low thermal stability (150 °C max.).

Exercise 5.12

A solution of an organometallic complex in a high-boiling solvent is applied to a porous inorganic support. The diffusion-controlled reaction takes place in the solvent film.

Chapter 6

Exercise 6.1

a) Aluminosilicates with ordered crystalline networks of composed interlinked AlO_4^- and SiO_4 tetrahedra and a channel structure containing water molecules and mobile, exchangeable alkali metal ions.

b) Number and strength of the acid centers; isomorphic substitution; metal doping.

Exercise 6.2

a) Reactant shape selectivity in the cracking of octanes: the linear alkane fits in the pores and is cleaved. The branched alkane does not enter the pores and therefore does not react. A mixture of various alkanes and alkenes is formed.

b) Product shape selectivity in the alkylation of toluene with ethylene: Only the slim *p*-ethyltoluene molecule can escape from the pores. The *ortho* isomer is too bulky and remains in the pores.

Exercise 6.3

– Wide range of variability
– Shape selectivity
– High thermal stability
– Strongly acidic zeolites
– Incorporation of many metals
– Bifunctionality (acid/metal)

Exercise 6.4

The zeolite catalyst leads to a shape-selective hydrogenation, in which the strongly branched starting material does not fit in the pores and is hardly hydrogenated. There is no steric hindrance with the conventional catalyst.

Exercise 6.5

Shape selectivity can occur when the starting materials, products, or transition state of a reaction have dimensions similar to those of the zeolite pores.

Exercise 6.6

Shape-selective reaction of methanol with ZSM-5 catalysts. Hydrocarbons that exceed the ideal size for gasoline are retained in the zeolite pores until they have been catalytically shortened enough to escape. The product spectrum of this process is far more favorable than that of the Fischer–Tropsch process.

Exercise 6.7

Exchange of the alkali metal ions in the channels with ammonium ions followed by heating to 500–600 °C, which leads to cleavage of ammonia and leaves behind protons.

Exercise 6.8

With narrow-pore acidic zeolites in the H form. The trimethylamine fraction can be lowered to less than 1 % as a result of shape selectivity.

Exercise 6.9

a) The acid form of ZSM-5 is sufficiently strong to form carbocations:

$$CH_2=CH_2 + H^+ \longrightarrow [CH_3–CH_2]^+$$

The carbocation can attack benzene. Deprotonation of the product then gives ethylbenzene:

$$[CH_3–CH_2]^+ + C_6H_6 \longrightarrow C_6H_5CH_2CH_3 + H^+$$

b) No, because there are no acid centers.

Exercise 6.10

The rate equation corresponds to the Eley–Rideal mechanism. The equation can be written as

$$r = k\theta_O c_T$$

where θ_O is the degree of coverage of the active centers by the olefin. The equation thus applies to the rate-determining step of the reaction of the adsorbed carbenium ion with toluene in the pore volume of the zeolite.

Exercise 6.11

$AlPO_4$ molecular sieves do not have ion-exchange properties.

Exercise 6.12

By dealumination with reagents such as $SiCl_4$: Al is removed from the lattice as $AlCl_3$; hydrothermal treatment with steam at 600–900 °C.

Chapter 7

Exercise 7.1

In the given sequence: f, f, t, f, t.

Exercise 7.2

Reaction time (min) **A**	30				45			
Temperature (°C) **B**	50		80		50		80	
Catalyst concentration (%) **C**	0.25	0.4	0.25	0.4	0.25	0.4	0.25	0.4
Experiment	(1)	c	b	bc	a	ac	ab	abc

Exercise 7.3

a) Effects and interactions:

A = 4.6   B = 3.6   AB = –0.6   C = 2.1   AC = 2.4   BC = 0.4   ABC = –1.4

The mean yield of 21.88% is theoretically obtained for an experiment with average experimental conditions:

Reaction time	25 min
Temperature	60 °C
Catalyst concentration	0.15%

Effect A means that the yield increases by 4.6% when the reaction time is increased by 5 min, i.e., the yield increases at a rate of 0.92%/min. In the same way, improvements of 0.73%/°C and 42.6%/% cat. can be calculated.

b) From Equation (7-14) we obtain

$$\sigma_{Eff} = \sqrt{1/8} \cdot 4.2 = 1.49$$

$$\text{Test quantity } z = \frac{\text{Effect}}{\sigma_{Eff}} = \frac{\text{Effect}}{1.49}$$

and therefore:

Effects			$z = \dfrac{\text{Effects}}{1.49}$	$z > c$?
A	=	4.6	3.09	yes
B	=	3.6	2.42	yes
AB	=	−0.6	0.40	no
C	=	2.1	1.41	no
AC	=	2.4	1.61	no
BC	=	0.4	0.27	no
ABC	=	−1.4	0.94	no

Result: The yield depends only on the reaction time and temperature, and not on the catalyst quantity, provided it is at least 0.1%.

Exercise 7.4

Result of the Plackett–Burman matrix:

	A	B	C	D	E (−)	F	G (−)	Yield
Σ	−0.613	−0.025	0.049	0.075	−0.003	−0.023	0.055	3.219/8
$\Sigma/4$	−0.153	−0.006	0.012	0.019	−0.0008	−0.006	0.014	= 0.4024

From Equation (7-15) we obtain

$$s^2 = \dfrac{0.0008^2 + 0.014^2}{2} = 9.85 \times 10^{-5}$$

Standard deviation $s = \sqrt{s^2} = 0.0097$

Determination of the significance of the effects by a t-test:

$$\text{Test quantity } t = \dfrac{\text{Effect}}{\text{Standard deviation } s}$$

We obtain

Effect A (temperature) = −0.153/0.0097 = −15.74
Effect D (solvent)   = 0.019/0.0097 =   1.93

Comparison with the values in the *t*-test table shows: only effect A is highly significant; D is only significant to 80% and hence practically unconfirmed. The other effects are meaningless.

Exercise 7.5

Experiment 1 represents the worst corner.

a) From Equation (7-16) we obtain

$$\vec{x}_{n+2} = \frac{2}{2} 114 - \left(1 + \frac{2}{2}\right) 37 \rightarrow x_1 = 40$$

$$\vec{x}_{n+2} = \frac{2}{2} 64.8 - \left(1 + \frac{2}{2}\right) 21 \rightarrow x_2 = 22.8$$

b) From Equation (7-17) we obtain

for x_1: $\vec{x}_{n+2+1} = \left(1 + \frac{2}{2}\right) 40 - \left(1 + \frac{2}{2}\right) 39 + 37$

$$= 80 - 78 + 37 = 39$$

for x_2: $\vec{x}_{n+2+1} = \left(1 + \frac{2}{2}\right) 22.8 - \left(1 + \frac{2}{2}\right) 21 + 21$

$$= 45.6 - 42 + 21 = 24.6$$

Coordinates of 5th experiment: $x_1 = 39$, $x_2 = 24.6$.

Chapter 8

Exercise 8.1

1. G
2. G, H
3. E
4. J
5. K
6. B, L
7. I
8. A
9. D
10. C, D
11. A
12. F

Exercise 8.2

C (B)

Exercise 8.3

– Liquid phase: low temperature, high concentration of starting materials, higher conversion, better heat transfer
– Disadvantages: higher pressure, slower mass transfer
– Gas phase: mild, fewer side products, but higher energy demand

Exercise 8.4

– Hydroformylation of propene to butanals (oxo synthesis); distillative separation.
– Aldol condensation of n-butanal to 2-ethylhexenal.
– Hydrogenation of the unsaturated aldehyde to 2-ethylhexanol.

Exercise 8.5

a) A. Dissociative chemisorption of H_2; π-olefin complex of propenol.
    B. Hydrogenation to the σ-alkyl complex.
    C. Complete reduction of the alkyl complex; desorption of the alcohol from the catalyst surface.
    D. Abstraction of H from the OH group of the alkyl complex; rearrangement.
    E. π-Complexation of the aldehyde; hydride complex.
    F. Desorption of the aldehyde; H remains dissociatively bound on the surface.
b) Isomerization of an α,β-unsaturated alcohol (enol) to an aldehyde.

Exercise 8.6

- CO insertion into the metal hydride complex **A** to give the formyl complex **B**.
- Hydrogenation of the formyl complex to the σ-alkyl complex **C**.
- CO insertion to give the acyl complex **D**.
- Hydrogenation of the acyl complex and water cleavage to give the ethyl complex **E**.
- CO insertion reaction followed by reduction; chain growth; longer chain alkyl complex **F**.
- Hydrogenative cleavage of the alkyl complex **F** to give the alkane **G** and the metal hydride complex **A**.

Exercise 8.7

1. Dissociative chemisorption of propene; π-allyl complex of Mo.
2. Nucleophilic attack by Mo lattice oxygen; insertion to give the σ-allyl complex.
3. Elimination (desorption) of acrolein; reduction to the dihydroxy complex.
4. Reoxidation by take up of oxygen from the gas phase; dehydration; formation of oxide ions; diffusion to the vacant lattice sites.

Exercise 8.8

1. Dissociative chemisorption of the aldehyde; aldehyde oxygen atom migrates to the vacant anion site; reduction of the metal; formation of OH^- by H abstraction from the aldehyde.
2. Intermediate state; coupling of the aldehyde carbon to two lattice oxide ions.
3. Oxidation of the intermediate state with two lattice oxygen atoms and take up of a hydrogen atom from lattice hydroxide to give the carboxylic acid; further reduction of the metal; formation of two vacant anion sites in the lattice.
4. Reoxidation of the lattice by atmospheric oxygen; the metal provides electrons for the formation of oxide anions, which are incorporated in the lattice.

Exercise 8.9

a) SCR process for denitrification of flue gases; important environmental process for power stations.

b) TiO_2
   WO_3/MoO_3
   V_2O_5
   300–400 °C

Chapter 9

Exercise 9.1

a) Catalyst residence time.

b)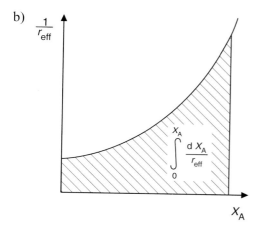

c) Ideal tubular reactor.

d) Besides temperature and concentration, r_{eff} also depends on macrokinetic factors (e.g., mass transport).

Exercise 9.2

Constant reaction rate, on average lower that in a tubular reactor, in which the highest reaction rate is reached at the beginning and then declines towards the end of the reactor.

Exercise 9.3

a) From Equation (7-7) we obtain:

$$-r'_A = \frac{\dot{n}_{A,0} - \dot{n}_A}{m_{cat}} \qquad \dot{n} = c \cdot \dot{V}$$

$$-r'_A = \frac{(c_{A,0} - c_A)\dot{V}_0}{m_{cat}} = \frac{(2-0.5)\cdot 1}{3} = 0.50 \text{ mol h}^{-1} \text{ g cat}^{-1}$$

$$-r'_A = k\, c_A^2$$

$$k = \frac{-r'_A}{c_A^2} = \frac{0.50}{0.5^2} = 2.00 \text{ L}^2 \text{ h}^{-1} \text{ g cat}^{-1} \text{ mol}^{-1}$$

b) From Equation (9-1)

$$\frac{m_{cat}}{\dot{n}_{A,0}} = \int_0^{X_A} \frac{dX_A}{-r'_A} \qquad -r'_A = k\, c_{A,0}^2 (1-X_A)^2$$

Introducing the kinetic expression as a function of conversion

$$\frac{m_{cat}}{\dot{n}_{A,0}} = \int_0^{X_A} \frac{dX_A}{k\, c_{A,0}^2 (1-X_A)^2}$$

Rearrangement and integration gives:

$$\frac{m_{cat}\, k\, c_{A,0}^2}{c_{A,0}\, \dot{V}_0} = \frac{X_A}{1 - X_A}$$

$$m_{cat} = \frac{X_A \dot{V}_0}{(1-X_A) k\, c_{A,0}} = \frac{0.8 \times 1000}{0.2 \times 2.00 \times 1} = 2000 \text{ g} = 2 \text{ kg}$$

c) Equation (9-2) applies; introducing the kinetic equation gives:

$$\frac{m_{cat}}{\dot{n}_{A,0}} = \frac{X_A}{-r'_A} \qquad -r'_A = k\, c_{A,0}^2 (1-X_A)^2 \qquad \dot{n}_{A,0} = c_{A,0} \dot{V}_0$$

$$\frac{m_{cat}}{c_{A,0} \dot{V}_0} = \frac{X_A}{k c_{A,0}^2 (1 - X_A)^2}$$

$$m_{cat} = \frac{X_A \dot{V}_0}{k c_{A,0} (1 - X_A)^2} = \frac{0.8 \times 1000}{2.00 \times 1 \times 0.2^2} = 10\,000 \text{ g} = 10 \text{ kg}$$

For a simple irreversible reaction, the ideal tubular reactor is always advantageous due to freedom from backmixing. Especially at high throughputs, backmixing has a strongly negative effect on the required reaction volume or catalyst mass.

Exercise 9.4

1. The partial pressures in the kinetic equation have to be expressed as functions of the conversion, and this achieved by means of a material balance. After conversion X:

Substance	n_i	x_i	$p_i = x_i\, P = x_i \cdot 30$ (bar)
T	$1 - X$	$\dfrac{1 - X}{11}$	$p_T = \dfrac{(1 - X)\,30}{11}$
H	$10 - X$	$\dfrac{10 - X}{11}$	$p_H = \dfrac{(10 - X)\,30}{11}$
B	X	$\dfrac{X}{11}$	$p_B = \dfrac{X \cdot 30}{11}$
M	X	$\dfrac{X}{11}$	$p_M = \dfrac{X \cdot 30}{11}$
	$\Sigma = 11$		

2. r is expressed as a function of conversion:

$$r = \frac{k \cdot 0.9 (1 - X)\,30 \cdot (10 - X)\,30}{11 \cdot 11 \left(1 + 0.9 \dfrac{(1 - X)\,30}{11} + 1.0 \dfrac{X \cdot 30}{11}\right)}$$

$$\frac{1}{r} = \frac{1 + 2.454\,(1 - X) + 2.727\,X}{0.00135\,(1 - X)\,(10 - X)}$$

3. The design equation (9-1) applies:

$$\frac{m_{cat}}{\dot{n}_{A,0}} = \int_0^X \frac{dX}{r} = A$$

By numerical integration using the Simpson rule, we obtain

X	$1/r$ (kg h kmol^{-1})
0	255.8
0.1	289.4
0.2	331.5
0.3	385.7
0.4	458.2
0.5	559.9
0.6	712.7

$$A = \frac{0.6}{3 \times 6} (255.8 + 4 \times 289.4 + 2 \times 331.5 + 4 \times 385.7 + 2 \times 458.2 \\ + 4 \times 559.9 + 712.7)$$

$A = 249.6$ kg h kmol^{-1}

4. Calculation of the feed

$$\dot{n}_{A,0} = \frac{2000 \times 1000}{8000 \times 92} = 2.72 \text{ kmol toluene h}^{-1}$$

5. Calculation of the catalyst mass

$m_{cat} = \dot{n}_{A,0} \cdot A = 2.72 \times 249.6 = 678.3$ kg

Exercise 9.5

Single-bed reactor:	isomerization of light gasoline
Tubular reactor:	methanol synthesis (low-pressure process)
Multibed reactor:	contact process (oxidation of SO_2 to SO_3)
Shallow-bed reactor:	combustion of ammonia to nitrous gases (Ostwald process)
Fuidized-bed reactor:	ammoxidation of propene to acrylonitrile

Exercise 9.6

The formaldehyde formed should be removed as quickly as possible from the active catalyst. The temperature in the catalyst bed should be as low as possible, and the temperature profile should be uniform.

This requires a catalyst with low porosity and high thermal conductivity.

A shallow-bed reactor is used.

Exercise 9.7

	Trickle-bed reactor	Suspension reactor
Temperature distribution	profile	uniform
Selectivity	low	high
Residence time behavior of liquid	ideal plug flow reactor	ideal stirred tank
Catalyst diameter	large, mm	small, μm
Catalyst effectiveness factor	$\ll 1$	≈ 1
Catalyst performance	low	high

Exercise 9.8

a) Ammoxidation of a methyl aromatic compound with an allylic double bond.
b) Fluidized-bed reactor (cf. SOHIO process).

Exercise 9.9

A) The microkinetics are decisive. There is no limitation by pore diffusion.

B) For the trickle-bed reactor, larger catalyst pellets must be used. This case lies in the region of pore-diffusion inhibition, which distorts the kinetics and allows only a very low degree of exploitation of the catalyst.

References

Textbooks and Reference Books on Homogeneous Catalysis [T]

[T1] Buddrus, J. (1980): Grundlagen der organischen Chemie, de Gruyter, Berlin. (Kap. 15: „Metallorganische Verbindungen")
[T2] Collman, J.P., Hegedus, L.S. (1980): Principles and Applications of Organotransition Metal Chemistry, University Science Books, Mill Valley, Calif. (USA).
[T3] Davies, S.G. (1982): Organotransition Metal Application to Organic Synthesis, Pergamon Press.
[T4] Elschenbroich, Ch., Salzer, A. (1989): Organometallics, 2. Aufl., VCH, Weinheim.
[T5] Falbe, J. (Ed.) (1980): New Syntheses with Carbon Monoxide, Springer-Verlag, Berlin–Heidelberg–New York.
[T6] Huheey, J.E. (1993): Inorganic Chemistry, 4th ed, Harper Collins College Publishers.
[T7] Keim, W. (Ed.) (1983): Catalysis in C_1-Chemistry, D. Reidel Publ. Comp., Dordrecht.
[T8] Jones, W.H. (1980): Catalysis in Organic Synthesis, Academic Press, London.
[T9] Kirk-Othmer (1992): Encyclopedia of Chemical Technology, J. Wiley, New York. (Vol. 5, 324: „Homogeneous Catalysis").
[T10] Kochi, J.K. (1978): Organometallic Mechanisms and Catalysis, Academic Press, London.
[T11] Masters, C. (1981): Homogeneous Transition-metal Catalysis – a gentle art, Chapman and Hall, London–New York.
[T12] Nakamura, A., Tsutsui, M. (1980): Principles and Applications of Homogeneous Catalysis, J. Wiley, New York.
[T13] Negishi, E.I. (1980): Organometallics in Organic Syntheses, J. Wiley, New York.
[T14] Parshall, G.W. (1980): Homogeneous Catalysis, J. Wiley, New York.
[T15] Pearson, A.J. (1985): Metallo-Organic Chemistry, J. Wiley, New York.
[T16] Pracejus, H. (1977): Koordinationschemische Katalyse organischer Reaktionen, Theodor Steinkopff, Dresden.
[T17] Sheldon, R.A. (1983): Chemicals from Synthesis Gas, Serie Catalysis by Metal Complexes, D. Reidel Publ. Comp., Dordrecht.
[T18] Shriver, D.F., Adkins, P.W., Langford, C.H. (1994): Inorganic Chemistry, 2nd ed, Oxford University Press.

Textbooks, Reference Books and Brochures on Heterogeneous Catalysis [T]

[T19] Atkins, P.W. (1994): Physical Chemistry, 5th ed, Oxford University Press.
[T20] Bond, G.C. (1987): Homogeneous Catalysis – Principles and Applications, Oxford Science Publ., Clarendon Press, Oxford. (Short introduction to heterogeneous catalysis).
[T21] Bremer, H., Wendlandt, K.P. (1978): Heterogene Katalyse – Eine Einführung, Akademie-Verlag, Berlin.

[T22] Campbell, I.M. (1988): Catalysis of Surfaces, Chapman and Hall, London–New York.
[T23] Fonds der Chemischen Industrie (1985): Katalyse, Folienserie, Frankfurt/M.
[T24] Gates, B.C. (1992): Catalytic Chemistry, J. Wiley, New York. (Textbook on catalysis, fundamentals and applications).
[T25] Gates, B.C., Katzer, J.R., Schuit, G.C.A. (1979): Chemistry of Catalytic Processes, Mc Graw Hill, New York.
[T26] Hagen, J. (1992): Chemische Reaktionstechnik – Eine Einführung mit Übungen, VCH, Weinheim.
[T27] Hauffe, K. (Ed.) (1976): Katalyse, de Gruyter, Berlin. (Selected Chapters on homogeneous, heterogeneous, and enzymatic catalysis of simple reactions).
[T28] Hegedus, L. (Ed.) (1987): Catalyst Design – Progress and Perspectives, J. Wiley, New York.
[T29] Hughes, R. (1984): Deactivation of Catalysts, Academic Press, London.
[T30] Jaffe, J., Pissmen, L.M. (1975): Heterogene Katalyse, Akademie-Verlag, Berlin.
[T31] Kirk-Othmer (1993): Encyclopedia of Chemical Technology, Vol **5**, 340. (Heterogeneous Catalysis)
[T32] Kripylo, P., Wendlandt, K.P., Vogt, P. (1993): Heterogene Katalyse in der chemischen Technik, Dt. Verlag für Grundstofindustrie, Leipzig–Stuttgart. (Chemistry and technology of the production of catalysts, and Chemistry and technology of heterogeneously catalyzed reactions).
[T33] Le Page, J.F. (1987): Applied Heterogeneous Catalysis – Design, Manufacture, use of solid catalysts, technip, Paris.
[T34] Muchlenow, I.P. (1976): Technologie der Katalysatoren, VEB Dt. Verlag für Grundstoffindustrie, Leipzig.
[T35] Richardson, J.T. (1989): Principles of Catalyst Development. Plenum Press, New York–London.
[T36] Römpps Chemie-Lexikon (1983). Francksche Verlagsbuchhandlung, Stuttgart.
[T37] Satterfield, C.N. (1980): Heterogeneous Catalysis in Practice, Mc Graw Hill, New York.
[T38] Schlosser, E.G. (1972): Heterogene Katalyse, Verlag Chemie, Weinheim.
[T39] Shriver, D.F., Atkins, P.W., Langford, C.H. (1994): Inorganic Chemistry, 2nd ed, Oxford University Press.
[T40] Trimm, D.L. (1980): Design of Industrial Catalysts, Elsevier, Amsterdam.
[T41] Ullmann's Encyclopedia of Industrial Chemistry, 5. Aufl. (1985). VCH, Weinheim. (Vol. A 5, 340; Heterogeneous Catalysis).
[T42] Wedler, G. (1970): Adsorption – Eine Einführung in die Physisorption und Chemisorption, Verlag Chemie, Weinheim.
[T43] Wedler, G. (1982): Lehrbuch der Physikalischen Chemie, Verlag Chemie, Weinheim.
[T44] White, M.G. (1990): Heterogeneous Catalysis, Prentice Hall, Englewood Cliffs, New Jersey. (Physisorption, Chemisorption).

[T45] Katalyse (1994): Topics in Chemistry. Heterogene Katalysatoren. BASF AG Ludwigshafen. (Brochure with some reviews on hetereogeneous catalysis of industrial relevance).

Chapter 1

[1] Baltes, J., Cornils, B., Frohning, C.D. (1975): Chem. Ing. Tech. **47** (12), 522.
[2] Emig, G. (1987): Chemie in unserer Zeit **21** (4), 128.
[3] Falbe, J., Bahrmann, H. (1981): Chemie in unserer Zeit **15** (2), 37.
[4] Folienserie des Fonds der Chemischen Industrie, Nr. 19 (1985): Katalyse.
[5] Godfrey, J.A., Searles, R.A. (1981): Chemie-Technik **10** (12), 1271.
[6] Hagen, J. (1992): Chemische Reaktionstechnik – Eine Einführung mit Übungen, VCH, Weinheim.

[7]   Mroß, W.D. (1985): Umschau **1985** (7), 423.
[8]   Riekert, L. (1981): Chem. Ing. Tech. **53** (12), 950.
[9]   Süss-Fink, G. (1988): Nachr. Chem. Tech. Lab. **36** (10), 1110.
[10] Ugo, R. (1969): Chim. Ind. (Milano) **51** (12), 1319.

Chapter 2

[1]   Berke, H., Hoffmann, R. (1978): J. Am. Chem. Soc. **100**, 7224.
[2]   Brown, S., Bellus, P.A. (1978): Inorg. Chem. **17**, 3726.
[3]   Calderazzo, F. (1977): Angew. Chem. **89**, 305.
[4]   Collman, J.P. et al. (1978): J. Am. Chem. Soc. **100**, 4766.
[5]   Hagen, J. (1977): Chemie für Labor und Betrieb **28** (4), 125.
[6]   Hagen, J., Tschirner, E.G., Fink, G., Lorenz, R. (1985): Chemiker-Ztg. **109**, 3.
[7]   Hagen, J. (1985): Chemiker-Ztg. **109** (2), 63 (Teil I). Chemiker Ztg. **109** (6), 203 (Teil II).
[8]   Hagen, J. (1987): Der Innovationsberater (DIB), 2/87, 1883. Rudolf Haufe Verlag Freiburg.
[9]   Hagen, J. (1988): Chemie für Labor und Betrieb, **39** (12), 605.
[10] Halpern, J. (1970): Acc. Chem. Res. **3**, 386.
[11] Henrici-Olivé, G., Olivé, S. (1971): Angew. Chem. **83** (4), 121.
[12] Herrmann, W.A. (1988): Kontakte (Darmstadt) **1988** (1), 3.
[13] Klein, H.F. (1980): Angew. Chem. **92**, 362.
[14] Pearson, R.G. (1977): Chem. Brit. **3**, 103; (1963): J. Am. Chem. Soc. **85**, 3533; (1966): Science **151**, 172.
[15] Shriver, D.F. (1970): Acc. Chem. Res. **3**, 231.
[16] Shriver, D.F. (1983): Chem. Brit. **1983** (6), 482; (1981): Am. Chem. Soc. Symp. Ser. **152**, 1.
[17] Stille, J.K., Lau, K.S.Y. (1977): Acc. Chem. Res. **10**, 434.
[18] Taube, R. (1975): Z. Chem. **15** (11), 426.
[19] Tolman, C.A. (1972): Chem. Soc. Rev. **1**, 337.
[20] Tolman, C.A. (1977): Chem. Rev. **77**, 313.
[21] Ugo, R. (1969): Chim. Ind. (Milano) **51** (12), 1319.

Chapter 3

[1]   Bach, H., Bahrmann, H., Gick, W., Konkol, W., Wiebus, E. (1987): Chem. Ing. Tech. **59** (11), 882.
[2]   Brunner, H. (1980): Chemie in unserer Zeit **14** (6), 177.
[3]   Casey, C.P. (Ed.) (1986): J. Chem. Educ. **63**, 188.
[4]   Cornils, B., Herrmann, W.A., Kohlpaintner, C.W. (1993): Nachr. Chem. Tech. Lab. **41** (5), 544.
[5]   Falbe, J., Bahrmann, H. (1984): Chemie in unserer Zeit, **15** (2), 37. Falbe, J. Bahrmann, H. (1984): J. Chem. Educ. **61**, 961.
[6]   Halpern, J. (1981): Inorg. Chem. Acta, **50**, 11.
[7]   Herrmann, W.A. (1991): Kontakte (Darmstadt), **1991** (3), 29.
[8]   Keim, W. (1984): Chem. Ind. XXXVI, 397.
[9]   Keim, W., in Graziani, M., Giongo, M. (Ed.) (1984): Fundamental Research in Homogeneous Catalysis **4**, 131. Plenum Press, New York.
[10] Keim, W. (1984): Chemisch Magazine Juli 1984, 417.
[11] Keim, W. (1984): Chem. Ing. Tech. **56** (11), 850.
[12] Parshall, G.W. (1978): J. Mol. Catal. **4**, 243.

[13]   Parshall, G.W. (1987): Organometallics **6**, 687.
[14]   Russell, J.H. (1988): Chemie-Technik **17** (6), 148.
[15]   Sheldon, R.A., Kochi, J.K. (1981): Metal-Catalyzed Oxidations of Organic Compounds, Academic Press, New York.
[16]   Waller, J.F. (1985): J. Mol. Catal. **31**, 123.
[17]   Weissermel, K., Arpe, H.-J. (1998): Industrielle Organische Chemie, 5. Aufl., Wiley-VCH, Weinheim.

Chapter 4

[1]   Beeck, O. (1945): Rev. Modern Physics **17**, 61.
[2]   Boudart, M., Djéga-Mariadassou, G. (1984): Kinetics of Heterogeneous Catalytic Reactions. Princeton Univ. Press, Princeton New Jersey.
[3]   Bradley, S.A., Gattuso, M.J., Bertolacini, R.J. (1989): Characterization and Catalyst Development. ACS Symp. Ser. **411**, 2.
[4]   Bradshaw, A.M., Hoffmann, F.M. (1978): Surf. Sci. **72**, 513.
[5]   Coughlin, R.W. (1967): Classifying catalysts, some broad principles. Ind. Eng. Chem. **59** (9), 45.
[6]   Delmon, B., Froment, G. (1980): Catalyst Deactivation. Elsevier, Amsterdam.
[7]   Dirksen, F. (1983): Chemie-Technik **12** (6), 36.
[8]   Emig, G. (1987): Chemie in unserer Zeit **21**, 128.
[9]   Erbudak, M. (1991): Swiss.Chem. **13** (11), 63.
[10]   Ertl, G. (1990): Angew. Chem. **102**, 1258.
[11]   Friend, C.M., Stein, J., Muetterties, E.L. (1981): J. Am. Chem. Soc. **103**, 767.
[12]   Hölderich, W., Mroß, W.D., Gallei, E. (1985): Arab. J. Sci. Eng. **10** (4), 407.
[13]   Hsiu-Wei, C., White, J.M., Ekerdt, J.G. (1986): J. Catalysis **99**, 293.
[14]   Huder, K. (1991): Chem. Ing. Tech. **63** (4), 376.
[15]   Jakubith, M. (1991): Chemische Verfahrenstechnik. VCH, Weinheim.
[16]   Klabunde, K.J., Fazlul Hoq, M., Mousah, F., Matsuhashi, H. (1987): Metal Oxides and their physico-chemical properties in Catalysis and Synthesis. In: Preparative Chemistry using supported reagents. Academic Press, London.
[17]   Kung, H.H. (1989): Transition metal oxides. In: Surface chemistry and Catalysis. Elsevier, Amsterdam.
[18]   Lamber, R., Jaeger, N., Schulz-Ekloff, G. (1991): Chem. Ing. Tech. **63** (7), 681.
[19]   Levsen, K. (1976): Chemie in unserer Zeit **10**, 48.
[20]   Lintz, H.G. (1992): Chemie in unserer Zeit **26**, 111.
[21]   Maier, W.F. (1989): Chem. Industrie 12/89, 52.
[22]   Maier, W.F. (1989): Einfluß der Katalysatorstruktur auf Aktivität und Selektivität von Hydrierreaktionen. In: Dechema-Monographien Bd. **118**, 243.
[23]   Maier, W.F. (1989): Angew. Chem. **101**, 135.
[24]   Moulijn, J.A., Tarfaoui, A., Kepteijn, F. (1991): Catal. Today **11** (1), 1.
[25]   Mroß, W.D., Kronenbitter, J. (1982): Chem. Ing. Tech. **54** (1), 33.
[26]   Mroß, W.D. (1984): Ber. Bunsenges. Phys. Chem. **88**, 1042.
[27]   Neddermeyer, H. (1992): Chemie in unserer Zeit **26**, 18.
[28]   Niemantsverdriet, J.W. (1993): Spectroscopy in Catalysis. VCH, Weinheim.
[29]   Noerskov, J.K. (1991): Prog. Surf. Sci. **38** (2), 103.
[30]   Polanyi, M., Horiuti, J. (1934): Trans. Faraday Soc. **30**, 1164.
[31]   Pulm, H. (1991): GIT Fachz. Lab. 9/91, 969.
[32]   Schäfer, H. (1977): Chemiker-Ztg. **101** (7/8), 325.
[33]   Schwankner, R.J., Eiswirth, M. (1985): Umschau **85**, 471.
[34]   Schwankner, R.J. (1989): Praxis d. Naturwiss.-Chemie 1/38, 2.
[35]   Stone, F.S. (1990): J. Mol. Cat. **59**, 147.

[36] Suib, S.L. (1993): Selectivity in Catalysis. In: ACS Symp. Ser. **517**, 1.
[37] Vannice, M.A. (1990): J. Mol. Cat. **59**, 165.
[38] Vannice, M.A. (1975): J. Catal. **37**, 449.
[39] van Santen, R.A. (1991): Surf. Sci. 251/252, 6.

Chapter 5

[1]   Delmon, B. et al. (Ed.) (1979): Preparation of Catalysts. Elsevier, Amsterdam.
[2]   Emig, G. (1977): Chem. Ing. Tech. **49**, 865.
[3]   Griebbs, H.R. (1977): Chemtech Aug. 1977, 512.
[4]   Gubicza, L., Ujhidy, A., Exner, H. (1988): Chemische Industrie 7/88, 48.
[5]   Hartley, F.R. (1985): Supported Metal Complexes. D. Reidel Publ. Comp., Dordrecht.
[6]   Hesse, D., Redondo de Beloqul, M.S. (1989): Einfluß der Porenstruktur auf das Umsatzverhalten von Supported Liquid-Phase-Katalysatoren. In: Dechema-Monographien Bd. 118, 305. VCH, Weinheim.
[7]   Higginson, G.W. (1974): Chem. Eng. **81** (20), 98.
[8]   Hölderich, W., Schwarzmann, M., Mroß, W.D. (1986): Erzmetall **39** (6), 293.
[9]   Kotter, M., Riekert, L. (1982): Chem. Eng. Fundam. Vol. **2** (1), 19.
[10] Kotter, M. (1983): Chem. Ing. Tech. **55**, 179.
[11] Luft, G. (1991): Chem. Ing. Tech. **63** (7), 659.
[12] Mroß, W.D. (1985): Jahrbuch 1985 der „Braunschweigische Wissenschaftliche Gesellschaft", 101.
[13] Panster, P. (1992): Chemie in Labor und Biotechnik **43**, 16.
[14] Schneider, P., Emig, G., Hofmann, H. (1985): Chem. Ing. Tech. **57**, 728.
[15] Scholten, J.J.F. (1985): J. Mol. Cat. **33**, 119.
[16] Schuit, G.C.A., Gates, B.C. (1983): Chemtech Sept. 1983, 556.

Chapter 6

[1]   Chen, N.Y., Garwood, W.E., Dwyer, F.G. (1989): Shape Selective Catalysis in Industrial Applications. Chemical Industries, A Series of Reference Books and Textbooks, Bd. 36. Marcel Dekker, New York, Basel.
[2]   Dyer, A. (1988): An Introduction to Molecular Sieves. J. Wiley, New York.
[3]   Hölderich, W., Gallei, E. (1985): Ger. Chem. Eng. **8**, 337.
[4]   Hölderich, W., Hesse, M., Näumann, F. (1988): Angew. Chem. **100**, 232.
[5]   Karge, H.G., Weitkamp, J. (Hrsg.) (1984): Zeolites as Catalysts, Sorbents and Detergent Builders. Elsevier, Amsterdam.
[6]   Kerr, G.T. (1989): Spektrum d. Wissenschaft **1989** (9), 94.
[7]   Tißler, A., Müller, U., Unger, K. (1988): Nachr. Chem. Tech. Lab. **36** (6), 624.
[8]   Unger, K., Kanz-Reuschel, B., Brenner, A., Wallau, M., Spichtinger, R. (1992): Labor 2000 (1992), 179.
[9]   Vedrine, J.C. (1982): Physical Methods for the Characterization of Non-Metal Catalysts. In: Surface Properties and Catalysis by Non-Metals. NATO ASI Series, 123. C.D. Reidel Publ. Comp., Dordrecht.

Chapter 7

[1] Agar, D., Bever, P.M., Wenert, D. (1988): Chem. Ing. Tech. **60** (9), 712.
[2] Baltzly, R. (1976): J. Org. Chem. **41** (6), 920.
[3] Bandermann, F. (1972): Statistische Methoden beim Planen und Auswerten von Versuchen. In: Ullmanns Enzyklopädie der technischen Chemie, 4. Aufl., Bd. 1, 294. VCH, Weinheim.
[4] Berty, J.M. (1983): Appl. Ind. Catalysis **1**, 41.
[5] Carlson, R. (1992): Design and optimization in organic synthesis. Elsevier, Amsterdam.
[6] Davis, L. (1992): Chemistry and Industry **1992**, 634.
[7] Deller, K. (1990): Chemische Prod. 1/2/90, 44.
[8] Dreyer, D., Luft, G. (1982): Chem.-Tech. (Heidelberg) **11** (8), 1061.
[9] Emig, G. (1977): Chem. Ing. Tech. **49** (11), 865.
[10] Greger, M., Gutsche, B., Jeromin, L. (1992): Chem. Ing. Tech. **64** (3), 253.
[11] Gut, G. (1982): Swiss Chem. **4**, 17.
[12] Hagen, J. (1975): Diss. RWTH Aachen.
[13] Hagen, J., Roessler, F., Zwick, T. (1993): Chemie-Technik 7/93, 76.
[14] Heidel, K. (1973): Fette, Seifen, Anstrichmittel **75**, 233.
[15] Herskowitz, M. (1991): Hydrogenation of Benzaldehyde to Benzylalcohol in a slurry and fixed-bed reactor. In: M. Guisnet et al. (Hrsg.). Heterogeneous Catalysis and Fine Chemicals II, 105. Elsevier, Amsterdam.
[16] Hoffmann, U., Hofmann, H. (1971): Einführung in die Optimierung. Verlag Chemie, Weinheim.
[17] Hofmann, H. (1975): Chimia **29** (4), 159.
[18] Ingham, J., Dunn, I.J., Heinzle, E., Prenosil, J.E. (1994): Chemical Engineering Dynamics — Modelling with PC Simulation. VCH, Weinheim.
[19] Kromm, K. (1994): Dipl.-Arbeit FH Mannheim — Hochschule für Technik und Gestaltung.
[20] Mahoney, J.A., Robinson, K.K., Myers, E.C. (1978): Chemtech Dec. 1978, 758.
[21] Petersen, H. (1992): Grundlagen der Statistik und der statistischen Versuchsplanung. ecomed, Landsberg.
[22] Plackett, R.L., Burman, J.P. (1946): Biometrica **33**, 305.
[23] Ramachandran, P.A., Chaudhari, R.V. (1980): Chem. Eng. Dec. 1, 74.
[24] Reh, E. (1992): GIT Fachz. Lab. **5**, 552.
[25] Retzlaff, G., Rust, G., Waibel, J. (1978): Statistische Versuchsplanung. Verlag Chemie, Weinheim.
[26] Schermuly, O., Luft, G. (1978): Ger. Chem. Eng. **1**, 222.
[27] Schneider, P., Emig, G., Hofmann, H. (1985): Chem. Ing. Tech. **57** (9), 728.
[28] Tarham, M.O. (1983): Catalytic Reactor Design. Mc Graw Hill, New York.
[29] Trimm, D.L. (1973): Chemistry and Industry, 3. Nov. **1973**, 1012.
[30] Zwick, T. (1992): Dipl.-Arbeit FH Mannheim — Hochschule für Technik und Gestaltung.

Chapter 8

[1] Anderson, J.B.F, Griffin, K.G., Richards, R.E. (1989): Chemie-Technik **18** (5), 40.
[2] Augustine, R.L. (1985): Catalytic Hydrogenation. Marcel Dekker, New York.
[3] Bröcker, F.J., Kaempfer, K. (1975): Chem. Ing. Tech. **47** (12), 513.
[4] Cerveny, L. (Ed.) (1986): Catalytic Hydrogenation. Elsevier, Amsterdam.
[5] Emig, G. (1977): Chem. Ing. Tech. **49** (11), 865.
[6] Emig, G. (1987): Chemie in unserer Zeit **21** (4), 128.
[7] Engler, B.H. (1991): Chem. Ing. Tech. **63** (4), 298.
[8] Ertl, G. (1990): Angew. Chem. **102**, 1258.

[9]   Fink, K. et al. (1992): Chem. Ing. Tech. **64** (5), 416.
[10] Godfrey, J.A., Searles, R.A. (1981): Chemie-Technik **10** (12), 1271.
[11] Guisnet et al. (Ed.)(1991): Heterogeneous Catalysis and Fine Chemicals II. Elsevier, Amsterdam.
[12] Hagen, J. (1987): Der Innovationsberater (DIB) 2/87, 1883. Rudolf Haufe Verlag, Freiburg.
[13] Herrmann, W.A. (1991): Metallorganische Chemie in der industriellen Katalyse: Reaktionen, Prozesse, Produkte. Teil 1: Kontakte (Darmstadt) **1991** (1), 22. Teil 2: Kontakte (Darmstadt) **1991** (3), 29.
[14] Hölderich, W., Schwarzmann, M., Mroß, W.D. (1986): Erzmetall **39** (6), 292.
[15] Kanzler, W., Schedler, J., Thalhammer, H. (1986): Chemische Industrie 12/86, 1188.
[16] Kotowski, W., Bekier, H. (1992): Chem. Tech. **44** (5), 163.
[17] Mroß, W.D. (1985): Umschau **1985** (7), 423.
[18] Mücke, M. (1975): Chem. Lab. Betr. **26** (1), 10.
[19] Schmidt, K.H. (1984): Katalysatoren für chemische Großsynthesen. Teil I: Chem. Ind. Okt. 1984, 572. Teil II: Chem. Ind. Nov. 1984, 716.
[20] Sleight, A.W., Linn, W.J., Aykan, K. (1978): Chemtech April 1978, 235.
[21] Weisweiler, W. (1989): Umweltfreundliche Entstickungskatalysatoren. In: Dechema-Monographien Bd. **118**, 81.

Chapter 9

[1]   Agar, D.W., Ruppel, W. (1988) Chem. Ing. Tech. **60** (10), 731.
[2]   Alper, E., Wichtendahl, B., Deckwer, W.D. (1980): Chem. Eng. Sci. **35**, 217.
[3]   Concordia, J.J. (1990): Chem. Eng. Prog. **86** (3), 50.
[4]   Deckwer, W.D., Alper, E. (1980): Chem. Ing. Tech. **52**, 219.
[5]   Falbe, J. (Hrsg.) (1978): Katalysatoren, Tenside und Mineralöladditive. G. Thieme, Stuttgart.
[6]   Fogler, H.S. (1992): Elements of Chemical Reaction Engineering. 2. Aufl., Prentice Hall, New Jersey.
[7]   Gianetto, A., Silveston, P. (Eds.) (1986): Multiphase chemical reactors. Hemisphere, Washington.
[8]   Gianetto, A., Specchia, V. (1992): Chem. Eng. Sci. **47** (13, 14), 3197.
[9]   Greger, M., Gutsche, B., Jeromin, L. (1992): Chem. Ing. Tech. **64** (3), 253.
[10] Herskowitz, M., Smith, J.M. (1983): AIChE J. **29**, 1.
[11] Jenck, J.F. (1991): Gas-liquid-solid reactrs for hydrogenation in fine chemicals synthesis. In: M. Guisnet et al. (Ed.): Heterogeneous Catalysis and Fine Chemicals II, 1. Elsevier, Amsterdam.
[12] Ramachandran, A., Chaudhari, R.V. (1980): Three-phase Catalytic Reactors. Gordon and Breach, New York.
[13] Tarham, M.O. (1983): Catalytic Reactor Design. Mc Graw Hill, New York.
[14] Trambouze, P. (1981): Chem. Ing. Tech. **53** (5), 344.
[15] Trambouze, P., Van Landeghem, H., Wauquier, J.P. (1988): Chemical reactors — design/engineering/operation. Editions Technip, Paris.
[16] Weiss, S. et al. (Hrsg.) (1987): Verfahrenstechnische Berechnungsmethoden, Teil 5. Chemische Reaktoren, Ausrüstungen und ihre Berechnung. VCH, Weinheim.

Chapter 10

[1] Asche, W. (1993): Chemische Ind. 11/93, 40.
[2] Creek, B.F. (1989): Chem. Eng. News **67** (22), 29.
[3] Deller, K., Focke, H. (1990): Chemie-Technik **19** (6), 21.
[4] Fonds der Chemischen Industrie (1989): Folienserie Nr. 12. Frankfurt/M.
[5] Hölderich, W., Schwarzmann, M., Mroß, W.D. (1986): Erzmetall **39** (6), 292.
[6] Roth, J.F. (1991): Industrial Catalysis: Period for a new Generation of Major Innovations. In: Catalytic Science and Technology, Vol. **1**, 1. Kodansha Ltd, Japan.
[7] Schmidt, K.H. (1984): Chem. Ind. 7/1984, 380.
[8] Schmidt, K.H. (1984): Katalysatoren für chemische Großsynthesen. Teil I: Chem. Ind. Okt. 1984, 572. Teil II: Chem. Ind. Nov. 1984, 716.
[9] Weng, L.T., Delmon, B. (1992): Appl. Catal. A: General **81**, 141.

Chapter 11

[1] Asche, W. (1993): Chemische Ind. 11/93, 40.
[2] Gallei, E.F., Neuman, H.P. (1994): Chem. Ing. Tech. **66** (7), 924.
[3] Grünert, W., Völker, J. (1992): Chem. Technik **44** (11/12), 395.
[4] Haber, J., Herzog, K. (1994): Heterogene Katalyse. Trends und Perspektiven. In: Topics in Chemistry. Heterogene Katalysatoren. BASF AG Ludwigshafen.
[5] Herrmann, W.A. (1991): Metallorganische Chemie in der industriellen Katalyse: Reaktionen, Prozesse, Produkte. Teil 1: Kontakte (Darmstadt) **1991** (1), 22. Teil 2: Kontakte (Darmstadt) **1991** (3), 29.
[6] Keim, W. (1984): Chem. Ind. 7/1984, 397.
[7] Keim, W. (1984): Chemisch Magazine Juli 1984, 417.
[8] Kochloefl, K. (1989): Chem. Ind. 8/89, 41.
[9] Kral, H. (1989): Chem. Ind. 8/89, 44.
[10] Roth, J.F. (1991): Industrial Catalysis: Period for a new Generation of Major Innovations. In: Catalytic Science and Technology, Vol. **1**, 1. Kodansha Ltd, Japan.
[11] Schmidt, K.H. (1985): Chem. Ind. 11/85, 762.

Index

σ acceptor   22
π-acceptor complex   24
acceptor
– function   134
– reaction   131 f.
acetonitrile
– adsorption   127
– decomposition   127
acid
– hard   37
acid-base
– catalyst   130
– – concept   154
– complex   157
– concept   24
– reaction   24 ff., 47
acidic catalyst   154 f., 161, 300
acidity function   156
acrolein
– synthesis   211 f., 250 f.
activation energy   4 f., 87 f. 101 ff.
– apparent   101
– catalytic   101
– true   101
active center   87, 89, 92 f., 101, 121
activity   4 ff., 102, 126 ff., 135, 139 f.
– kinetic   4
– loss   190 f.
– series   138 f.
additive   165, 187, 189
Adkins catalyst   209
adsorbate   87, 196 f.
adsorption   83, 86 ff., 100, 195, 197
– chemical   86
– coefficient   102
– competitive   125, 172, 238
– cooperative   125
– dissociative   105, 133
– enthalpy   89, 103, 112
– equilibrium   88 f.
– – law of mass action   89
– heat   86 f., 102, 104, 112
– hydrocarbons   102
– hydrogen   104 f., 112

– isotherm   88
– measurement   114
– mixed   90, 94
– molecular   114
– multipoint   117
– physical   86
– single-point   117
ageing   180
aggregation state   8
alcohol synthesis   130
Aldox process   220 f.
alkyl migration   34
alkylation   130, 300 f.
– aromatics   243
π-allyl complex   23, 39, 52 f., 315 ff.
σ-allyl complex   23
allyl radical   106
aluminophosphate   240
aluminosilicate   9, 154, 157 ff., 348
ammonia synthesis   108, 111, 122, 130, 211, 299 f., 304 ff., 331
– catalyst   124, 304 f.
– mechanism   304 f.
– promoter   176, 304
– steps   305
ammoxidation   300 f., 316 ff.
– methane   301, 332
– propene   301, 316 ff., 333
– – mechanism   316
– variants   318
Andrussow process   332
anionic complex   24, 38
anti-Markownikov addition   50
aromatization   153
– alkane   153, 175
– – promoter   175
Arrhenius equation   4
Arrhenius law   85
asymmetric catalysis   351
Auger electron spectroscopy   202
autoclave   60
automobile exhaust
– catalyst   345, 347 f.
– purification   212, 302 f.

π backbonding   38, 55, 176
backbonding   23
Balandin   112
band model   131 f., 140
base
– hard   37 f.
– soft   38
basic catalyst   154 f., 160 ff.
basicity   162
benzaldehyde
– hydrogenation   281
benzene
– alkylation   234, 242, 301
– ethylation   238
– selective oxidation   212
BET
– equation   195
– surface   194, 196
bifunctional activation   40
bimetallic cluster   137
binary oxide   153
biocatalyst   8 f.
bismuth molybdate catalyst   315 f.
blank
– effect   272
– reaction   262
– variables   273
π bonding   22 f.
boron zeolite   239
Brønsted acid   28, 107
– chemisorption   107
– center   155 ff., 235
– – zeolite   235
bubble column   338 f.
– reactor   341
bulk catalyst   208
butynediol
– synthesis   336

calcination   210
carbonyl complex
– ligand reactions   37
carbonylation   12, 31, 33 f., 47, 52, 58, 67
– 1-decene   60
– methanol   53, 59, 67, 72 f. 215, 241, 341
– α-olefin   50
carboxylation   35
catalysis
– definition   1
– enantioselective   78
– enzymatic   5
– future development   351
– hard   47 f.

– history   2 f.
– importance   2
– shape-selective   225, 229
– soft   47, 50
catalyst
– acidity   160
– activity   4, 180
– bifunctional   171, 240 f., 300
– bimetallic   186, 315
– bleeding   12
– blocking   186
– classification   8, 129
– complex intermediate   1
– concept   139
– – heterogeneous catalysis   100
– – homogeneous catalysis   44
– definition   1
– demand   346, 348
– development   249 ff., 257, 348, 353, 356
– economic importance   345
– energy generation   355
– hard   49
– impregnated   208, 210 f., 213
– lifetime   8, 178, 180, 186, 251
– loading   211
– losses   181, 189 f.
– mode of action   4
– multifunctional   215, 221
– planning   249, 251 ff.
– poisons   181 f.
– precipitated   208
– production   207 ff.
– pyrophoric   211
– regeneration   178, 180 f. 186 f., 191
– screening   5, 258 ff.
– – selective hydrogenation   258
– selection   256 f.
– shapes   207 ff.
– solid-state   8
– stability   4, 8, 178
– testing   249, 254, 256
– turnover number (TON)   121
catalytic cracking   333
catalytic cycle   1 f., 30, 45
Catatest plant   285 f., 309
cationic complex   24, 38
CH activation   352
– alkanes   352
chain reaction   75
chemical catalyst   345, 347
– application   347

chemisorption   84, 86 ff., 101, 105, 167, 193, 197
– alkene   106
– associative   103, 170
– capability   135, 167
– CO   107
– dissociative   12 f., 90, 103 ff., 170
– enthalpy   112 f.
– gases   102
– heat of   88
– heterolytic   107
– on metal   102
– molecular   103 ff.
– processes   139
– semiconductors   145 f.
– specific   197
– term   92
Chevron process   80
chirality   77
chromium
– complex   220
– supported catalyst   320
Claus process 299 f.
clay   154
CO adsorption   124
– IR spectrum   124
CO conversion   209
CO hydrogenation   40, 107, 130, 133 f., 162, 168 ff., 176 f., 299, 301
– catalytic activity   168
– potassium promoter   177
– selectivity   168
CO oxidation   95, 122, 125 f., 150
– metal oxide catalyst   151
cobalt catalyst   68 ff., 76
coke formation   158, 160, 168
coking   180 f., 184, 186 f., 234
_ complex   19 f., 106
complex formation   21, 26
– Lewis acid   26
condensation   300
– acetone   163
conduction band   132, 134, 140 ff.
conductivity   140 ff.
– electrical   130 f.
conductor   131
cone angle   18 f.
constraint index   232
contact process   331, 345
copper   103, 135 f.
– alloys   135 ff.
– catalyst   73 f., 307 f., 312
– – supported   170

– chromate   209
– complex   59
coupling reaction   47
coverage
– degree of   89 f., 95 f., 124
cracking   9, 130, 232, 242 f., 300, 302
– alkanes   238
– catalyst   157, 237
– – zeolite Y   237
– crude oil   159
– cumene   155
– n-heptane   161
– heptane isomers   232
– n-hexane   232, 238
– hexenes   235
– n-octane   238
– processes   155
cumene dealkylation   236
cyclization
– heptane   122
– hexane   122
cyclooligomerization   23, 32

deactivation   1, 180 ff., 234
– kinetics   190 f.
– mechanism   181 f., 189 f.
– model   189 f.
– rate   190
decomposition   35 f.
– ammonia   124
– ethanol   151 f.
– – semiconductor oxides   152
– formic acid   124, 130, 136
– hydrogen peroxide   136
– isobutane   155
– liquid hydrocarbons   212
– methanol   136
– N_2O 149 f.
dehydration   130, 151 f., 300
– alcohols   155
– butanols   233
– ethanol   158 f.
dehydroaromatization
– olefins   251 ff.
– – catalyst planning   251, 255
dehydrochlorination   168
dehydrocyclization   265
dehydrogenation   122, 130, 135 f., 151 ff., 300
– alcohol   161 f.
– butanes   153, 301
– cyclohexane   118, 124, 136 f., 168 f.
– cyclohexanone   168

– ethylbenzene   209, 301, 332
– methanol   332
dehydrohalogenation
– alkyl halide   162
– catalyst   162
denitrification   303
denitrogenation   153
DENOX process   318
deoxygenation   153
deposit   186
DESONOX process   303
desorption   83, 101
– heat   112
desulfurization   153, 303, 336
development cycle   354
dewaxing   232, 240, 242 f.
differential circulating reactor   260 ff.
diffusion   83, 194, 289
– coefficient, effective   284
– rate   92
Dimersol process   67
dispersion   121, 127 f., 197 f.
– degree of   121, 168 f., 197
– metal   127
dispersity   119, 128
disproportionation   131
– methylcyclopentene   163
– m-xylene   234
dissociation   104
– energy   105
dissymmetry   47 f., 51
σ-donor   25 f., 30, 63
– softness   52
donor
– function   134
– reaction   131 f., 136
L-DOPA   65, 67, 77
doping   143, 150

economic importance   345
electron density   29
16/18-electron rule   19, 44, 58
electron spectroscopy for chemical analysis
   see ESCA
electron transfer   131 f.
electronic
– effect   22, 131, 167 f.
– factor   129 f., 133
Eley-Rideal mechanism   96 f., 125
elimination   33, 35 f.
α-elimination   36 f.
ensemble effect   175
entropy   103

environmental catalyst   345
– automobile   345, 347
– industrial   345, 347
environmental protection   302 f.
enzyme   8 f.
– catalysis   351
epoxidation   47
– olefin   128
– propene   49
– selective   48
error-propargation law   271
ESCA   200 ff.
– application 202
– spectrum   200 f.
ester hydrolysis   47
ethylene
– adsorption   106, 123
– complex   20, 109
– – IR spectrum   109
– dimerization   50 f.
– hydrogenation   96, 112, 121 ff., 135, 152 f., 218
– – catalyst   124
– oligomerization   67, 78 ff.
– oxidation   54, 73, 96, 121 f., 176, 220, 241, 300 f., 314, 332, 341
– – promoter   176
– oxychlorination   300 f.
ethylene glycol   12
ethylene oxide   96
2-ethylhexanol
– production   220

factorial design   267 f.
– evaluation   268, 270
– oxo synthesis   269
– significance   270 ff.
– test plan   267
fat hardening   301
Fermi level   132 ff.
film diffusion   84 f.
– region   84 f.
– resistance   84
fine chemicals   308, 336
– production   308
Fischer-Tropsch synthesis   7, 40
fixed catalyst   214
fixed-bed reactor   328, 330, 334, 337
flue gas
– denitrogenation   347
– desulfurization   319
– purification   302 f., 318 f.
fluidized-bed reactor   330, 332

formic acid   153
– cleavage   153
– decomposition   110
Freundlich equation   88 f.
Friedel-Crafts reaction   47
fuel cell   356

gallium zeolite   239
gas-phase reaction   83, 92 f., 101, 260, 329
– activation energy   101
– bimolecular   93, 95, 97
– heterogeneous-catalyzed   260
– individual steps   83
– monomolecular   90
– reaction rate   92
gas-solid reactor   330
geometric effect   108, 117
Gibb's free energy   103 f.
glycol   7
grafted catalyst   218
Gulf process   80

H_2/D_2 exchange   114
– hydrocarbons   114
Haber-Bosch process   304
Halcon process   67
heterogeneous catalysis   10, 83 ff.
– future development   353
heterogeneous catalyst   8 ff., 193
– characterization   193 ff.
– concept   100
– industrial use   299
– production   207 ff.
– shape   207 ff.
heterogenization   215
heterolytic process   66
heteropolyacid   155, 357
high-pressure IR apparatus   60 f.
homogeneous catalysis   5, 10 f., 44 ff., 17
– concept   44
– future development   351
– key reactions   18, 44
– mechanism   10
– reactors   340 f.
homogeneous catalyst   8 ff. 57, 214 ff.
– characterization   57
– immobilization   214
homolytic process   66
HSAB concept   20 f., 35 ff., 47, 50, 53 f.
hybrid catalyst   214
hydration   130, 300
– olefin   155
hydride elimination   12

β-hydride elimination   35 f., 49 f.
– nitrile   301
hydrocarbon
– synthesis   130
– H_2/D_2 exchange   114
hydrocarboxylation   53, 67
hydrocracking   240, 242 f., 300, 302, 331, 336
– bifunctional catalysis   243
hydrocyanation
– butadiene   55, 66 f.
hydrodenitrogenation   128
hydrodesulfurization   121, 128, 265, 299, 302
– catalyst promoter   178
hydroformylation   31, 45 f., 67 f., 175 f., 220, 275 f.
– ethylene   175
– 2,4-hexadiene   275 f.
– ligand influence   63
– mechanism   68
– olefin   52, 59 f., 220, 341
– optimization   275 f.
– 1,3-pentadiene   273
– – Plackett-Burman plan   273
– – reaction parameters   273
– propene   68, 70 f., 341
hydrogenation   27, 33, 47, 306 ff.
– acetone   172 f.
– acetylene   98, 114 f., 336
– – mechanism   115
– activtiy   135
– aldehyde   65, 137, 153, 301, 336
– adiponitrile   336, 339
– – technology   339
– alkene   49, 52, 62, 112 f., 130, 138 f., 162, 265
– – mechanism   113
– alkyne   52, 185
– allene   336
– amide   49
– aromatics   130
– asymmetric   65, 67, 77
– benzaldehyde   281 ff.
– – conversion   287
– – modeling   281, 285
– – selectivity   287
– benzene   65, 67, 122, 124, 301, 332, 339
– benzoic acid   340
– butadiene   52
– butynediol   336
– butyraldehyde   138 f.
– catalyst   114, 168, 219
– chloronitrobenzene   168, 339

410 Index

- CO_2 130
- crotonaldehyde 138 f., 172
- – selectivity 172
- cyclohexene 52, 98, 340
- diene 49, 52 f., 162
- ester 301
- 2-ethylanthraquinone 339
- ethylene 112, 121 f., 152 f.
- fat 306 ff.
- fatty acid 336, 339
- fatty ester 339
- ketone 122, 336
- lactone 287 ff.
- linoleic ester 340
- C_{12}-C_{22} nitriles 340
- 2-nitrobenzonitriles 258 ff., 277 f.
- – test plan 278
- oil 339
- poisoning 184
- selective 114, 153, 185, 299, 308 f., 336
- – – ester 309
- shape-selective 241
- triglyceride 306 ff.
- α,β-unsaturated carboxylic acids 49

hydrogenolysis 310
- activity 137
- catalyst 219
- cyclohexane 122
- ethane 122, 136 f., 168 f.
- methylcyclopentane 122

hydrolysis 130
hydrorefining 300
hydrosilylation 27

immobilization 214
- homogeneous catalyst 214
immobilized catalyst 8
impregnation 209 f.
industrial processes 65 ff., 304 ff.
inhibitor 182
inorganic chemicals 299
insertion 33 ff., 49
- acetylene 33
- carbene 35
- CO 33 f., 40
- olefin 33
integral reactor 262, 265 f., 285
- advantage 266
- data evaluation 266
interaction
- catalyst-additive 165
- catalyst-support 165

- hard-hard 48
- soft-soft 49
ion scattering spectroscopy 202 f.
ion-exchange catalyst 333
ionic catalyst 130
IR spectroscopy 108 ff., 114, 127, 200
- application 200
iridium complexes 114
iron catalyst 124, 304
iron oxide catalyst 209
ISIM program 290
- dynamic modeling 291
- lactone hydrogenation 290
isolator 130 f., 144, 150, 154 f.
isomerization 131, 300
- alkane 302
- 1-alkene 50 f.
- butene 162
- dichlorobutene 66 f.
- double-bond 50
- – mechanism 50
- n-hexane 137
- hexane 122
- isobutane 122
- light gasoline 331
- 2-methyl-3-butenenitrile 66
- n-pentane 161
- xylene 302, 234, 242 f.
isotopic labelling 57 f.

jet loop reactor 262 ff., 338 f.

key reactions 58
- homogeneous catalysis 18
kinetic measurement 57, 60
kinetics 58, 260
- heterogeneous catalysis 86

lactone
- hydrogenation 287 ff.
Langmuir
- equation 88 f., 194
- adsorption isotherm 89 f., 145
Langmuir-Hinshelwood mechanism 93 ff. 125 f., 151, 159
lattice
- defect 119
- distance 118, 123
- plane 119 f.
- types 118 f.
LEED 123 ff., 199 f.
- application 200

Lewis acid 39
– center 155 ff.
– hard 39
Lewis base 39
– center 158
– hard 39
– soft 39
Lewis basicity 162
ligand 17
– π acceptor 23, 30, 38
– alkyl phosphine 23
– allyl 23
– basicity 62
– cone angle 18
– coordination 18
– DIPAMP 78
– σ donor 38
– effect 30
– exchange 18, 20 f.
– hard 48
– influence 29 f.
– – in hydroformylation 63
– ionic 17
– multidentate 216
– neutral 17
– reaction 37
– soft 47
Lindlar catalyst 185
liquid
– feed 291
– holdup 291, 293
loop reactor 337 f., 341
low-energy electron diffraction see LEED
Lurgi process 312

macrokinetics 84, 91 f.
macropores 193 f.
manganese anomaly 112
Markownikov addition 50
Mars-van Krevelen mechanism 148
mass transfer
– coefficient 293 f.
– influence 293
– limitation 293
– resistance 284, 338
membrane reactor 357
mercury porosimetry 193 f., 196
mesopores 193 f.
metal alkyl
– complex 52
– mechanism 50
metal allyl mechanism 50
metal basicity 24 ff., 30

metal catalyst 299
– poisons 182 ff.
– rules 139
– supported 303
metal cluster catalyst 12
metal complex 130
metal oxide 130, 148 f., 154
– catalyst 300
metal-olefin
– bonding 22
– complex 20 f.
– – π bonding 22
– – softness 22
metallic bonding 134 f.
metals 131 ff.
– classification 102 f.
metathesis 37, 66 f.
methane
– catalytic combustion 355 f.
– oxidation 355
– oxidative coupling 355
– steam reforming 299 f. 355
methanization 7, 122, 300
– catalyst promoter 177
methanol oxidation 144, 148, 153, 301
– promoter effect 177
– selective 218
methanol synthesis 7, 95, 153, 175 ff., 262 ff., 299, 301, 311 ff., 331 f.
– copper catalyst 311 f.
– high-pressure process 311
– kinetic modeling 262
– low-pressure process 311
– mechanism 311 ff.
– promoter 175, 177
– – effects 177
methanol-to-gasoline (MTG) process 234, 242 f.
– pentasil catalyst 243
methanol-to-olefin (MTO) process 242 f.
methyl migration 59
microkinetics 84, 86, 91, 289
micropore 194
Miller indices 119
Mittasch 311
mixed oxide 153, 155, 160
– acid stregth 160
Mobil-Badger process 243
model
– analysis 278 f.
– equation 278, 281
model-space representation 279

modeling   58, 290
– hydrogenation   277 f., 281
– kinetic   281
– statistical   281
molecular sieve   230
– aluminophosphate   240
molybdenum catalyst   189
molybdenum oxide catalyst   108
– chemisorption   108
– immobilized   218
Monsanto process   53, 65, 67, 72 f., 77
– catalytic cycle   72
– L-DOPA   65, 67, 77
multibed reactor   331
multiphase reaction   263
– kinetic study   265
– test reactor   263
multiple-tube reactor   312, 331
multiplet theory   117

nickel catalyst   78 ff., 109, 122, 168 f., 189, 309
– supported   109, 209, 211, 309
– surface   123, 127
nitrogen oxide   318
– selective reduction   318
noble metal catalyst   133, 209, 307 ff., 348 f.
– additive   137 f.
– production   209

off-gas
– catalytic afterburning   346
– purification   302 f., 345, 347
olefin
– carbonylation   50
– cyclization   254
– – mechanism   254
– dimerization   49, 67, 254
– – mechanism   254
– disproportionation   37
– hydroformylation   215
– hydrogenation   50, 215
– insertion   33, 52
– isomerization   36, 50
– metathesis   218, 300
– oligomerization   47, 50, 67, 241
– oxidation   215
– polymerization   36, 49, 67, 300 f., 320 f.
oligomerization   66, 302
– butadiene   67
optimization   267, 274, 278, 280 f.

organic chemicals   299
Ostwald   1
– nitric acid process   9, 189, 299 f., 332
oxad reaction   13
oxidation   47 f., 130, 301
– activity series   149
– aldehyde   48
– ammonia   9, 97, 189, 299 f.
– benzene   301, 314
– n-butane   314
– butene   301, 314
– catalyst   48, 149
– – selective   154
– cyclohexane   67, 75 f.
– – mechanism   76
– ethylene   122, 220, 301, 314, 332, 341
– hydrocarbons   147
– mechanism   148
– methanol   144, 148, 153, 301
– naphthalene   301, 333
– propene   301, 314
– selective   128, 147 f., 211 f., 239, 314
– selectivity series   149
– SO_2 299 f.
– toluene   341
– o-xylene   301, 333
– p-xylene   341
oxidative addition   12, 21, 26 ff., 47 f., 54
oxidative coupling   32
– methane   162
oxirane process   49
oxo process   65
oxo synthesis   9, 12, 61, 68 ff., 215, 220, 241, 269, 341
oxychlorination   220, 300 f.
– alkene   220

palladium alloys   137
palladium catalyst   73, 114, 168 f., 309
– poisoned   185
– supported   109, 128, 139, 170, 259
passivation   211
petroleum catalyst   345, 347 f.
phase change   188
phenol
– ethylation   238
Phillips catalyst   320 f.
Phillips process   220
Phillips triolefin process   66
phosphate   154
phosphine ligand   24 f., 62, 69 f., 215
– chiral   77
– donor character   24

photosensitization 241
– oxygen 241
phthalic anhydride 212
– production 212
physisorption 86f., 104f., 195
– isotherm 195
Plackett-Burman plan 272
– hydroformylation of 1,3-pentadiene 273
platinum catalyst 108, 126
– crystallite agglomeration 199
– shell type 211
– supported 169, 172
plug flow reactor 328f.
poisoning 12, 157, 174, 180ff.
– acid center 185
– desired 185
– halogens 184
– metals 182ff.
– mode of action 183
– nitrogen compounds 185
– semiconductor oxides 185
– solid acids 185
polycondensation 130
polyethylene 320
– high-density 320
– low-density 320
polymer support 214, 217
polymerization 66, 130
– butadiene 163
– diene 67
– ethylene 320f., 341
– mechanism 321
– olefin 67, 155, 160, 320f.
– α-olefin 220
– propene 161
– propylene 220, 320
– supported catalyst 320
polypropylene 220, 320
polysiloxane catalyst 218f.
– organofunctional 218
pore
– diffusion 83ff., 211, 340
– – region 84f.
– distribution 193
– radius 193
– shape 193
– size 194
– size distribution 193f.
– structure 92, 193
– volume 193f.
porosity 117
porphyrins 186f.

potential diagram 104
process improvement 354
product selectivity 231, 233
promoter 174ff., 184, 251
– bromide 54
– chloride 54
– cesium 176
– chemical effect 177
– cobalt 178
– electronic effect 176f.
– ensemble effect 175
– function 174f., 178
– molybdenum 178
– potassium 176, 178
– type 174
propene 314ff.
– ammoxidation 314, 316ff.
– dimerization 163
– oxidation 250, 314ff.
– – catalyst 314
– – mechanism 315
protonation 25

quench reactor 331

rate constant 4f.
rate equation 5, 92ff.
raw material costs 355
reactant selectivity 230f., 233
reaction
– kinetics 88
– order 90f.
– parallel 7
– rate 4f., 84, 91, 94ff.
– – effective 85, 91, 328
– – function of partial pressure 94
– – relative 111
– sequential 7
– steric effects 122
– structure-insensitive 121, 125, 172
– structure-sensitive 121, 124
reactor 327ff.
– calculations 98
– choice 339
– design 257, 281, 327, 330
– homogeneous catalysis 340f.
– length 292f.
– modeling 327
– simulation 257, 281
redox
– catalyst 108, 130
– mechanism 47, 130
– reaction 26ff., 47f.

reduction
- adiponitrile   336
- CO_2 97
- N_2O 96
reductive
- amination   310
- elimination   21, 26, 30 f., 47 f.
refinery processes   300, 302
refining   336
reforming   137, 232, 240, 300, 302
- catalyst deactivation   186 ff.
- n-hexane   171
- methane with CO_2 173 f.
- naphtha   178, 188 f., 331
Reppe   66
- alcohol synthesis   53
- reaction   67
residence time   6, 125, 287 ff.
restricted transition state selectivity   231, 234 f.
rhodium catalyst   12, 68 ff., 77, 215
- cluster   12
- complex   217
- DIPAMP ligand   78
- supported   169 f.
Rhône-Poulenc/Ruhrchemie process   12, 70, 341
ring opening   122
- cyclopropane   122
ring-opening metathesis   66
- cyclooctene   66
- norbornene   66
Roelen   66, 68
Rosenmund reaction   185

scatter of tests   269 ff.
- error variance   270
- standard deviation   270
SCR process   144, 189, 242 f., 318 ff., 347, 356
- catalyst   318
- mechanism   319
- variants   319
secondary ion mass spectrometry *see* SIMS
selective catalytic reduction *see* SCR process
selectivity   4, 6 f., 114, 127 f., 351
- promoter   357
selectoforming process   232
semiconductor   130 f., 140 ff.
- chemisorption   145 f.
- doping   141, 143 f., 150
- excitation energy   141
- hydrogen adsorption   146

- intrinsic   140 f.
- oxidation catalyst   146 f.
- oxide   141 ff., 169
- - poisoning   185
- - reactions   147
- - as support material   153
- oxygen adsorption   146
- properties   143 f.
- n-type   140 ff.
- p-type   140 ff.
shallow-bed reactor   332
shape selectivity   230, 232
shell catalyst   208, 211 ff., 340
- properties   212
Shell higher olefin process, SHOP   12, 66 f., 78 ff.
Shell hydroformylation process   63
Shell oxo process   66
Shell process   69
significance test   269 ff.
silica gel   154
silicalite   239
silicoaluminophosphate   240
silver catalyst
- supported   200 f., 211, 300
simplex method   274 ff.
SIMS   203 f.
- application   204
simulation   281
single crystal   118
- surface   119, 121
- - model   121
single-bed reactor   331
sintering   167 f., 172, 178, 180 f., 188 f., 211
Smidt   73
SMSI *see* strong metal-support interaction
SOHIO process   316 ff., 333
solid acid   185, 235
- poisoning   185
solid-state catalyst   131
solvent screening   259 f.
sorptometer   197
space velocity   5 f.
space-time yield   5 f.
spectroscopy   57 ff.
- in-situ   57 f.
- IR   57 ff.
- NMR   57 f.
- UV   58
steric effect   117, 122
stirred-tank reactor   328 f., 338 f., 341
- cascade   338, 340

strong metal-support interaction 167
– effects 171 ff.
superacid 155, 237
support
– application 165
– effect 153
– form 165, 167
– functionalized 216
– inorganic 217
– interaction 167, 170, 173
– loading 167
– material 108, 165, 167, 170, 208, 211, 357
– – impregnation 209
– particle size 167
– pore structure 167
– selection 166
– semiconductor type 171
– surface 166
supported catalyst 8 f., 153, 165, 209
– activity 166
– costs 166
– production 210
– regenerability 166
– selectivity 166
supported liquid phase catalyst, SLPC 216, 220, 333
supported solid phase catalyst, SSPC 216, 220
surface 83 f., 89 f., 101, 124, 194
– acidity 155 f.
– analysis 119, 126, 198, 200, 204
– characterization 117, 128, 198
– composition 200
– coverage 87
– irregularity 87
– reaction 289, 293
– specific 196
– structure 118, 127 f.
– titration 196
suspension reactor 285, 288, 334, 336 ff.
synthesis gas 7, 12, 130, 173, 299 f., 311, 352
– reactions 7

target function 274
target quantity 274 f., 278 ff.
– isolines 279 f.
test planning 267, 276 f.
– APO program 276 ff.
– factorial 267
– hydrogenation 277 f.
– statistical 267, 276 f.
test reaction 232

test reactor 5, 249, 260 ff., 285
– kinetic measurement 260
texture 193
thermal process 188
three-phase reactor 328, 333 f., 337
Tischtschenko reaction 162
titanium
– complex 22
– silicalite 239
Tolman 19, 44 f.
toluene
– alkylation 233
– demethylation 302
– disproportionation 155, 233, 242 f. 302
– – selective 233
– methylation 233
transition metal 132 f.
– catalyst 11 f., 17, 103, 112, 351
– – immobilized complex 333
– catalytic activity 132, 135
– complex 340
– electron density 132
– oxides 318
– work function 132
transition state 103, 105
transmission electron microscopy 198 f.
– application 198
trickle-bed reactor 281, 284 f., 287 ff., 334 ff.
– ISIM program 290
– modeling 287 ff.
– plug-flow model 288
tubular reactor 265, 328 f.
– design equation 265
turnover number, TON 5
two-phase reactor 328 f., 340
two-phase technology 70 f.

Ugo 47

valence band 132, 134, 140 ff.
valence bond theory 133
valence structure theory 131
vanadium molybdenum oxide 212
variable space 277 ff.
– experimental results 277, 279
vinyl chloride
– synthesis 211
volcano plot 110 f., 123, 133

Wacker process 54, 73 ff., 215, 241, 341
– mechanism 74
Wacker-Hoechst process 67, 75
waste-gas purification 144

Wilkinson's catalyst   19, 52, 77
work function   132 f.

Yates scheme   269

zeolite   154 f., 186, 194, 225 ff.
– A   226, 228
– acidity   229, 235 f., 239
– advantage   229
– application   242
– bifunctional catalyst   240 f.
– Brønsted acid center   237 f.
– β-cage   225
– catalytic properties   229
– composition   225
– cracking   232, 235, 238
– faujasite   227 f., 236
– framework   239
– isomorphic substitution   239
– Lewis acid center   237 f.
– metal-doped   240 f.
– organic syntheses   244
– pentasil   227 ff., 244
– pore size   228, 230, 232
– production   228
– shape selectivity   230, 232
– Si/Al ratio   236 ff.
– sodalite   225 f., 228
– structural unit   226
– structure   225
– Y   227 ff., 234, 237, 243
– ZSM-5   232, 237 f., 356
Ziegler-Natta
– catalysis   67
– catalyst   320 f.